The Illusions of Time

Valtteri Arstila • Adrian Bardon
Sean Enda Power • Argiro Vatakis
Editors

The Illusions of Time

Philosophical and Psychological Essays on Timing and Time Perception

palgrave
macmillan

Editors
Valtteri Arstila
University of Turku
Turku, Finland

Sean Enda Power
University College Cork
Cork, Ireland

Adrian Bardon
Wake Forest University
Winston-Salem, NC, USA

Argiro Vatakis
Panteion University of Social and Political Sciences
Athens, Greece

ISBN 978-3-030-22050-1 ISBN 978-3-030-22048-8 (eBook)
https://doi.org/10.1007/978-3-030-22048-8

Cover illustration: Yurii Perepadia © 123RF.com

This Palgrave Macmillan imprint is published by the registered company Springer Nature Switzerland AG.
The registered company address is: Gewerbestrasse 11, 6330 Cham, Switzerland

Contents

List of Contributors

Valtteri Arstila Department of Philosophy & Turku Institute for Advanced Studies, University of Turku, Turku, Finland

Maria Balcells Department of Philosophy, Bucknell University, Lewisburg, PA, USA

Adrian Bardon Department of Philosophy, Wake Forest University, Winston-Salem, NC, USA

Karin M. Bausenhart University of Tübingen, Tübingen, Germany

Marc Buehner School of Psychology, Cardiff University, Cardiff, UK

Natalja Deng Underwood International College, Yonsei University, Seoul, South Korea

Andrea Desantis Département Traitement de l'Information et Systèmes, ONERA, Salon-de-Provence, France
Integrative Neuroscience and Cognition Center, CNRS, Université Paris Descartes, Paris, France
Institut de Neurosciences de la Timone, CNRS, Aix-Marseille Université, Marseille, France

Sylvie Droit-Volet CNRS, UMR 6024, Université Clermont Auvergne, Clermont-Ferrand, France

Mark A. Elliott School of Psychology, National University of Ireland Galway, Galway, Republic of Ireland

Anne Giersch INSERM U1114, Pôle de Psychiatrie, Centre Hospitalier Régional, Université de Strasbourg, Strasbourg, France

Christoph Hoerl Department of Philosophy, University of Warwick, Coventry, UK

Timothy L. Hubbard Arizona State University, Tempe, AZ, USA
Grand Canyon University, Phoenix, AZ, USA

Tijana Jokic School of Philosophy, Psychology and Language Sciences, University of Edinburgh, Edinburgh, UK

Luke A. Jones Division of Neuroscience and Experimental Psychology, School of Biological Sciences, University of Manchester, Manchester, UK

Kristie Miller Department of Philosophy, School of Philosophical and Historical Inquiry and The Centre for Time, University of Sydney, Sydney, Australia

Nadia Paraskevoudi Brainlab, Cognitive Neuroscience Research Group, Department of Clinical Psychology and Psychobiology, University of Barcelona, Barcelona, Spain
Institute of Neurosciences, University of Barcelona, Barcelona, Spain

Ingmar Persson University of Gothenburg, Gothenburg, Sweden
The Oxford Uehiro Centre for Practical Ethics, Oxford, UK

Eric Pfeifer Catholic University of Applied Sciences Freiburg, Freiburg, Germany

Sean Enda Power Department of Philosophy, University College Cork, Cork, Ireland

Rodika Sokoliuk Centre for Human Brain Health, University of Birmingham, Birmingham, UK

Kristoffer Sundberg University of Oslo, Oslo, Norway

Rolf Ulrich University of Tübingen, Tübingen, Germany

Rufin VanRullen Centre de Recherche Cerveau et Cognition, Université Paul Sabatier, Toulouse, France

Argiro Vatakis Cognitive Systems Research Institute, Athens, Greece
Department of Psychology, Panteion University of Social and Political Sciences, Athens, Greece

Marc Wittmann Institute for Frontier Areas of Psychology and Mental Health, Freiburg, Germany

List of Figures

Introduction

Illusion and error have always been important points of entry for the study of mind. Philosophers since Plato have looked at sense-perception to elucidate the difference between knowledge and opinion, as well as the corresponding metaphysical difference between the world of appearance and transcendent reality. René Descartes thought that understanding positive and seemingly inextricable illusions like phantom limb syndrome were key to understanding mind-body interaction. We learn a great deal about how the brain generates 3-D images out of visual inputs by studying the mechanics of, say, the Necker cube illusion and the Müller-Lyer illusion. Our understanding of brain function has been heavily influenced by the study of the effects of specific brain injuries or lesions on sense-perception, personality, and cognition.

Immanuel Kant thought that the determination of timing and time-order is the key to mind's fundamental operations. He identified the organization of our own experience in time as the prerequisite for any coherent experience. Our experience is not a simple reflection of external reality. We experience a world in time—of ongoing processes and enduring structures—but our actual experience is only directly of the 'present' moment. Until the raw data of experience is temporalized, there can be no pattern or coherence in our experience; for example, successive experiences of a static state of affairs (as in circumnavigating a house, and thereby experiencing its four sides successively) need to be distinguished in kind from successive experiences relating to an event (as in experiencing the extended event of a ship leaving a dock, noting the different positions of the ship successively). As William James later put it, 'A *succession of feelings, in and of itself*, is not a feeling of succession'. The unified, coherent *temporality* of experience is a construction. The mechanics of the construction of temporal experience are exposed through systematic illusions regarding the passage of time, illusions of duration, time-order illusions, and related temporal illusions involving memory, representation, and cognition.

Judgments about the timing of experiences vary a great deal, depending on circumstances, content of experience, and the mental or emotional state of the perceiver. The study of psychological time includes the examination of the effects of unexpected experiences, emotional arousal, and altered states of consciousness, among other variables, on experienced duration. Judgments about time-order and synchrony are bound up, in complex and mutually influencing ways, with judgments about objective causal relations. In assessing (or mistaking) the order of events, the brain is faced with the challenge of integrating the data of different sensory modalities. Researchers studying perception and cognition must confront the binding problem—the puzzle of the deeply mysterious process by which the brain unifies all the different elements and modalities of input into a single, spatiotemporally unified picture of the world around us. Timing and time-order judgments lie at the center of the binding problem; in addressing it, experimental psychologists and neuroscientists face the philosophical problem of how to clearly reify aspects of the relation between objective time and psychological time.

This collection of original essays examines what temporal illusions can teach us about cognition and perception. Here, leading philosophers, psychologists, and neuroscientists study time in mind via the many ways our perception of time and temporality can diverge from reality.

Part I: The Passage of Time

An ancient question about the passage of time is whether, given what we know about physical reality, the passage of time itself should be considered an illusion. Researchers since the fifth-century BCE philosopher

Parmenides of Elea have strongly suspected that there is a dramatic gap between the perception of change and its reality; models in twenty-first century, post-classical physics typically do not include any notion of a privileged present moment, nor any time-value corresponding to what we might think of as the passage of time. Should we thus conclude that the sense of change, and the difference between past and future so pertinent to emotion and judgment, is a fiction?

Our best physical theories point toward a 'block' universe, including neither a privileged present nor an objective, dynamic transition of events from future to present to past. Yet many find it indisputable that in some respect it *seems* as though time passes. This has led some to argue that—on the assumption that time appears to pass—the passage of time must therefore be some sort of illusion. Philosopher Natalja Deng rejects the idea that there is an illusory phenomenology of temporal passage—not because time objectively passes and thus our experience of passage is veridical, but because there is, in fact, no phenomenology of passage. She explains the difference between the (alleged) experience of the *passage* of time and the experience of temporal *succession*.

Philosopher Kristie Miller agrees with Deng that we do not experience the passage of time and investigates how it is possible to form false beliefs about how things actually seem to us. How and why could one come to mistake or misdescribe one's own temporal phenomenology? Human languages almost always include metaphors alluding to movement through time. Is language itself responsible for our misconceptions about the nature of our own experience, or is this kind of linguistic expression itself a reflection of some deeper, adaptive cognitive mechanism?

Philosopher Maria Balcells argues that there is a way we can understand our experiences of the block universe as dynamic. Balcells compares the experience of veridical motion with motion illusions like the phi phenomenon. She argues that each can be described as stemming from discrete, static inputs. The controversy over how we could be subject to some sort of pervasive illusion that time passes is dissolved when we understand that there is nothing non-veridical about the dynamic experience of a so-called static universe.

One important area of research on time perception concerns the familiar way time can, subjectively, seem to pass more quickly under some conditions than others. Under some predictable conditions, timing judgments—that is, judgments about the duration of events—can be way off. Why is this, and what does it tell us about how the mind works? Psychology researcher and neuroscientist Luke A. Jones helpfully sorts out some different categories of timing judgments. Jones distinguishes between judgments about the duration of an experience and 'passage of time judgments', or one's alterable sense of rapidity of the passage of time. He notes that there is a different story to be told about retrospective judgments regarding timing, as opposed to prospective ones. Jones surveys research supporting these distinctions and points the way to future research on timing and time perception.

Part II: Duration

The questions of what time is and how we know about it are questions with a long history. Aristotle thought of time as merely the number of change, a kind of measure of change. St. Augustine noted that we commonly talk about differences in length between different times: some times are short, others long. He also wondered which change was being measured; for example, was time the measure of the movement of the stars? St. Augustine puzzled about simply calling time some measure of change. How, he wondered, could we measure time? Whatever time is, it is either past, present, or future. He thought the present had no temporal extent—it was merely a point; if so, in the present, there is no time to measure: we do not measure different lengths of time, all within the present. He also thought we could not measure the past and future: nothing past or future is in the world that we can go out and measure, as we might measure the distance between, say, a mountain and the sea; it is not even hidden in the world. Instead, what is past is gone and what is future is yet to come. So, how and what can we be measuring when we measure time?

For both Aristotle and St. Augustine, the aspect of time they are focusing on is not temporal passage, temporal order, or a moment of time. They focus on the temporal extent or *duration* of time. The nature of duration and our knowledge of it are still discussed. One way into such discussions is to consider illusions of duration. Illusions of duration are experiences in which some event appears to have a duration but does not have that duration. Examples seem easy to come by. Absorbed in work, bored at a talk, and running from a tiger: frequently, we spend time doing something, feel it lasts a certain length of time (or *duration*), only to discover it lasted a much longer or shorter time.

One task in cognitive science is to explain such experiences. Ulrich and Bausenhart examine an important example, the *oddball effect*. The oddball effect occurs when subjects are presented with several stimuli, each for the exact same clock duration (the amount of time measured by a clock); however, one stimulus—the 'oddball'—seems to last longer. The authors provide a comprehensive review of studies devoted to establishing and understanding the origin of this phenomenon. They overview the various experimental methods used to investigate the effect, potential limitations, and caveats associated with these methods. They also consider various studies that examine the generality of the oddball effect and explanations for the occurrence of this effect. They conclude that the current most plausible explanation assumes two independent mechanisms that exert opposite effects on the perceived duration: (a) low-level repetition suppression and (b) high-level stimulus expectations.

Duration is an extension or amount; it is a quantity of time. Another role illusions of duration play is in the investigation of other important kinds of experience. Persson uses the difference between felt duration and actual duration and how we retrospectively reason with both, to argue that for even the simplest physical pleasure and pain, their intrinsic value is not quantifiable, unlike their duration. Their intrinsic value is not quantifiable because their intensity is not quantifiable. Sensory intensities do not have even a rough position on the same scale in which their physical stimuli have a more precise position, as the duration of sensations have a rough location in time in which their stimuli have a more precise location. Consequently, there is an argument that demonstrates the non-transitivity of the relation of sameness as regards sensory intensity but is not applicable to their duration. This non-transitivity rules out the existence of a numerical scale along which sensory intensity can be even roughly plotted.

For there to be illusions of duration, two conditions must be met: there must be an actual duration and there must be an experience of duration. If we do not know the amount of time, then we cannot use it to tell if the experienced duration is different to it, and so if there's an illusion. Similarly, if we do not know what duration is experienced, then, even if we know the actual time, we cannot know if the experience is different to it, and so there's an illusion.

So, an important question about illusions of duration is how to measure experienced duration, and what it is we measure. One possibility is that we measure duration through some internal clock. The cognitive scientist Droit-Volet reviews the recent studies on emotion and time perception and passage of time awareness, and how they question the models of internal clock. She notes that time seems to exist as a reality independent of us, as a physical feature of an objective world that we can measure with an internal clock. However, numerous studies have shown how easily our time estimates can be distorted by our emotions. Under the influence of emotion, time often seems to speed up or slow down. Thus, the duration is a pure product of our emotions and of the upheavals they produce in our bodies and minds. This is the paradox identified by Droit-Volet and Gil (2009): if we possess a sophisticated mechanism for measuring time, why are our time estimates so variable?

Duration need not only play a role in temporal illusions that are obviously of duration—involving obvious experiences of duration. It can also play a role in other temporal illusions. If St. Augustine is correct, when we measure duration, we do not measure something even *hidden* in the present. However, perhaps a hidden duration may explain other kinds of illusion. One possibility is for illusions around our experience of events in the physical world. There is a duration between the events we seem to experience and the associated sensory stimuli and neural activity of that experience. This duration leads to an argument that philosophers of time call the *time-lag argument*.

The philosopher Sundberg notes that time-lag argument forces us to acknowledge that all perceptions are perceptions of the past. While the spatial distance between perceiver and perceived typically is obvious, the temporal distance usually remains hidden from the perceiver. This temporal distance provides interesting challenges for perceptual realism. If objects and events in the world are presented to us *as* simultaneous with the experience, then the experience is illusory. If we deny that the past exists, then all perception is, like a hallucination, perception of what no longer exists. This chapter argues that there are in fact several time-lag arguments and that their importance for philosophical theories of perception has been underestimated. It is argued that they provide important means of evaluating recent theories about perception.

Another way to investigate illusions of duration is to consider them in a variety of circumstances, for example, meditative states and altered states of consciousness. The cognitive scientists Wittmann et al. note review work they have recently done on varying durations. They propose that felt time is modulated together with the awareness of the bodily self. For inter-individual differences in the perception of time, they showed that individuals with higher levels of self-reported impulsivity felt more irritated in a waiting situation and in turn overestimated its duration. Regarding intra-individual differences, they showed that silence after body-related depth relaxation led to longer duration estimates than the same period of silence following a seminar. Their studies concerning longer duration in the minute range are paradigmatic because ecologically valid situations are sensitive to inter- and intra-individual variations in subjective time.

Finally, duration is an amount or measure of change. So, with experiences of duration, we might ask, which change is measured in our experience of duration? We need to know this if we are to know an experience is an illusion. If we do not know the change measured, then how can we hold that the experience of that measure is wrong, and so an illusion?

The philosopher Power argues that although we commonly make mistaken judgments about duration, we ought not to conceive of these judgments as illusions of duration. First, it is not enough for an illusion that experience merely contributes to mistaken judgments. To be an illusion, the experience itself must be mistaken. Judgments from experience can be mistaken because of one's background beliefs, which can be mistaken independent of the experience's accuracy. Second, there is rarely a reason to explain mistaken judgments of duration from experiences of duration in terms of mistaken experiences. A better explanation is that one mistakenly interprets experienced duration in terms of an inappropriate standard—that of standard time. Whatever it is that we might be experiencing in experiencing duration, there is little reason to think that it should be described and evaluated in terms of standard time.

These two claims lead to the position that there is rarely (if ever) reason to explain mistaken judgments of duration, even driven by experiences of time, as being due to illusions of duration. Power suggests an alternative account of such experiences of duration.

Part III: Simultaneity and Temporal Order

An illusory aspect of timing is that of the percept of synchrony in everyday multisensory events when physical synchrony is the exception rather than the rule. This illusory temporal percept has intrigued many different disciplines. Kant thought this sort of phenomenon is critical to understanding the mind. In experimental psychology, tasks measuring the percept of simultaneity and temporal order have been utilized in order to define when and how two events or sensory inputs are simultaneous or successive in time and, in the case of succession, what is the exact event/modality order. These measures allowed for the definition of a threshold for simultaneity (where multiple inputs can be fused into a unified percept or the point where the perceiver cannot successfully define the event/input order) and succession (where the minimum time required to perceive two events/inputs separated in time is present). These thresholds turn out to be variable and dependent on many factors.

Paraskevoudi and Vatakis present an overview of the recent research findings and developments on the illusory perception of multisensory event synchrony. Although many issues still remain largely unknown, evidence suggests that synchrony perception is modulated by stimulus features, the individual's age and prior experience, and exhibits abnormalities in pathology. Paraskevoudi and Vatakis review the factors modulating our ability to align sensory inputs in time and summarize recent findings on the behavioral and neural correlates of perceptual synchrony.

Understanding event order and successiveness provide insight into our ability to perceive time continuity. Phenomenologists have long reported a breakdown of the temporal structure of consciousness in patients with schizophrenia, with a disruption of the sense of time continuity. Anne Giersch, in her chapter, summarizes the models in phenomenology and experimental psychology that have been proposed to explain how we attain a sense of time continuity. She also describes a series of experimental results that reveal timing difficulties in patients with schizophrenia, both at unconscious and conscious levels, with a surprising high time resolution at the unconscious level. Finally, she discusses how the idea of closed-loop systems and close interaction between non-conscious and conscious mechanisms may help to understand the underpinnings of the feeling of time continuity.

It is yet unclear how the perception of simultaneity and successiveness relates to the present moment and event causality. Desantis and Buehner discuss the relation between time perception and causality. Traditionally, it has been considered that causality is partly inferred from temporal relations between events with people being more likely to report two events as causally linked if they follow each other closely in time rather than if they are separated by a long delay. Desantis and Buehner review and discuss a complementary literature proposing that time perception is in turn guided by people's assumptions about the causal connection between events. This implies that time perception is shaped by the combination of sensory and internal information such as our prior causal expectations. Finally, a discussion is devoted to how Bayesian models of uncertainty reduction may offer an exploratory hypothesis to understand how causality structures time perception.

Part IV: Cognition and Representation of Temporal Phenomena

The common factor of the chapters in the last section is that they all tackle the issue of temporal illusions on a context that goes beyond the perception of temporal phenomena. The section begins with Mark A. Elliott's chapter, which addresses the nature of time as a psychological construct. In particular, Elliott is interested in the immediate experiences, especially of humans, and how such experiences relate to the serial event structure and the sense of the passage of time. In the end, he concludes that psychological time is best understood as a measure of dynamical psychological systems.

Christoph Hoerl suggests that some of the purported cases of temporal illusions are not temporal illusions. Though his reasons differ from those discussed in the following chapter by Hubbard, Hoerl proposes that temporal binding, which is the case he focuses on, is not a temporal illusion because it is not an illusion in the first place. This follows from the relational view of perception he endorses. In this view, experiences do not present us with a content of how things are in the world, as the discussion on illusions assumes, but they put us in contact with the external world.

Many temporal illusions appear to be intrinsically linked with the perception of spatial properties. For example, the kappa effect demonstrates how the distance a stimulus moves influences the perceived duration of its movement, while the looming effect shows that subjects underestimate the time of contact of an approaching object. Whether all these so-called spatiotemporal illusions are indeed temporal illusions or merely spatial illusions is an open question. Often the answer depends on the way in which they are explained. These phenomena, factors that modulate the effects, and several explanations for them are reviewed in Timothy L. Hubbard's chapter.

Many perceptual phenomena exhibit some sort of rhythmic nature. Rodika Sokoliuk and Rufin VanRullen review these phenomena and argue that the intrinsic rhythmic activity in the brain has two kinds of effect on visual perception. In the case of rhythmic perception, the neural oscillations modulate the visibility of the stimuli. That is, whether the stimuli are perceived or not depends on the suitable match between the time when the stimuli are presented and the phase of ongoing neural oscillations. In the case of discrete perception, the timing of the stimuli is affected. For instance, two non-simultaneous stimuli can be perceived as simultaneous or non-simultaneous depending on the phase of neural oscillations during which they are presented.

Finally, Valtteri Arstila addresses the question as to whether the subjective time order reflects the order in which experiences are processed in the brain or the order in which the events occur in the world. The first position is often called the brain-time view and the latter the event-time view. However, Arstila argues that there are two different versions of each view and that, theoretically, the two possibilities are not mutually exclusive.

Part I

The Passage of Time

1

One Thing After Another: Why the Passage of Time Is Not an Illusion

Natalja Deng

Is the passage of time an illusion? This has meant many different things to different people. As rightly noted by Gruber, Smith, and Block (2018), both 'passage of time' and 'illusion' invite a multitude of interpretations. Moreover, it's a question that cries out for interdisciplinary efforts (such as this volume), since numerous disciplines—including physics, cognitive science, developmental psychology, evolutionary biology, and philosophy—are potentially relevant. It's also a very timely question (no pun intended). For example, time has recently taken center stage at the frontiers of physics. Physicists working toward a theory of quantum gravity have had to re-examine some very basic assumptions about time, such as whether it exists, and if it does, whether it is fundamental or in some sense emergent (Callender, 2010).

The title of this chapter might lead one to expect a theory of what temporal passage is, and an argument for the view that it's not an illusion. In fact, my aims are a bit more narrowly circumscribed. First, I'll describe what 'passage' stands for in current philosophical usage. For there to be

N. Deng (✉)
Underwood International College, Yonsei University, Seoul, South Korea

V. Arstila et al. (eds.), *The Illusions of Time*,
https://doi.org/10.1007/978-3-030-22048-8_1

an illusion of passage, one would have to answer 'Yes' and 'No', respectively, to the following two questions: '(a) Does time seem to pass?', and '(b) Does it pass?'. Assuming for present purposes that the answer to (b) is indeed 'No', I'll suggest that the answer to (a) is also 'No'. The upshot will be that there is no illusion of passage: All we perceive is one thing after another.

1 'The Passage of Time'

The phrase 'the passage of time' (or 'the passing of time', 'temporal passage', or 'temporal flow') has a particular meaning in philosophical debates about the nature of time. Passage is a (putative) feature that time is supposed to be able to have or lack, while in any case existing. That is, we're not here concerned with the possibility that time may not exist, nor with the possibility that it may not be fundamental. (So this is not directly about those basic questions arising from within current research in physics.) Rather, the question is whether or not time, a real feature of the world, itself has the feature of being such that it passes. This question may well sound a little strange at first. Ordinary usage may not allow much leeway between time existing and time passing. But philosophical reflection has produced a distinction between two ways that time could in principle be, namely dynamic (such that it passes) or non-dynamic. To illustrate, consider a few ideas for what time's passing might consist in, according to different dynamic views of time: Time would pass in this sense if only one time was (ever) real, so that times came into and went out of existence constantly; or, if only the past and the present were real, so that reality as a whole grew constantly; or, if future events constantly became less and less future, until they became present and then more and more past. All these views have in common that they take the metaphors about time being like a river, exhibiting some kind of sui generis movement of its own, very seriously. On these views, time is not much like space, at least not in this respect. Space doesn't pass, after all.

All these dynamic views (versions of the A-theory) are opposed by the non-dynamic, or block universe view (also called the B-theory). The block universe view denies that time does anything like passing in this sense. At least in this respect, time is no different from space.

Though it's the product of philosophical reflection, the distinction between dynamic and non-dynamic views of time is very intuitive at root. After all, the metaphors that drive the distinction are commonplace in many languages (though the details reveal many fascinating differences as well, see Sect. 4). We commonly speak as if we were moving through time, or as if time and events were moving past us. Is time really like this, or does it lack the dynamic features the metaphors point to? It's not surprising that something like this question has left traces throughout the history of philosophy. The British idealist philosopher J. M. E. McTaggart brought it into sharp focus (arguing that for time to exist it would have to be dynamic, but it can't be, so it doesn't exist), but its history stretches back at least to Heraclitus and Parmenides.

Its history is also intertwined with the history of time in physics. The block universe view has often been associated with time in relativity theory. One not so good reason for this is the four-dimensional nature of relativistic spacetimes. On a spacetime diagram, past, present, and future are seemingly already there. There is no dynamism, just the block universe. This isn't a good reason to associate the B-theory with time in relativity theory, mainly because there are spacetime formulations of non-relativistic physics too. But there are other reasons. In particular, all the dynamic views seem to rely on there being a 'now' that is the same everywhere in space. But relativity teaches us that there is no objective notion of distant simultaneity—no notion of the same time at different spatial locations. What's simultaneous with what depends on one's state of motion. So this makes trouble for any dynamic view that says that what's dynamic is some global present, ever moving on, or constituting the edge of what exists, or the only time that exists. By elimination, the block universe view emerges looking vindicated. Philosophers and physicists alike have expressed the sense that relativity theory shows time to be very much unlike what the metaphors suggest. Hermann Weyl famously remarked that '[t]he objective world simply is, it does not happen. Only to the gaze of my consciousness, crawling upwards along the life line of my body, does a section of this world come to life as a fleeting image in space which continuously changes in time' (Weyl, 1949). Admittedly, this almost sounds like a dynamic view of time again, but only almost: Time itself is non-dynamic; 'only to the gaze of my consciousness' does temporal reality present itself as dynamic. Similarly, Arthur Eddington said that '[e]vents do not happen; they are just there and we come across them' (Eddington, 1920). Albert Einstein himself went so far as to claim spiritual significance for this view of time. Shortly after the death of his friend Michele Besso, he wrote: 'Now he has also gone ahead of me a little in departing from this peculiar world. This means nothing. For us believing physicists, the division between past, present and future has only the significance of a stubbornly persistent illusion' (Einstein, 1972).

2 'Illusion'

This brings us to the term 'illusion'. Roughly, I'll take this to mean a mismatch between how things seem, and how they are. In this case, the mismatch would be between time's being non-dynamic and yet seeming dynamic. Thus, the question at hand is whether two things are the case: Time doesn't pass (in the sense of the A versus B debate), and yet it seems to (in that same sense). (From now on 'passage' will denote passage in the sense of the A versus B debate.) Let's call the corresponding questions '(a) Does time seem to pass?' and '(b) Does time pass?'. And let's assume for present purposes that time doesn't pass, and concentrate on (a): Does it seem to? that is, do we have passage phenomenology?

Is passage something one should even expect to show up in experience? Some are skeptical about this. Suppose that the passage of time turns out to be the 'birthing' of new elements of a causal set, i.e. discrete spacetime points. This is a suggestion recently made by some of the physicists working on causal set theory, which is intended to be a stepping stone toward a theory of quantum gravity. Isn't the growth of causal sets the last thing one should expect ordinary experience to be responsive to? Since when does our phenomenology track the most fundamental layer of reality as revealed by our best physical theories? How likely is it that when I glance at my watch, I somehow intimate the stochastic growth process constituting spacetime itself? On the other hand, those physicists writing about this seem to be motivated precisely by the link to experience. For example,

Rafael Sorkin writes, '[S]equential growth [...] provides an objective correlate of our subjective perception of "time passing" in the unceasing cascade of birth-events that build up the causal set' (Sorkin, 2007). More generally, as the Weyl and Einstein quotes already suggest, it's somewhat natural to think that a non-dynamic temporal reality is at odds with how things seem. There's something striking about the claim that time doesn't pass, and it's natural to suppose the reason is that it seems to.

Forget passage for a moment and just consider our experience of time as such. That is, just consider our experiences as of temporal duration, of events as occurring in a particular order, and as following on from or succeeding one another. ('As of x' is meant to leave it open whether the experience is veridical or illusory, that is, one can have an experience *as of* x without there being any x.) In order to understand the question at hand, namely '(a) Does time seem to pass?', one has to keep in mind that both sides to this dispute can and typically do allow that we have temporal experiences. Whether we have experiences as of time passing or not, we definitely have experiences as of succession, order, and duration. However, there are those who answer 'Yes' to (a) think that we only have experiences as of succession, order, and duration, *by* having experiences as of time passing, or *because* we have experiences as of time passing. For them, it's the impression of dynamicity that *enables* or *underlies* the impression of succession, order, and duration. Some speak of 'animated change' in this connection (Paul, 2010). So the question is, do we experience temporal features of the world only by experiencing time's passing? Does our temporal phenomenology involve passage phenomenology? Or just impressions of succession, order, and duration?

So far, I've talked of 'experiences', 'impressions', and 'phenomenology'. But when it comes to time, there is one rough distinction that it's useful to make at the outset. This is the distinction between our perceptual experiences (or perceptions) and experiences in a broader sense (let's call these EXPERIENCES). Our (perceptual) experiences of time unfold over small timescales. They include seeing the second hand move around the clock, hearing a succession of notes, or feeling a raindrop run down one's neck. Our EXPERIENCES unfold over a wide range of timescales longer, all of them longer than the timescales relevant to (perceptual) temporal experiences. The following are all examples of temporal

EXPERIENCES: Noticing that the hour hand has moved around the clock, reflecting on how a city has changed over the years, and feeling as if time somehow passes more quickly as one gets older. These EXPERIENCES are quite varied, and it would be no easy task to say more precisely what distinguishes them from (perceptual) experiences, or how the two categories are related. But (perceptual) experiences of time are a good starting point. Does seeing the second hand move involve seeing the passing of time?

3 Why We Don't Perceive Time's Passing

With this intended meaning of (a) firmly in mind, it would be reasonable to worry that it won't be easy to decide on an answer. But here is a consideration that (indirectly) speaks in favor of 'No'.

Consider what happens when one answers 'Yes', and when one combines that with the claim that time does not pass (answering 'No' to (b)). This is the view that the passage of time is an illusion. It lends itself to a methodology whereby one looks to insights in cognitive science work on perceptual illusions and sees whether they can be applied here. One influential proposal of this kind is due to philosopher Laurie Paul. Paul asks us to consider cases of illusory perception of motion, such as films, time-lapse photography, or old-fashioned flipbooks. Another central example for Paul is (what in the philosophical literature has become known as) the 'color phi' experiment. If a subject is presented with appropriately spaced and timed flashes of differently colored dots on opposite sides of a screen, it can seem to them as though a single dot is continuously moving back and forth and changing color mid-trajectory. So, here we have static inputs of the form 'red flash left' and 'green flash right', leading to dynamic outputs, namely the impression of a single dot moving back and forth, changing color. The idea is that something analogous might be underlying our sense of passage. Although the inputs are 'static', passage-less facts, like 'object O has property P1 at t1' and 'object O has property P2 at t2', what we actually perceive is an animated change in O, in which temporal passage is the key ingredient. In a nutshell, Paul's proposal likens time in the block universe to something like a film, giving rise to illusory passage phenomenology.

I think this proposal fails in an interesting way. Recall that all parties to the debate agree that there is change and succession in the block universe, and that our temporal phenomenology in a block universe includes experiences as of change and succession. The question is how to explain the additional illusory element of animation underlying this phenomenology. But on closer inspection, the proposed explanation suggests that there wasn't this additional element in the first place. Consider again the analogy between the 'static' block universe facts, and the static inputs in the color phi experiment. In the latter case, the brain's limited powers of discrimination lead to the gaps between images not being perceived. The problem is that facts about what happens when in the block universe aren't 'static' in the same sense: These facts are not themselves in time. Moreover, there need not be any gaps in the block universe. Whether or not time is continuous is orthogonal to the A versus B debate. But in order for the analogy to work, we have to pretend that there is discreteness in the block universe, or rather a 'gappiness' of the kind found in the color phi phenomenon.

Suppose there was. Then, plausibly, what we would have explained would be an illusion of continuity—of continuous motion and persistence. In a gappy block universe, the brain produces experiences as of non-gappiness. But if that's all that needed explaining, then there wasn't an additional illusion of 'animation' (A-theoretic passage) to our temporal phenomenology in the first place. Only experiences as of continuous motion, of order, and duration and succession.

Note that none of this bears on whether our temporal phenomenology is in some other respect illusory. In particular, what I've argued is compatible with our perception of motion and other change involving an illusion of continuity (though it's also compatible with this not being the case).[1] It's worth noting this, because some of the same cognitive science work that Paul draws on has been presented as evidence for claims of that

[1] Here's what would have to be the case for our perceptions of change to involve an illusion of continuity: (i) our perceptions of change would have to be as of continuous change and persistence, so that the world would seem continuous, *and* (ii) that impression would have to be mistaken, for example, because time would be discrete rather than continuous (or something similar).

kind. Consider, for example, the position defended by Gruber et al. They distinguish between a upper-level and a lower-level 'flow of time', by which they mean two distinct aspects of our temporal phenomenology. The upper-level component they describe as a false cognition, and as having (what I'm calling) A-theoretic content: 'There is an upper-level component, the past/present/future, which entails a moving present' (Gruber et al., 2018, p. 133). The lower-level component they describe as 'the experience of events flowing or happening, such as occurs when a volcano erupts' (Gruber et al., 2018, p. 133); '[i]t involves visual, auditory, olfactory, and other sensory modalities' (Gruber, Bach, & Block, 2015, p. 9). Since the volcano example is presumably generic, by the 'lower-level component of the flow of time', they just mean our perception of change, rather than a perception of passage. (Recall that all sides to the dispute I'm focusing on typically agree that our temporal phenomenology involves perceptions of change, but that only one side takes these to involve passage phenomenology.) Now, Gruber et al. argue that the lower-level component is illusory, because it is the 'perceptual completion that provides continuity to discrete observation' (Gruber et al., 2018, p. 134). They go on to cite the phi phenomenon as part of the evidence for this claim, and they also make a connection with Paul's work: 'These views as to the illusory nature of motion are in good agreement with philosophical views. Paul (2010) argued that, in general, the flow associated with motion (a change of location of a persisting object) is merely an effect of the brain' (Gruber et al., 2018, p. 135). The point to note is that none of my arguments bear on whether we should think that our perception of change involves an illusion of continuity. My topic is whether temporal phenomenology is illusory in a different respect, namely in that it involves passage phenomenology. Incidentally, this is also the topic of Paul's proposal (as well as that of Christoph Hoerl's and Simon Prosser's writings in this area, who are also cited in this context by Gruber et al. (2015, pp. 10, 12; 2018, p. 128)). And this is so in spite of the fact that, as we've seen, Paul makes use of a lot of the same experimental results as Gruber et al., and in spite of the fact that she builds her proposal on an analogy with illusory perceptions of continuous motion.

We've seen one (indirect) reason to think we don't have passage phenomenology. Here are some more. Recall that in this discussion 'passage' means an updating of which time is 'special'; for example, which time is the one time that exists, or the latest time that exists, or the one time that is objectively present. Thus, in order for it to seem like time passes, one time and its contents should seem 'special'. Does it? Suppose it's Tuesday today. Do your Tuesday experiences have a special phenomenal character? Suppose you spent all of today, and all of yesterday, in a completely red room (Skow, 2015, p. 203), meditating while looking at the red walls. And suppose you're able to produce perfectly accurate descriptions of your phenomenology. Will your descriptions of your experiences on Monday and on Tuesday differ in some way? Do the Tuesday experiences feel somehow more vivid or more lustrous or in some other way 'special'? Presumably not: Instead, the descriptions will be exactly the same. Presentness makes no difference to how things look or feel or sound.

Of course, perceiving something is different from remembering or anticipating it. So, in that sense, there is a connection between presentness and experience. At each time, one perceives only some things and not others. But this is a kind of 'specialness' that is compatible with the block universe: Each time is 'present' and therefore 'special' at itself, just like each spatial location is 'here' at itself. No illusion there.

What, then, do we perceive, if not the passing of time? We perceive temporal succession, temporal order, and temporal duration. Our temporal phenomenology includes all of that. We are perceptually aware of changes and of persistence (things existing over time). These perceptual experiences present the world in a way that is perfectly compatible with the block universe view. If this is right, then perceptual contents are not best described using notions such as 'past, 'present', or 'future'. When we perceive a succession of events, we don't experience some things as present while experiencing other things as past (and perhaps still other things as future). Rather, perceptual contents are best described using notions like 'before' and 'after'. When we perceive a succession of events, we simply experience one event as happening after the other.

This way of thinking about temporal experiences may enjoy additional support from another direction. In particular, it may be well aligned with certain independently plausible positions in a different philosophical debate about the nature of temporal experience. This other debate has a long and even more illustrious history than the one we're concerned with,

with key historical protagonists including William James and Edmund Husserl. There is some disagreement about how best to set it up (see e.g. Phillips, 2014). Descriptions of what's at stake range from 'how is temporal experience possible?', to 'how does the temporal structure of experience relate to the temporal structure of what is experienced?', to simply 'what is the nature and content of temporal experience?'. Interestingly, some approaches to these questions center on the idea that an experience of something temporally extended is itself temporally extended. The experience itself takes time to unfold—in fact it takes as much time as the process that it is an experience of. Arguably, this makes it easier to understand the nature of temporal experience. We no longer have to ask how it is that multiple individual experiences, each succeeding the other, can add up to an experience of succession. Instead, we recognize that the fundamental experiential unit is itself temporally extended, and use this to explain how there can be an experience of a temporally extended content. For present purposes, what's interesting about these approaches is that they may be particularly well aligned with the view that we don't perceive time as passing. Christoph Hoerl argues that if the experience takes as long as what's experienced, then 'there is nothing that falls within the experience to which the notion 'a short while ago' could apply, because there is nothing experienced as having happened at a time that pre-dates the onset of the experience' (Hoerl, 2009). Temporal perceptual contents are then not temporally view-pointed, in the way they would be if we experienced the passing of time.

4 Do We Think Time Passes?

In the growing philosophical literature on negative answers to (a) (Does time seem to pass?), a common thread is that if you answer 'No', you have a lot of explaining to do. In particular, you owe an explanation for why people were ever attracted to the view that time passes (why people tend to answer 'Yes' to (b)), and you owe an explanation for why people tend to think they experience time as passing (why people tend to answer 'Yes' to (a)).

Kristie Miller and co-authors explore different possible relations between these explanations (Miller, Holcombe, & Latham, forthcoming). Perhaps one and the same mechanism explains both (our view that time passes and our view that we experience time as passing). Or maybe something makes us think time passes, and we sub-personally infer from this belief that we experience time's passing. Either way, the proposed explanation of why we think time passes is roughly this. First, we conceptualize time as having a direction, due to the many temporally asymmetric phenomena are all around and within us. This is reflected in the fact that many natural languages are direction-friendly, in the sense that they make it easy to express a conception of time as directed. Moreover, our lives are temporally ordered series of temporal perspectives, each of which includes mental representations of decisions at earlier times and deliberations about decisions at later times (Ismael, 2010, 2012). To allow us to express how things are from each temporal perspective, languages need to be minimally passage-friendly: They need to include tense, or at least deictic adverbials like 'tomorrow' or 'yesterday'. Some are even substantially passage-friendly, in that they include moving time ('Christmas is approaching') or moving ego metaphors ('He's nearing his 70th birthday'). There is much evidence that language influences the way we think. So these facts about natural languages can in turn explain why we tend to think time passes.

This is a fascinating explanation. Nonetheless, in what remains I'll suggest that one can take on board what's most convincing in it without assuming so uniform an explanandum as that 'we' tend to think time passes. This assumption is, as Miller et al. note, widespread in this area of philosophy. But there are good reasons not to make it.

To begin with, as their article shows indirectly, there is fascinating variation in how different languages treat time. Not all languages are 'substantially passage-friendly'. Yet, it's precisely these moving time and moving ego metaphors that drive the A versus B debate. A-theories posit temporal passage in a way that is supposed to make these metaphors literal; the block universe view (B-theory) says time doesn't pass, in that same sense. Meanwhile, the universal 'minimal passage-friendliness', which characterizes all languages, barely goes beyond 'direction-friendliness'. The former allows one to express degrees of pastness or futurity with respect to the present utterance (whether via tense or not); the latter 'make it easy to distinguish what has been the case from what is the case, and what will be the case'. This too is revealing, because direction-friendliness can't by itself dispose one toward an A-theoretic conception (on which time passes), for the simple reason that the directionality issue is different from the passage issue. It's perfectly possible to think of time in the block universe as having a direction, namely from earlier to later.

This leads to the main reason for caution: The A versus B debate is far more metaphysically involved and theoretical than its close connection to the metaphors make it seem. Using the moving time metaphors available in one's language is not the same as being committed to a particular metaphysical conception of time. Of course, Miller et al. are in fact suggesting that the former causes the latter. But does the latter actually obtain, in addition to the former? A lot of philosophical work goes into formulating both the (A-theoretic) idea that the metaphors could be getting at some literal truth about time, and the (B-theoretic) idea that time could exist without passing in this sense. To get at the subtle differences between these requires much metaphysical engineering. It's true that once one is in the business of pursuing these views (e.g. when doing metaphysics), A-theoretic ideas tend to have some intuitive appeal. But that's not to say that human beings in general walk around with tacit A-theoretical models in their heads.

'Is the passage of time an illusion?' has meant many things to many people. Metaphysicians of time are one relevant group with a specific, captivating, and elusive meaning in mind. When the question is understood this way, the answer is that there is no illusion of passage.

Acknowledgments This chapter was partly written while I was supported by the Yonsei University Future-Leading Research Initiative 2018 (2018-22-0100).

References

Callender, C. (2010). Is time an illusion? *Scientific American, 302*, 58–65. https://doi.org/10.1038/scientificamerican0610-58

Eddington, A. (1920). *Space, time, and gravitation: An outline of the general relativity theory*. Cambridge: Cambridge University Press.
Einstein, A. (1972). *Einstein, A. & Besso, M. correspondance, 1903–1955* (P. Speziali, Trans.). Paris: Hermann.
Gruber, R., Bach, M., & Block, R. (2015). Perceiving two levels of the flow of time. *Journal of Consciousness Studies, 22*(5/6), 7–22.
Gruber, R., Smith, R., & Block, R. (2018). The illusory flow and passage of time within consciousness: A multidisciplinary analysis. *Timing and Time Perception, 6*, 125–153.
Hoerl, C. (2009). Time and tense in perceptual experience. *Philosophers' Imprint, 9*(12), 1–18.
Ismael, J. (2010). Temporal experience. In C. Callender (Ed.), *Oxford handbook on time* (pp. 460–482). Oxford University Press.
Ismael, J. (2012). Decision and the open future. In A. Bardon (Ed.), *The future of the philosophy of time* (pp. 149–169). Routledge.
Miller, K., Holcombe, A., & Latham, A. (forthcoming). Temporal phenomenology: Phenomenological illusion versus cognitive error. *Synthese*. Published online 23 February 2018. https://doi.org/10.1007/s11229-018-1730-y
Paul, L. (2010). Temporal experience. *Journal of Philosophy, 107*(7), 333–359.
Phillips, I. (2014). The temporal structure of experience. In D. Lloyd & V. Arstila (Eds.), *Subjective time: The philosophy, psychology, and neuroscience of temporality*. MIT Press.
Skow, B. (2015). *Objective becoming*. Oxford: Oxford University Press.
Sorkin, R. (2007). Relativity theory does not imply that the future already exists: A counterexample. In V. Petkov (Ed.), *Relativity and the dimensionality of the world*. Springer.
Weyl, H. (1949). *Philosophy of mathematics and natural science* (Based on a translation by Olaf Helmer). Princeton: Princeton University Press.

2

Does It Really Seem to Us as Though Time Passes?

Kristie Miller

1 Temporal Phenomenology

Let us call our experience of temporal properties and relations our *temporal phenomenology*. This phenomenology includes our experiencing events to be in a temporal order, of a temporal duration, and separated by a temporal distance. It includes our experience of events as occurring in succession. It may include our experiencing time as having a direction (from earlier, to later). Perhaps it also includes our experiencing the past as somehow fixed and immutable, the present as especially salient and real, and the future as as-yet unsettled (though I will remain silent on whether it does). It is also often said that we experience time *as passing*. This is often described as an experience of future events coming ineluctably toward us, and then reaching us and becoming present, and then receding into the past, and is known as the phenomenology as of temporal passage. (I return to the 'as of' shortly.) The following are some characterizations of this phenomenology.

K. Miller (✉)
Department of Philosophy, School of Philosophical and Historical Inquiry and The Centre for Time, University of Sydney, Sydney, Australia

V. Arstila et al. (eds.), *The Illusions of Time*,
https://doi.org/10.1007/978-3-030-22048-8_2

> We are not only aware of the passage of time when we reflect on our memories of what has happened. We just *see* time passing in front of us, in the movement of a second hand around a clock, or the falling of sand through an hourglass, or indeed any motion or change at all. (Le Poidevin, 2007, p. 76, emphasis added)
>
> the flow of time, or passage, as it is known, is given in experience, that it is as indubitable an aspect of our perception of the world *as the sights and sounds that come in upon us*, even though it is not the peculiar property of a special sense. (Schuster, 1986, p. 695, emphasis added)

In what follows, I focus on just one aspect of our rich temporal phenomenology: our phenomenology as of temporal passage. That phenomenology is usually taken to be perceptual, in that it seems to us to represent the way the *world* is (as opposed to, say, imaginings, which seem a certain way to us, but don't seem to be about the way the world is). Wherever we have perceptual phenomenology, we can ask a range of questions, including:

(a) Phenomenal content question: What are those experiences *like*—what is the *content* of the experiences in question? (i.e. what do those experiences say the world is like?)
(b) Mode of presentation question: Is there some special way that some contents of experiences are presented to us?
(c) Cognitive mechanisms question: What cognitive mechanisms are responsible for generating those experiences? (i.e. what brain processes cause these experiences, and which brain processes subserve, or realize, those experiences?)
(d) Evolutionary explanation question: What evolutionary explanation is there for our having those experiences?
(e) Veridicality question: Are those experiences veridical? (i.e. is the world the way those experiences say it is?)

Philosophers have been interested in all five questions with regard to our temporal phenomenology generally, and our phenomenology as of temporal passage specifically.

Most philosophers have begun by considering the phenomenal content question. Since we want to be able to describe the content of mental states without prejudging the question of whether that content is veridical, we talk about a mental state having content *as of X*, by which we mean that the mental state represents that things are *as though* X is the case. This leaves it open that X might, or might not, in fact be the case. In the case of temporal phenomenology, the mode of presentation question arises when we wonder whether the very same phenomenal contents can be presented to us as future, present, or past, or whether, instead, it is part of the content of an experience that said experience is presented to us as being past, present, or future. By and large, I am going to try and sidestep the mode of presentation question and just focus on the remaining four questions: That is more than enough to be getting on with!

Philosophers have typically answered the phenomenal content question by assuming that we have experiences that are *as of* time passing. Whether these are also experiences *of* time passing then depends on whether time passes, and whether its passing is appropriately connected to our having those experiences: It depends on the answers to the cognitive mechanisms, evolutionary explanations, and verticality questions. It is, then, these three questions that have been these philosophers' predominant focus.

2 The Veridicality Question

Having answered the phenomenal content question by supposing that we have phenomenology as of temporal passage, philosophers frequently attempt to answer the veridicality question before they answer either the cognitive mechanism question or the evolutionary explanation question. That makes some sense. We would expect to search for somewhat different cognitive mechanisms to explain our phenomenology as of passage on the assumption that said phenomenology is veridical (because time passes) as opposed to the assumption that it is not veridical (because time does not pass). After all, based on the former assumption, our phenomenology is responsive to something in the world and is tracking that thing. Based on the latter assumption, our phenomenology is illusory. Since we would expect a mechanism that is successfully tracking something to behave rather differently from an illusion-generating mechanism, determining which kind of mechanism we are looking for, before we look for it, makes some sense.

Moreover, we would expect a somewhat different evolutionary explanation for the presence of a mechanism that successfully tracks the presence of temporal passage, than for a mechanism that generates an illusory phenomenology as of passage. In either case, of course, we will assume that the mechanism confers evolutionary benefits. But in the latter case we might expect the generation of such *pervasive* illusory phenomenology to be an unintended by-product (a spandrel) of the selected-for mechanism (rather than merely the output of a mechanism which, due to various evolutionary trade-offs, issues in occasional mistaken outputs).

So, although this approach is by no means the only reasonable one, it makes some sense to answer the veridicality question *before* one attempts to answer the cognitive mechanism or evolutionary explanation questions. Given this, we find philosophers predominantly divided into two groups.

The first group of philosophers holds that time does not pass, and hence concludes that our phenomenology *as of* passage is not veridical: It is not a phenomenology *of* passage at all. We can call this view the *Phenomenal Illusion Thesis*. I consider that view in Sect. 2.2. Philosophers endorse this view largely (though not entirely) because of evidence from physics showing that temporal passage does not appear in contemporary physics. This is generally agreed, amongst physicists (and many philosophers), to be strong evidence that time does not pass.[1] By contrast, there are those who suppose that time does pass, sometimes in part because it allows us to answer the veridicality question in the affirmative. These philosophers conclude that physics must be mistaken, or at least, incomplete. This view allows us to say that *typically* our phenomenology is

[1] Price (1997)and Callender (2017).

veridical: It seems as though time passes because, in fact, it does.[2] We can call this the *Veridical Passage Thesis*. I consider this view in Sect. 2.1.

2.1 The Veridical Passage Thesis

According to the veridical passage thesis, it seems to us as though time passes because we are having veridical experiences of the passage of time.[3] The explanatory challenge for this view is to say *how* we track time's passage, and thereby to answer the cognitive mechanism question (as well as to say how we came to track its passage, and thus to answer the evolutionary explanation question).

On the face of it, the assumption that these experiences are (typically) veridical ought to make answering these two questions relatively methodologically straightforward: Neuroscience, psychology, and evolutionary biology are fairly good at determining how it is we come successfully to track things in our environment. It's worth noting that matters are a *little* more complicated in this case, since if time passes, it passes everywhere, at every time, for all of us, and its passage cannot be manipulated. So it will not be possible to experimentally manipulate the presence of temporal passage (or its rate of passing) in order to see what differences in the brain (say) are correlated with said manipulation. Nevertheless, setting aside this complicating factor, the explanatory task here is the familiar one of explaining how we come to have cognitive mechanisms that track something in the external world.

2.2 The Phenomenal Illusion Thesis

By contrast to the veridical passage thesis, according to the phenomenal illusion thesis, our experiences as of time's passing are *never* veridical

[2] Defenders of this view include Craig (2000, 2001), Gale (1968), Hestevold (1994), Maudlin (2002), Maudlin (2007) (though what Maudlin means by time's passing is somewhat different from what other authors mean) Norton (2010), Prior (1968), Schuster (1986), Smith (1994), and Zimmerman (2005, 2008).

[3] Leaving it open that *sometimes* our phenomenology might be illusory, or otherwise mistaken.

(Callender, 2008; Le Poidevin, 2007; Dainton, 2011, p. 405).[4] Instead, we are subject to a pervasive phenomenal illusion as of time's passing. If that is so, we can proceed to try and answer the cognitive mechanism and evolutionary explanation questions by trying to determine which evolutionary process, and which mechanism, could be responsible for such an illusion. Since being subject to a pervasive illusion is not likely to itself enhance fitness, we can expect (and indeed, this is what we find) that these accounts will try to explain the presence of this illusion as a by-product of something that is fitness enhancing: most likely the having of other veridical experiences. The details of that story, as it has hitherto been spelled out, vary. But in very broad brushstrokes, these accounts appeal to the idea that our experiences as of temporal passage are either the product of our motion phenomenology (Le Poidevin, 2007, p. 76; Paul, 2010), or the product of our motion phenomenology in combination with our phenomenology of change (Paul, 2010, p. 346). In either case, details of the story are controversial along two dimensions. First, it is controversial exactly what the connection is between motion and motion/change phenomenology and a phenomenology as of temporal passage Second, it is controversial just what motion and change phenomenology *themselves* consist in. For instance, on this second issue, there is debate about how we come to have a phenomenology as of motion at all, rather than simply a phenomenology as of something being in a sequence of positions. The most common views about how motion phenomenology (and others besides) arises hold that the contents of the experiences presented to us *as present* consist in a temporally extended interval in which the contents appear as temporally structured. This is either because said experiences (and contents) are in fact temporally extended (as per *extensionalist* models such as those defended by Dainton (2011)) or because, although the experience is not temporally extended, the contents of experiences are temporally extended (as per *retentionalist* models—see Gallagher (2003)). By contrast, there has, of late, been a resurgence of interest in snap-shot models, according to which neither

[4] Prosser (2007, 2012, 2013) defends the conditional claim that were there passage, our phenomenology would be illusory because we would be failing to track said passage. That is not to defend Phenomenal Illusionism, since it is consistent with the claim that, in fact, we do have experiences as of passage at all.

the experiences themselves, nor their contents, are temporally extended (see Le Poidevin, 2007; Arstila, 2016, 2018). This latter have been motivated by recent empirical work which suggests that past phenomenal contents are not available to us in any significant manner. In many ways, however, it ought not matter just which of these proposals we adopt in order to make sense of temporal phenomenology in terms of motion or change phenomenology

To that end, a number of recent proposals have spelled out the idea that our phenomenology of change plays a crucial role in producing our phenomenology as of passage. Ismael, for instance (Ismael, 2012), argues that each of us has a temporally embedded point of view: a representation of time relativized to a particular moment in a psychological history. The temporally embedded perspective is an individual's representation of its history; its memories of memories, its anticipations, its memories of anticipations, and so on. From each temporally embedded perspective, there is an asymmetry in how we represent, and come to know about, past and future decisions: We know about past decisions by *remembering* them, we know about future decisions by *deliberating* about what to do. In addition, there is a *temporally evolving view*; an ordered set of embedded perspectives, such that at each embedded perspective more of the past has been represented, and some of what were previously open future decisions are represented, later in the ordering, as fixed past decisions. It is the presence of this temporally evolving view, argues Ismael, that is responsible for our phenomenology as of passage: for the existence of this view explains why it seems to us as though future events are coming ever closer, and past events receding, why it seems as though we are moving through time, or time is moving over us.

A related proposal casts elements of Ismael's view within a hierarchical Bayesian inference theory of perception (Hohwy, Paton, & Palmer, 2015).[5] On this view, conscious perception is determined by whatever hypothesis about the cause of sensory input has the highest posterior probability—the 'winning' perceptual hypothesis. A perceptual hypothesis only counts as the winning one for a short period. As soon as our

[5] Though for somewhat different purposes, Hohwy et al. aim to explain why our temporal phenomenology is *as of* passage, though they use the phrase *temporal flow* instead of *temporal passage*.

perceptual system settles on a winning hypothesis, it begins decreasing the probability of that hypothesis as the best account of our current sensory input, because the external world constantly changes, and the best perceptual hypothesis to explain the incoming sensory stimuli at one time is less likely to be the best hypothesis about incoming sensory stimuli at later times. Our perceptual system is, at each time, placing the winning perceptual hypothesis head-to-head with alternative hypotheses about what is happening simultaneously with that very hypothesis.

According to Hohwy et al., our phenomenology as of passage is caused by the perceptual system predicting that the world is a changing place so that at each time the perceptual system generates new alternative perceptual hypotheses that next best explain the current incoming sensory input. The idea is that the constant creation of new hypotheses (which are either vindicated, or not) about what is happening at successive moments generates a phenomenology of motion and change, and generates the sense that we are heading toward the future (or it is heading toward us). One might worry that such a view cannot accommodate its being the case that we have the relevant temporal phenomenology even if *nothing* perceptually changes. That, in turn, might give you reason to think that, at best, internal psychological changes must do some (or perhaps all) of the work of explaining why we have this phenomenology. Hohwy et al., however, might suggest that even in the absence of perceptual changes, the perceptual mechanism will still be updating in this manner; it is just that the input it received will be the same. Since the mechanism is *attuned* to expect change, it will still update in this manner, by decreasing the probability of the current hypothesis; it is just that it will turn out to have been wrong to do so.

3 The Cognitive Error Thesis

More recently, however, some philosophers have suggested that we ought to reverse the order in which we attend to our four questions. These philosophers argue that we ought to begin with the working assumption that *typically*, our temporal phenomenology—*whatever its content*—is veridical. That is, we should give very low credence to the idea that some aspects of our temporal phenomenology are *always* in error. Of course, this is not to say that we couldn't find out that they are always in error. Rather, it is to suppose that this is a very unlikely outcome, and that, as such, we should think it unlikely that the phenomenal illusion thesis will be vindicated. So, at this stage of investigation, either we should accept the veridical passage thesis as a working hypothesis, or we should accept, as a working hypothesis, that it does *not* seem to us as though time passes. Since many philosophers think we have independent reason to think that time does not pass (given to us from physics), this gives us reason to investigate the second of these options. Let's call this second option the *Cognitive Error Thesis* (Hoerl, 2014; Torrengo, 2017; Braddon-Mitchell, 2013; Deng, 2013; Bardon, 2013; Baron, Cusbert, Farr, Kon, & Miller, 2015; Miller, Holcombe, & Latham, 2018).[6,7] It is the view that we do not have phenomenology as of passage; instead, we falsely *believe* that we have phenomenology as of passage.

Since defenders of this view think we don't have phenomenology as of passage, let's call the phenomenology we do have, our *purported passage phenomenology*. While the phenomenal illusionist has to explain how (cognitive mechanisms question) and why (evolutionary explanation question) we are subject to a pervasive phenomenal illusion, the cognitive error theorist has to explain how, and why, we are subject to a pervasive false belief about the content of certain of our phenomenal states. In what follows, I gesture toward a recent proposal for addressing both how, and why, we come to have such false beliefs.

3.1 The Cognitive Mechanism Question

Miller, Holcombe, and Latham (2018) suggest two broad explanatory strategies on behalf of the cognitive error theorist: the inferentialist and

[6] Torrengo's view, known as the phenomenal modifier view, may not be best thought of as a version of the cognitive error thesis. It is the view that our temporal experiences do not have a representational content as of time passing. So in this sense, it is a denial of the veridical passage thesis or the phenomenal illusion thesis. However, Torrengo holds that our experiences are phenomenally modified, in that the way we experience things is 'passagey'.

[7] Following Baron et al. (2015), this view is sometimes also known as veridicalism, since it holds that our phenomenology has veridical, not illusory, content; it is just that said content is not as of passage.

misdescriptionist strategies. Each of these offers somewhat different answers to the cognitive mechanism and evolutionary explanation questions. While Miller et al. do not point to *specific* mechanisms, they offer a higher-level account of what they think said mechanisms must be *doing*; that account, in turn, can be parlayed into a mechanistic account by discovering which mechanisms subserve the processes they describe.

Usually one would expect to explain our having a belief that time passes in terms of our having a phenomenology as of time passing. We believe the former on the basis of the latter. Since the cognitive error theorist denies that we have a phenomenology as of passage, she/he *must* deny this. He/she has to provide some independent explanation for our believing that time passes, which does not appeal to our phenomenology.

Inferentialists hold that the direction of explanation goes in the other direction: from our believing that time passes, to our believing that our phenomenology has content as of passage. Misdescriptionists hold that there is no direct explanatory connection between our belief that time passes and our belief about our own phenomenology; instead, there is a common cause both of our belief that time passes and our belief about the content of our phenomenology.

According to the misdescriptionist, the common cause both causes us to believe that time passes and to *mis*describe our phenomenology. Our misdescribing our phenomenology causes us to come to believe that its content is as of passage. Miller, Holcombe, and Latham suggest that this common cause is the presence of passage-friendly language (more on this in Sect. 3.2).

To get a sense of how such a view might go, consider a different case: our phenomenology as of the sun rising. The sun doesn't rise; instead, the earth moves relative to the sun. Phenomenal illusionists will say that we have phenomenology as of the sun rising, and that phenomenology is systematically illusory. Cognitive error theorists will say that our phenomenology is *not* as of the sun rising at all. Instead, it is (presumably) phenomenology as of the earth moving relative to the sun. But in English (at least), we use a host of expressions that imbed terms such as 'sunrise' and 'sunset', and we describe our purported sun-rising phenomenology using these expressions. That is, we *misdescribe* our phenomenology.

Having done so, we come to believe that its content represents the rising of the sun when in fact it merely represents the relative movement of the earth and sun. Of course, in this case, it's easy to see why we came to use expressions like 'sunrise' and 'sunset': We had, in the past, explicit beliefs that the sun rose and set! So while the cognitive error theorist wants to say something similar about our temporal phenomenology as about our sun-related phenomenology, she/he will need some account of why we come to use the kinds of expressions we do, which does not appeal to standing beliefs about there being temporal passage. I return to this issue shortly. First, however, a question arises as to whether, *even if we had an account of how we come to use such language* which does not appeal to existing beliefs about there being temporal passage, there being such language would result in us misdescribing our experiences.

In fact, there is reason to think the misdescriptionist is on firm ground here. Miller et al. suggest that the misdescriptionist appeal to a suite of evidence regarding the connection between the way linguistic communities speak about time, and write and read language, and the way that they conceptualize time (Boroditsky, Fuhrman, & McCormick, 2011; Fuhrman et al., 2011; Chen, 2007; Boroditsky, 2001; Fuhrman & Boroditsky, 2010; Casasanto & Bottini, 2014). This evidence suggests that *if* we misdescribe our experiences using passage-friendly language, then we could indeed come to believe that those experiences are experiences as of passage.

By contrast, according to the inferentialist, we form the belief that our purported passage phenomenology is as of passage on the basis of a sub-personal inference from one belief to another. What are these two beliefs? One is the belief that time passes. The other is the belief that the *reason* we believe that time passes is on the basis of our temporal phenomenology. Why think we have this second belief? Well, our purported passage phenomenology is perceptual: It involves the presentation as of mind-independent features of the world (distinct from features of our experience) in such a way that our phenomenology seems to be immediately responsive to the things in the world that it represents. Given that we believe that time passes, then, it's natural to think that we believe that on the basis of having that phenomenology. After all, in general, our perceptual beliefs are formed as a result of having perceptual phenomenology.

So suppose that we have these two beliefs, the belief that time passes and the belief that the *reason* we believe that time passes, which is on the basis of our temporal phenomenology. Then the inferentialist holds that we make a sub-personal inference on the basis of holding these two beliefs: We infer that our purported passage phenomenology is as of passage. Why would we infer this? Well, if our purported passage phenomenology is the reason we believe that time passes, then it is very plausible that the content of that phenomenology is as of passage. Otherwise, how could our having said phenomenology be a reason for us to believe that time passes? Hence we come to attribute to that phenomenology the content as of time passing.

Miller et al. point toward cases in which, plausibly, sub-personal inferences generate mistakes about our own reasons. For instance, they note that Morrot, Brochet, and Dubourdieu (2001) gave a panel of 54 wine tasters white wine, dyed red. The scent of the resulting ambiguous stimuli was described by the wine tasters using terms typically used to describe red wine. That is, the wine tasters believed that their wine phenomenology was olfactorily as of red wine. Plausibly, the wine tasters believed that the wine was red on the basis of their visual perception, and because of that they came to *believe* that their olfactory wine phenomenology was as of smelling red wine. Hence the subjects believed that their olfactory phenomenology was (at least part of) the *reason* for their beliefs about the kind of wine they smelled. Subjects falsely believed they smelled red wine. On the basis of this, Miller et al. hypothesize that they made a sub-personal inference that their olfactory phenomenology had the content as of smelling of red wine, and hence they came to believe that their phenomenology was as of red-wine smelling. This inferential pattern is just like that proposed by the inferentialist.

Both inferentialists and misdescriptionists need to show that we have mechanisms that could do the work ascribed to them. More taxingly, both need to provide an account of *why* we come to have a belief that time passes, if that belief is not generated by our purported passage phenomenology. Doing so goes some way toward answering the evolutionary explanation question (though only some way) and it is to this that I now turn.

3.2 The Evolutionary Explanation Question

Miller, Holcombe, and Latham suggest that we believe that time passes because of *passage-friendly* language. According to the misdescriptionist, the presence of passage-friendly language is the common cause of both our belief that time passes and our misdescription of our purported passage phenomenology. According to the inferentialist, the presence of passage-friendly language is the cause of our belief that time passes, and we then infer, on the basis of this belief and our belief that our purported passage phenomenology is the reason for our belief that time passes, to the belief that the content of our phenomenology is as of passage.

But why would we have passage-friendly language in the absence of passage, or of passage phenomenology? Miller et al. appeal to a broadly evolutionary story about the function of language to suggest that we would. They argue that any language needs the resources to express the way things are from the standpoint of each embedded perspective. To do so, a language must mark the location and perspective of each embedded perspective, and distinguish past from future events at that location. The most straightforward way to do this is via grammatical tense and deictic (indexical) adverbs like 'tomorrow' or 'yesterday' and nominal temporal landmarks such as 'next Christmas' alongside the use of a marker such as 'present'. In fact, all languages denote presentness in one way or the other (Gell, 1992), and although there are languages without tense (such as Chinese), all languages have some set of deictic adverbials that indicate gradations of pastness and futurity of events with respect to the time of utterances (Sinha & Gardenfors, 2014). Miller et al. call such languages *minimally passage-friendly* and note that all actual languages are minimally passage-friendly. They then suggest that languages can also be *substantially passage-friendly* (or not) depending on the extent to which they include one (or both) of what are known as *moving time* or *moving ego* metaphors (Sinha & Gardenfors, 2014). The moving time metaphor includes a whole suite of expressions which suggest that time itself moves, while the moving ego metaphor includes a suite of expressions which suggest that the ego moves through time. In both cases, the relevant expressions employ motion verbs such as 'his death is *approaching* (moving time metaphor)' or 'he is *nearing* his death' (moving ego metaphor). These metaphors appear in language as an easy way of expressing the 'nested' nature of the embedded perspective, wherein at later times what was previously (i.e. earlier) represented as open for deliberation is represented as fixed and decided.

Hence, the idea is that both the inferentialist and the misdescriptionist can appeal to this strategy for explaining why it is that we have passage-friendly language, a strategy that does not, in turn, appeal to our believing that there is temporal passage (or to there being temporal passage).

Of course, there is more work to be done here. Veridical passage theorists might argue that the best explanation for our having these kinds of embedded perspectives is there being temporal passage. Perhaps, for instance, the reason that later perspectives have memories of earlier perspectives, rather than the other way around, has something to do with temporal passage. Nothing I have said here countervails that reasoning. Nevertheless, I see no reason to suppose this is true. As long as temporally asymmetric phenomena such as causation, memory, knowledge, and records more generally, can be explained without appeal to temporal passage (as many suppose they can) then there being *nested* embedded perspectives of this kind can be broadly thought to be the result of there being these temporally asymmetric phenomena. It's also worth noting that phenomenal illusionists also suppose there to exist these embedded perspectives, so considerations such as these do not militate in favor of that view over cognitive error theory: At best, they militate in favor of veridical passage theory. But that view, of course, faces its own problems in understanding what passage could be, within the constraints of extant physical theories.

4 Conclusion

The veridical passage thesis, phenomenal illusion thesis, and cognitive error thesis all face significant explanatory challenges. The aim of this chapter has been to outline the cognitive error thesis—a relatively new view in town—and present it as a viable contender alongside the other more common views.

References

Arstila, V. (2016). The time of experience and the experience of time. In B. Mölder, V. Arstila, & P. Øhrstrøm (Eds.), *Philosophy and psychology of time* (pp. 163–186). Cham: Springer International Publishing. https://doi.org/10.1007/978-3-319-22195-3_9

Arstila, V. (2018). Temporal experiences without the specious present. *Australasian Journal of Philosophy, 96*(2), 287–302.

Bardon, A. (2013). *A brief history of the philosophy of time.* (p. 95). Oxford University Press.

Baron, S., Cusbert, J., Farr, M., Kon, M., & Miller, K. (2015). Temporal experience, temporal passage and the cognitive sciences. *Philosophy Compass, 10*(8), 56–571.

Boroditsky, L. (2001). Does language shape thought? English and Mandarin speakers' conceptions of time. *Cognitive Psychology, 43*, 1–22.

Boroditsky, L., Fuhrman, O., & McCormick, K. (2011). Do English and Mandarin speakers think about time differently? *Cognition, 118*, 123–129.

Braddon-Mitchell, D. (2013). Against the illusion theory of temporal phenomenology. In T. Sato, S. Sugimoto, & T. Sakon (Eds.), *Proceedings of the CAPE International Workshops, CAPE Studies in Applied Philosophy and Ethics Series, Vol 2.* (pp. 211–222). CAPE Publications.

Callender, C. (2008). The common now. *Philosophical Issues, 18*(1), 339–361.

Callender, C. (2017). *What makes time special?* Oxford University Press.

Casasanto, D., & Bottini, R. (2014). Mirror reading can reverse the flow of time. *Journal of Experimental Psychology: General, 143*, 473–479.

Chen, J. Y. (2007). Do Chinese and English speakers think about time differently? Failure of replicating Boroditsky (2001). *Cognition, 104*, 427–436.

Craig, W. L. (2000). *The tensed theory of time: A critical examination.* Dordrecht: Kluwer Academic Publishers.

Craig, W. L. (2001). Wishing it were now some other time. *Philosophy and Phenomenological Research, 62*(1), 159–166.

Dainton, B. (2011). Time, passage, and immediate experience. In C. Callender (Ed.), *The Oxford handbook of philosophy of time.* Oxford: Oxford University Press.

Deng, N. (2013). On explaining why time seems to pass. *Southern Journal of Philosophy, 51*(3), 367–382.

Fuhrman, O., & Boroditsky, L. (2010). Cross-cultural differences in mental representations of time: Evidence from an implicit non-linguistic task. *Cognitive Science, 34*, 1430–1145.

Fuhrman, O., McCormick, K., Chen, E., Jiang, H., Shu, D., Mao, S., & Boroditsky, L. (2011). How linguistic and cultural forces shape conceptions of time: English and Mandarin time in 3D. *Cognitive Science, 35*, 1305–1328.

Gale, R. M. (1968). *The language of time*. Virginia: Humanities Press.

Gallagher, S. (2003). Sync-ing in the stream of experience: Time-consciousness in Broad, Husserl, and Dainton. *Psyche, 9*.

Gell, A. (1992). *The anthropology of time: Cultural constructions of temporal maps and images*. Berg.

Hestevold, H. S. (1994). Passage and the presence of experience. *Philosophy and Phenomenological Research, 50*(3), 537–552.

Hoerl, C. (2014). Do we (seem to) perceive passage? *Philosophical Explorations, 17*(2), 188–202.

Hohwy, J., Paton, B., & Palmer, C. (2015). Distrusting the present. *Phenomenology and the Cognitive Sciences, 15*(3), 315–335. https://doi.org/10.1007/s11097-015-9439-6

Ismael, J. (2012). Decision and the open future. In A. Bardon (Ed.), *The future of the philosophy of time* (pp. 149–169). Routledge.

Le Poidevin, R. (2007). *The images of time*. Oxford: Oxford University Press.

Maudlin, T. (2002). Remarks on the passing of time. *Proceedings of the Aristotelian Society, 102*(3), 237–252.

Maudlin, T. (2007). *The metaphysics within physics*. Cambridge University Press.

Miller, K., Holcombe, A. O., & Latham, A. J. (2018). Temporal phenomenology: Phenomenological illusion vs cognitive error. *Synthese*. https://doi.org/10.1007/s11229-018-1730-y

Morrot, G., Brochet, F., & Dubourdieu, D. (2001). The color of odors. *Brain & Language, 79*, 309–320.

Norton, J. (2010). Time really passes. *Humana Mente*, (13), 23–34.

Paul, L. (2010). Temporal experience. *Journal of Philosophy, CVII*(7), 333–359.

Price, H. (1997). *Time's arrow & Archimedes' point: New directions for the physics of time*. Oxford: Oxford University Press.

Prior, A. (1968). *Papers on time and tense*. Oxford: Oxford University Press.

Prosser, S. (2007). Could we experience the passage of time? *Ratio, 20*, 75–90.

Prosser, S. (2012). Why does time seem to pass? *Philosophy and Phenomenological Research, 85*, 92–116.

Prosser, S. (2013). Passage and perception. *Noûs, 47*, 69–84.

Schuster, M. M. (1986). Is the flow of time subjective? *The Review of Metaphysics, 39*, 695–714.

Sinha, C., & Gardenfors, P. (2014). Time, space, and events in language and cognition: A comparative view. *Annals of the New York Academy of Sciences. Issue: Flow of Time, 40*, 1–10.

Smith, Q. (1994). The phenomenology of a-time. In L. N. Oaklander & Q. Smith (Eds.), *The new theory of time* (pp. 351–359). New Haven and London: Yale University Press.

Torrengo, G. (2017). Feeling the passing of time. *The Journal of Philosophy, 114*(4), 165–188.

Zimmerman, D. (2005). The A-theory of time, the B-theory of time, and 'taking tense seriously'. *Dialectica, 59*(4), 401–457.

Zimmerman, D. W. (2008). The privileged present: Defending an "A-theory" of time. In J. Hawthorne & D. W. Zimmerman (Eds.), *Contemporary debates in metaphysics. Theodore sider* (pp. 211–225). Oxford: Blackwell Publishing Ltd.

3

The Dynamic Block Universe and the Illusion of Passage

Maria Balcells

1 Introduction

It seems clear that ordinarily our experiences of change are different from our experiences of variation; watching a balloon being inflated and deflated is very different than looking at a mountain, low and green on one side, tall and rocky in the middle, and low and green again on the other side. Putting aside the qualitative differences of the objects themselves, there is still something very different about the dynamic character of the two. The mountain, after all, does not *really* change—it's just sitting there! The balloon, on the other hand, is growing—*really* growing—not just growing depending on where I happen to be looking or walking.

This difference between the way we experience change and spatial variation has been the crux of many arguments against the four-dimensional block universe model of space and time, or B-theory, in which time is extended like space, and all states of affairs exist equally, eternally, and "statically" in this varied tapestry of space and time. The charge is that

M. Balcells (✉)
Department of Philosophy, Bucknell University, Lewisburg, PA, USA
e-mail: mlb050@bucknell.edu

V. Arstila et al. (eds.), *The Illusions of Time*,
https://doi.org/10.1007/978-3-030-22048-8_3

while the B-theorist is capable of accounting for an object being in different states at different times (B-change), this alone does not account for *genuine* or *real* change[1] (A-change). *Genuine* or *real* change is typically cashed out as being some kind of change of the momentary state itself, perhaps in terms of its *presentness*.[2] This additional feature would be something like a moving present, which passes over these successive states and brings them into a shared and objective present, animating them into existence and extinguishing them into the past. The B-theory's failure to incorporate a feature that would account for *real* change is considered a failure to properly and completely describe time.

However, for B-theorists, this difference in experience and the supposed necessity of an additional feature are not enough to justify such a problematic addition. Thus, there have been many attempts to reconcile our experiences of change with the B-theory and show that the temporal and causal properties contained within the block universe are sufficient for explaining change, our experience of change, and which elements of experience (if any) are supposed to be indicative of passage. Nick Huggett, for instance, holds that there is, in fact, *no* experience of passage, and argues, "[i]nstead what are experienced are the characteristic aspects of temporal perception—duration, causal connection, change and motion—and these are misidentified as percepts of passage." (Huggett, 2014, p. 9). Under his view, our belief that we experience passage is a cognitive error that could, in principle, be corrected. L.A. Paul, however, holds that passage is a central part of our experience and argues that our experience as of[3] animated change is bound up with our experience as of passage and utilizes visual illusions of motion, such as color-phi, to make the case that just as we fabricate animation in the case of apparent motion, so too do we fabricate animation in all of our experiences of motion, thus our experiences as of passage are illusory. Natalja Deng, on the other hand,

[1] Many have argued that genuine change does not merely involve being in a different state at different times, for that would only give you temporal variation. Rather, genuine change must involve some change in the state of affair itself, namely a change in its presentness. See McTaggart, J. E. M. (1908). The Unreality of Time. *Mind, 17*(68), 457–474.

[2] Most famously, C.D. Broad considers this the spotlight of the present.

[3] Paul (2010) uses the phrase "as of" to indicate that there is no object of experience that could be identified as passage.

identifies temporal passage with the temporal succession contained within B-theoretic models and both she and Christoph Hoerl have argued against Paul's claim that our experiences of animated change ought to be considered illusory.[4] I share the view that these experiences are non-illusory and that *real* or *genuine* change is exactly that kind of change as described by the B-theory. Underlying my view is a sentiment expressed by Lee Siegal in his book, *Net of Magic*, which captures a resistance in reducing certain phenomenology to all-too ordinary facts about the world:

> 'I'm writing a book on magic,' I explain, and I'm asked, 'Real magic?' By 'real magic,' people mean miracles, thaumaturgical acts, and supernatural powers. 'No,' I answer. 'Conjuring tricks, not real magic.' 'Real magic,' in other words, refers to the magic that is not real; while the magic that is real, that can actually be done, is not real magic. (Siegal, 1991, p. 425)[5]

What many people mean by *real* magic is the kind of thing that just does not and could not exist in this world, while the conjuring tricks that are actually being done are considered impoverished versions of magic. Similarly, we are inclined to think of *real* change as impossible A-change, while the B-change that actually exists in the world is considered an impoverished version of change. A closer inspection of how something so natural and part of this world can produce in us the kind of experience that it does will, hopefully, allow us to understand B-change as less impoverished than it initially seems.[6]

What many in the anti-illusion camp have pointed out is that the construction of animation involved in our experiences of both apparent and genuine motion does not warrant the move to calling all experiences of motion illusory; construction alone does not disqualify an experience from being veridical and there are elements within the B-theory that can

[4] Insofar as Deng and Hoerl argue that there is no phenomenology of passage of the type described in A-theoretic terms, they are not at odds with Huggett with respect to this point.

[5] Daniel Dennett (2003) uses this quote by Lee Siegel to make a similar point regarding consciousness.

[6] Although I have drawn on this analogy with magic, I still resist the idea that the world is producing in us illusions of temporal passage. In the case of the magic trick, there is a disconnect between what the magician has made you perceive and the reality of the situation—the magician is trying to deceive you—and importantly, the magician can only deceive you by taking advantage of the perceptual mechanisms that usually work very well in accurately tracking the way the world is.

distinguish these illusions from those genuine experiences. The B-change in the world no more produces illusions in us than light waves and reflectance patterns produce illusions of color experience, or objects that project differently sized areas on the retina produce illusions of depth. Certainly, there is a lot of fancy footwork that goes on between the reality and the experience—the signals leaving the objects, perhaps in the form of light waves or sound waves, the reception of this information by the specialized sensory organs, and the various different specialized neural systems responsible for processing different aspects of the incoming information sending the information out to yet other neural systems for processing. Some of this information will be integrated with information from other processes and passed along yet again, perhaps even back through some of the previous systems. Eventually, and if needed, I will come to have an experience (although not necessarily a unified experience) of the world that reflects, at least in part, the way the world actually is.

In what follows, I will examine the claim that our experience of animated motion is an illusion. Since, in an illusion, there is something that we are getting wrong about the way the world is, I will consider two points of contrast. First, I will look at how we are getting it, that is, I will consider the entry point into our visual motion processing system in experiences of both apparent and genuine motion. En route to this discussion, I will review some of the latent assumptions we naturally make about temporal representation and conscious experience, an issue Paul and others have discussed. Second, I will consider what it would mean for us to get it right. Both Deng and Hoerl have argued that we can make a meaningful distinction between experiences of apparent and genuine motion by focusing, at least in part, on the continuous nature of genuine motion that is absent when we have an experience of apparent motion. However, I would like to reevaluate this particular distinction and consider another reason why we cannot extend the illusory nature of apparent motion to all cases of experienced motion. In particular, I focus on the idea that illusions only make sense against a backdrop of veridicality, so in order to call something an illusion, we must have a sense of what it would mean for us to get it right. I will argue that with no viable account of a veridical experience of B-change, the animated character of such experiences should not be considered illusory.

2 The Experience of Dynamic Change and Dynamically Changing Experiences

When it comes to mental representations of time, it is intuitively held that what you see is what you get. Arthur Prior expresses this intuition when he states, "I believe that what we see as a progress of events *is* a progress of events, a *coming to pass* of one thing after another, and not just a timeless tapestry with everything stuck there for good and all."(Prior, 1998, p. 104). Here, he espouses the view that the vehicle of our representation of time resembles or is structurally similar to time, itself. He correlates an *experience of progress* with a structurally similar *progression of experiences*, a position also found in the snapshot view of temporal experience. Further, he claims that the underlying problem for the B-theorist is of explaining "the *appearance* of time's passage: for appearing is itself something that occurs in time." (104)

However, while appearing is something that occurs in time, the vehicle of that appearance need not be extended in time and flow-like. The tendency to think that mental representations of temporal properties convey those properties by themselves exhibiting those properties manifests in the demands we put on the B-theory for a proper representation of time. Imagine a square moving across the page from left to right over a certain period of time. The square would be in a series of positions at various times (Fig. 3.1). Now, imagine watching the square pass by. Once we account for a slight delay in the time it takes the light signal to leave the square and be processed, then intuitively one might think that the mental representations would have a similar structure (Fig. 3.2). If this structure is modeling the general way we

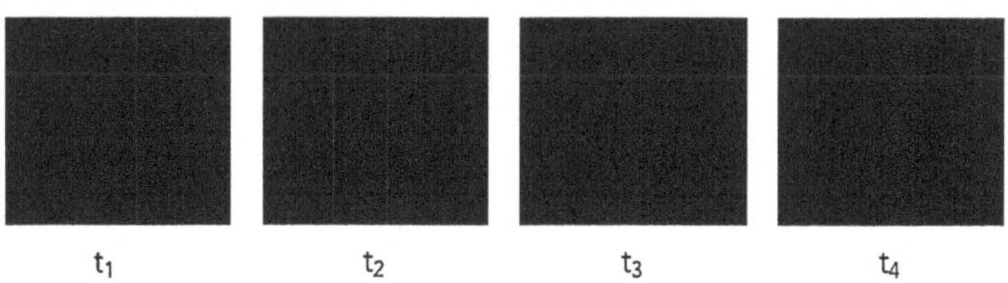

Fig. 3.1 A square occupying different positions on the page at different times

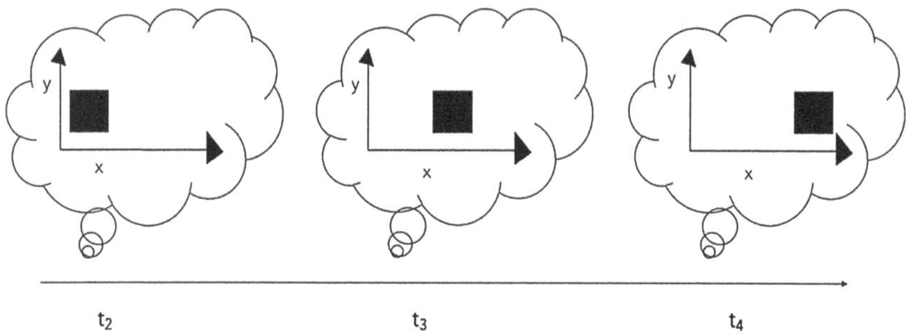

Fig. 3.2 The mental representation of the square occupying different positions on the page at different times

experience the world, then just as one may be tempted to think that genuine motion of the square could only occur if the spotlight of the present were to pass over each stage of the square's path, so too, we are tempted to think our experience could only have the animated character it does by virtue of the spotlight of the present passing over and presenting to us each of our representations of the square at its different stages. In this snapshot model of temporal experience, information crossing the threshold and making it onto the stage of consciousness becomes strikingly similar to events making it onto the lens of the present; the "eye" of our central observer is akin to the spotlight of the present, so it is no wonder we may consider an A-theoretic account of change to be the more intuitive view since it mirrors a more intuitive view of our own consciousness.

This model is considered problematic for a number of reasons involving various temporal properties including duration, succession, and motion and, as many have noted, resistance to this model is key in forming a coherent model of B-theoretic experience. In Dennett and Kinsbourne's paper, "Time and the Observer" (1992), they focus on the visual experience of apparent motion,[7] the phenomenon of experiencing

[7] Specifically, they refer to the phi-phenomenon; however, this term has since been shown to be misapplied in this case, although "color-phi" is still typically (and in this paper) used despite the misnomer. See Steinman, Pizlo, and Pizlo (2000) for a more thorough discussion.

motion when presented with a sequence of static images. In controlled experimental settings, when two small dots, separated by up to 4 degrees of visual angle are flashed sequentially, about half a second apart, the subjects experience a single moving dot. Further, in the color-phi experiment, when the flashed dots are of two different colors, subjects not only perceive the two dots as a single dot in motion, but they reported that the dot suddenly changed colors midway to its destination (the position of the second dot). The flash of a single dot does not appear to move, nor does it appear to change colors. It is only when experienced in conjunction with an appropriately timed and placed second flash that motion is perceived. Barring backward causation, we are faced with the problem of how to account for the later dot causing earlier experiences of the intermediary positions. In order to maintain a picture of temporal representation like in Fig. 3.2, Dennett and Kinsbourne argue that we have two options: either resort to the brain "rewriting history" by supplanting the two stimuli of the stationary dots with a false memory of a single moving dot, which changes colors midway (the Orwellian account), or have the information "wait in the wings" for more information to come in before being performed for consciousness (the Stalinesque account). The former case fails to properly preserve the model of experience in Fig. 3.2, but instead provides us with a false memory of experiences that could be modeled this way.[8] For the latter account to work, the delay needed for all the information to come in would be much longer than what is experimentally reported.

When we reject the snapshot model of temporal experience, we can also loosen the ties that an A-theoretic account of change has on how we understand our experiences of change and motion. Looking at the entry point into the visual processing system and the neural processes involved can further serve this purpose and also provide a background for considering whether our experience of motion is an illusion.

[8] The Orwellian account is not entirely at odds with the view of temporal representation whereby time is *not* its own representation. Rather, the problematic aspect of the Orwellian account comes from its denial that the "rewritten history" is part of conscious experience and its placement of the rewrite into memory instead.

3 Getting It: Visual Motion Processing in Experiences of Real and Apparent Motion

If we take the case of vision, the general form of basic motion processing can be described as follows. There are pairs of photoreceptors on the retina set up to detect movement. Any individual receptor cannot detect movement, it only detects a change in illumination. However, a pair of photoreceptors can detect movement. Two receptors, A and B, are set up to detect, say, rightward movement. They do so by sending their signals to a third comparator neuron, C. The signal sent from B to C is direct while the signal sent from A to C contains a short time delay (Fig. 3.3). This group of receptors works such that when the signal from the object (the change in illumination) first hits A and then hits B while moving rightward, the two signals are received by C simultaneously (or close to it) and this comparator neuron outputs a signal indicating movement is detected (Fig. 3.4).[9] This signal is then sent along to integrate with other signals with information about form, color, texture, and so on. It is in this way that if a car

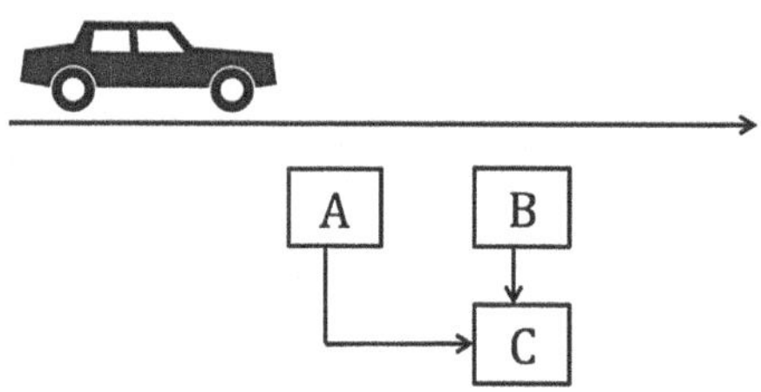

Fig. 3.3 Two receptors, A and B, detect the rightward motion of the car. They do so by sending their signals to a third comparator neuron, C. The signal sent from B to C is direct while the signal sent from A to C contains a short time delay. This group of receptors works such that when the signal from the object (the light/dark edge of the car) first hits A and then hits B while moving rightward, the two signals are received by C simultaneously (or close to it) and motion is detected

[9] While still working with signals only indicating changes in illumination, in primates, A and B are actually intermediate neurons gathering information from groups of photoreceptors rather than individual receptors.

Fig. 3.4 The mental representation of dynamic, smooth, continuous motion. Although I have represented the output signal from the comparator neuron, C, as a unified, multi-modal presentation, this is only to roughly depict a momentary representation of motion. At this point in the processing, many aspects would not be included in the information, such as it being a car, its particular shape, its color, and so on

were to move continuously rightward, the light/dark boundary of the car would hit photoreceptor A and then hit photoreceptor B and movement would be detected by the comparator neuron, C (Mather, 2006a). Even though the car is moving continuously and smoothly, the brain only gets information regarding position in discrete packages, those from the A and B photoreceptors. However, this is all the brain needs in order to produce an experience of dynamic movement. With this model of early stage motion processing (the neural substrate), we can easily see how we come to have illusory experiences of movement.

In the experience of apparent motion, the individual photoreceptors are triggered, not by an object continuously moving along a path, but by the series of disconnected frames. If we consider again our basic rightward motion detector, our first frame will depict a car in a certain position and the following frame will depict the car in a different position, to the right of its original position (Fig. 3.5). Photoreceptor A will again detect a change in illumination and, at a slightly later time, photoreceptor B will detect a change in illumination. Again, both photoreceptors will send those signals to comparator neuron C and motion will be detected (Fig. 3.6).

At this entry point into the visual processing system, there does not appear to be any difference between real and apparent motion, although

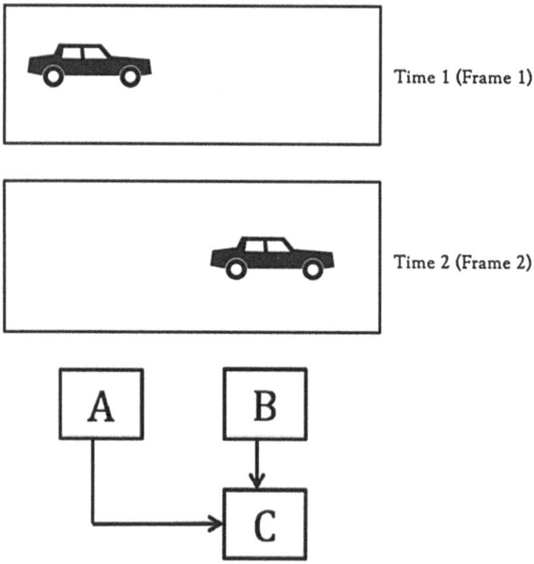

Fig. 3.5 The group of receptors for detection motion (see Fig. 3.3) applied to an apparent motion case. The images presented to the photoreceptors in this case are distinct frames, rather than those of a single continuously moving car

Fig. 3.6 The mental representation of dynamic, smooth, continuous motion, which is the same in this apparent motion example as it is in the continuous motion example depicted in Fig. 3.4

at higher orders of motion processing,[10] there certainly is. The higher order motion detection typically involves integration of information from various different processes including a variety of other motion detectors specific to direction and velocity, as well as information regarding texture, color, shape, and so on. These integrations do not happen serially, but in parallel and feed back into each other. For example, object shape is in part determined by the information regarding direction and velocity from early stage motion detection. The information about shape will then re-integrate with information regarding motion in order to produce the experience of an object moving. Failure to integrate information about shape and motion due to lack of information or competing information from other systems can produce, for instance, the phi-phenomenon, which is an experience of movement, but not movement of an object.[11]

Given that the neural substratum for visual motion detection is the same for the experience of both real and apparent motion, it may be tempting to conclude that our experience of real motion is just as much an illusory fabrication of the mind as our experience of apparent motion. I would like to resist this move, however, and explore why we ought instead to take these experiences of real motion to be veridical experiences. One way that B-theorists can distinguish illusory cases from genuine experience of motion is to note that in the illusory case there is an absence of continuous succession. For instance, Hoerl points out that a B-theorist could cash out the difference between the illusion involved in color-phi and ordinary experiences of real motion by saying that "in color-phi, it appears that there is a continuous locational variation across time, where in fact there are just two stimuli at two discrete locations" (Hoerl, 2014, p. 192). Similarly, Deng cites the presence of a continuous succession of stages to differentiate apparent motion from genuine

[10] Traditionally, there is thought to be two distinct systems involved in processing motion, short range and long range. Cavanagh and Mather, however, argue that experimental evidence points to there being a single system, but with lower and higher orders of processing, depending on the stage of the processing and the information being integrated (Cavanagh & Mather, 1989).

[11] Wertheimer calls this "pure" motion, which is produced when the flashes are timed sufficiently close together. See Steinman et al. (2000).

experiences of motion (Deng, 2013, p. 377). And even psychologists typically cash out their difference in B-theoretic-friendly terms: "[i]n natural images of real scenes, moving objects change position in the image in a smooth, continuous manner…[f]or obvious reasons, this kind of movement is called 'real movement'" (Mather, 2006b, p. 2).

However, that the B-theoretic account of the world can identify real motion with the continuous succession of positions does not, to me, provide us with enough to resist the claim that all experiences of motion are illusory. While I disagree with her ultimate conclusion, I am sympathetic to a point Paul makes in her essay, "for normal humans, there may never be any phenomenal difference between our experience of a series of instantaneous objects that are appropriately spatiotemporally spaced and qualitatively similar and our experience of a moving, changing, persisting object with the same qualitative and locational variation as the series." (Paul, 2010, pp. 25–26). My agreement with Paul here has less to do with the phenomenal similarity of the two cases described and more to do with their metaphysical similarity. If we change the above description of the first case to "a series of instantaneous *temporal stages* that are appropriately spatiotemporally spaced and qualitatively similar," and take "appropriately" to mean continuously,[12] then this starts to look remarkably similar to a four-dimensionalist description of "a moving, changing, persisting object." One would still need to add in the necessary persistence conditions to distinguish distinct stages of a single object from distinct objects, but continuous succession does not strike me as being sufficient for this distinction. If what it really means for an object to move is for a series of distinct temporal stages to occupy distinct, but continuous positions in space and time, then it is understandable (although still incorrect, I think) to hold that our experience is getting something wrong.[13]

As Deng and Hoerl point out, there is still a fundamental difference in the world between typical cases of apparent motion and experiences of real motion: the inputs in cases of apparent motion are themselves static

[12] No doubt, by "appropriate" Paul means sufficiently close in space and time (up to 4 degrees of visual angle and up to about half a second apart), but the lower end of this spectrum would limit zero.

[13] A thorough and closely related discussion of our experience of passage and our experience of objects as enduring is taken up more fully in Prosser (2012), Deng (2013), and Hoerl (2014).

and unchanging, while for real motion, the instantaneous temporal stages are not, since the determinations of "static" or "dynamic" are only applicable over intervals, not to individual instantaneous stages or facts. However, while continuous succession may be what separates the two in reality, the visual processing system is finite and, as discussed in above, can only take in discrete bits of information; it never gets the continuous succession. In discussing Paul's argument, Deng says "[w]e do not experience [*O* has property *P1* at *t1*] for a short time, and then [*O* has property *P2* at *t2*] for a short time, like we (perhaps pre-consciously) experience [red flash left] for a short time, and the [green flash right] for a short time," (Deng, 2013, p. 377) however, in whatever sense we preconsciously get the "static" information in the case of color-phi, we get that same kind of information in experiences of real motion.

4 Getting It Right: Defining the Backdrop of Veridicality

It is not surprising that the phenomenology of an illusion would be the same as in a veridical experience, after all, that is much of what it is to be an illusion. However, since the entry-level mechanisms involved in both cases are the same and in neither case is a continuum of intermediary positions utilized, I think that it is important to give a fuller account of why we should not consider ordinary motion perception to be illusory. In order to resist saying that we are getting something wrong in motion perception, we must be able to say what it would mean to get it right. In the case of color-phi, it appears as though there is a single moving dot that is changing colors halfway through its path. To get it right would be to see the two distinct dots, one red and the other green, and see the space between them. In the case of ordinary motion perception, what would it mean to get it right?

We tend to fall back onto that Cartesian Theatre intuition that a dynamic experience of motion involves each position of the path (a slice) to be represented in experience in a sequential manner and presented to the mind's eye in that order, and further, that each individual, discrete, static slice presented to consciousness must be infused with a dynamic change that is not present in the world, namely A-theoretic change. If this kind of thing is not present and the world is truly B-theoretic, we are inclined to think that these slices ought instead to appear as discrete, static slices. For a fuller description of this kind of veridical experience, consider the following passage:

> Suppose, further, that there were a being who did not experience the beta phenomenon and instead only experienced quicker and quicker sequences of still images, without motion or change. Suppose this being had an unlimited mental capacity and powers of discrimination; then, in the limit, the sequence would be experienced as a continuum of still images, yet still with no *motion* or *change* of any kind. (Prosser, 2012, p. 112)[14]

Here we have a being very unlike a human being, with its "unlimited mental capacity and powers of discrimination." It is unclear whether a worldly creature could even have such powers. In fact, it is doubtful that such a creature would have visual perception anything like what we typically think of as visual perception. For one, vision requires light to interact with an eye. Light is, itself, of some finite size as are the cells within an eye that interact with the light. A creature with unlimited powers of discrimination would need to have either infinitely small photoreceptors or an infinite number of photoreceptors in the eye in order to make such discriminations, both of which are impossible. Seemingly, a creature like the one described above would not be a creature with visual perception like ours, and most likely nothing we could call visual perception at all. The kind of experience outlined above does not seem a likely candidate for veridical experience of B-change since it is not a plausible account of an actual visual experience.

Considering the difficulties in cross-species comparison of experience, we may instead turn to a possibility for veridical experience of B-change

[14] In his paper, Prosser argues that, in our experiences of motion and change, we represent the objects undergoing such change as enduring, an A-theoretic form of persistence. This, in itself, is a rich topic that requires a much fuller treatment than can be done here, so this passage is only being used to discuss what a plausible description of a veridical experience of B-theoretic change might look like.

present in some humans: akinetopic experience. In cases of people with akinetopsia, a kind of motion blindness typically caused by damage to the MT (Medial Temporal)/V5 region of the brain, those suffering from extreme cases of the condition do not have a visual experience of dynamic motion. The reported experience is more like the experience of difference or *that* change has occurred; those with akinetopsia "described the perceptual experience of a moving target as if the visual stimulus remained stationary, but appeared at different successive positions" (Zihl, Von Cramon, & Mai, 1983, p. 314). One subject with gross akinetopsia reported that she could not see the coffee *flow* as she poured it into a cup, rather the liquid appeared be frozen in space. The cup would appear partially full and then in the next moment the cup would be overfilled. Here, we truly have an experience of a static world. So, akinetopic experience may be a possible candidate for veridical experience of B-change; those with akinetopsia have been freed of this persistent illusion of motion.[15]

Although one could take the position that akinetopic subjects have a veridical experience of B-change, it would be tough considering the difficulties those with akinetopsia have in successfully navigating the world. Given that akinetopic experiences and the fictional experience laid out by Prosser are not viable options for veridical experiences of B-change, we lack a backdrop for saying that our dynamic experience is getting it wrong. Thus, we should not consider our dynamic experiences of motion to be illusory, but rather just the particular way that we experience B-change.

5 Conclusion

Upon examining the neurological processes involved in early motion detection, we see that experiences of both apparent and real motion come about from discrete input, either in the form of discrete images being

[15] One may object that akinetopic experience importantly lacks the intermediary positions of the objects; the experience has gaps where there are none in the world. The absence of the intermediary positions in experience ought not be too troubling, though, since an anti-illusionist would concede that I can certainly experience, for instance, the real motion of an animal as it runs behind a slatted fence, obscuring its intermediary positions.

flashed or in the form of the discrete, discontinuous slices that constitute smooth, continuous motion for the B-theorist. In both cases, we need to reconcile the tension between the "static" input and the dynamic experience. Illusion seems to be the only option, but illusions must be understood against a backdrop of veridicality; if we are getting it wrong, there must be a way to get it right.

Our dynamic experiences are often thought to be inherently A-theoretic, which puts the B-theorist in the impossible position of explaining A-theoretic experience in a B-theoretic world; it seems that the only story we could tell of veridical experience of B-change is one that is "static." However, upon examination, possible stories describe either an experience of a non-worldly being (or at least a non-human being) or an experience we take to be quite inaccurate. Without a viable alternative to what it would be like to veridically experience B-change, our own dynamic experience of motion ought to be stripped of its A-theoretic framing and considered the most likely candidate.

Finally, if B-theorists take seriously their commitment to a B-theoretic world, and take seriously that change and motion can be cashed out in B-theoretic terms, then this should be considered *real* change in the fullest sense, a sense which regards B-change as dynamic and does not resign itself to the misnomic label of "static" simply because it lacks the impossible qualities of A-change. We ought to consider B-change to be dynamic and responsible for our dynamic experiences of motion. I return back to the sentiment expressed in Lee Siegal's paradoxical quote about magic. If that which we think of as truly *real* does not and cannot exist in this world, then we must consider that which can and does exist to be the real thing.

References

Cavanagh, P., & Mather, G. (1989). Motion: The long and short of it. *Spatial Vision, 4*(2/3), 103–129.

Deng, N. (2013). On explaining why time seems to pass. *The Southern Journal of Philosophy, 51*(3), 367–382.

Dennett, D. (2003). Explaining the "magic" of consciousness. *Journal of Cultural and Evolutionary Psychology, 1*(1), 7–19.

Dennett, D., & Kinsbourne, M. (1992). Time and the observer. *Behavioral and Brain Sciences, 15*, 183–247.
Hoerl, C. (2014). Do we (seem to) perceive passage? *Philosophical Explorations, 17*(2), 188–202.
Huggett, N. (2014). Skeptical notes on a physics of passage. *Annals of the New York Academy of Sciences, 1326*(1), 9–17.
Mather, G. (2006a). *Foundations of perception*. Hove: Psychology Press.
Mather, G. W. (2006b). Motion perception, psychology of. In L. Nadel (Ed.), *Encyclopedia of cognitive science*. John Wiley. https://doi.org/10.1002/0470018860.s00581
McTaggart, J. E. M. (1908). The unreality of time. *Mind, 17*(68), 457–474.
Paul, L. A. (2010). Temporal experience. *Journal of Philosophy, 107*, 333–359.
Prior, A. N. (1998). Some free thinking about time. In P. van Inwagen & D. Zimmerman (Eds.), *Metaphysics: The big questions* (pp. 104–107). Malden: Blackwell Publishing.
Prosser, S. (2012). Why does time seem to pass? *Philosophy and Phenomenological Research, 85*(1), 92–116.
Siegal, L. (1991). *Net of magic: Wonders and deceptions in India*. Chicago: University of Chicago Press.
Steinman, R. M., Pizlo, Z., & Pizlo, F. J. (2000). Phi is not beta, and why Weirtheimer's discovery launched the gestalt revolution. *Vision Research, 40*, 2257–2264.
Zihl, J., Von Cramon, D., & Mai, N. (1983). Selective disturbance of movement vision after bilateral brain damage. *Brain, 106*, 313–334.

4

The Perception of Duration and the Judgment of the Passage of Time

Luke A. Jones

A common complaint of any academic is that their field is misunderstood by related disciplines or by the general public. Perhaps nowhere more so than in the field of timing and time perception is this complaint justified. Confusion and misinformation abound even within time psychology itself, let alone in general psychology or the wider community. This essay is not a council of despair, but rather an opportunity to bring together for the general reader a schema for understanding some of the key types of timing judgment, the extent to which they are understood, and what the relationships between them might be. This chapter will first outline some of the types of questions that people raise about our relationship with time, before discussing the three main types of timing judgment that have been investigated and their models, and then returning to the questions and everyday phenomena to see how well they are addressed. Finally, I will discuss how although this categorization is useful, there are some real outstanding problems about how these judgments relate to each

L. A. Jones (✉)
Division of Neuroscience and Experimental Psychology, School of Biological Sciences, University of Manchester, Manchester, UK
e-mail: Luke.Jones@manchester.ac.uk

V. Arstila et al. (eds.), *The Illusions of Time*,
https://doi.org/10.1007/978-3-030-22048-8_4

other, and the extent to which our models really apply to our everyday experience of time.

Anyone who works in the field of time perception will agree that the topic of time, one's relationship with it and the way in which it can seem distorted, is a subject that people are intrinsically interested in. There seems to be something about our relationship with time and the fact that we feel it either dragging of flying by that is inherent to our experience of being human and self-aware. Simply observing the types of questions that repeatedly come up is revealing, for example:

Why does time fly when I'm having fun and crawl when I am bored?

Why Does Christmas seem to come around quicker every year?

How am I able to tell whether one note is longer than another?

Putting aside these types of questions, and simply observing some of things that people say about time is equally fascinating, and it is often paradoxical, or difficult, to cut through to what is actually meant by the speaker. For example, people will say:

It's been a long week!

That (one hour) lecture seemed like 4 hours long!

I can't believe it's October/Christmas/Thursday already!

I will return to each of these questions and statements later, but even the casual reader will see that there is a lot to unpack here, and these examples show that there is more than one type of timing judgment, or relationship to time, that people are making and expressing.

1 Prospective Timing

The type of timing that has been most comprehensively examined in the laboratory is the perception of duration. That is to say, making estimates of a stimulus duration (typically a tone, or a light, but sometimes a vibration), making a judgment about the duration of one stimulus compared to another, producing a given duration by performing some action, or reproducing a given duration. Most commonly, this perception of duration is conducted in a *prospective* manner, that is, the participants are made aware before the stimulus is presented that they will be required to make some action or decision based on its duration (as opposed to retrospective timing which will be discussed later).

It turns out that people are remarkably good at this type of timing. For example, people can distinguish a difference in duration between two tones that is as small as 1/10th of a second (100 ms: 1 second = 1000 ms); and under controlled laboratory conditions, this can drop to 10 ms or even less (Jones, Poliakoff, & Wells, 2009). The obvious question is how is this done? The key problem here is that we often refer to people having a sense of time, or time 'perception', but it is not a sense or a perception in the normal meaning of those terms. Unlike vision or hearing, there is no external source of energy called time that is transduced by a sensory organ into a bioelectrical signal that can be interpreted by the brain; in contrast to light/photos that are transduced by the retina, or vibrations that are transduced by the hair cells in the cochlea.

This has led psychologists to draw on a hidden internal process that is performing this timing, that is, the proposal that we possess some form of internal clock. In fact, if one considers real clocks, they to do not measure time directly, rather they count regular events: the swings of a pendulum or the oscillations of a quartz crystal. The simplest clock model consists of a pacemaker that generates and emits pulses or ticks, and an accumulator/counter that stores the ticks via a switch that is under attentional control to start and stop timing. This type of clock (with the addition of memory stores and a decision process) forms the basis of the model of Scalar Expectancy (timing) theory (SET: Gibbon, 1977; Gibbon, Church, & Meck, 1984). Even this simple clock model produces testable predictions in behavior. Because the accumulation process is linear, we would expect a linear relationship between elapsed duration and perceived duration. There is evidence for this. If we double the duration to be timed, then the mean estimate will also double; people can accurately express one duration as a percentage of another duration; people can perform mathematical operations on durations such as averaging three durations together and using the average as a standard to make other discriminations (Wearden & Jones, 2007). If timing was not linear, accurate performance on these tasks would not be possible.[1]

Other evidence for this internal clock comes from the fact under some circumstances its speed can be altered. An auditory click train (a series of regular clicks, typically at 5 Hz for around 5 seconds) presented before a stimulus to be timed has the effect of that stimulus being perceived as longer in duration than one presented without (Treisman, Faulkner, Naish, & Brogan, 1990; Penton-Voak, Edwards, Percival, & Wearden, 1996).[2] The skeptical reader may be thinking this does not prove any 'speeding up' effect, but rather that people are biased to estimate durations preceded by some noise as longer, or that they are including a small part of the end of the click train with the duration to be timed. However, if this were the case, then it would not matter how long the stimulus to be timed was; it would always be overestimated by the same additive amount. In fact, it turns out the effect is multiplicative, that is to say, the longer the duration to be timed, the greater the relative overestimation. Therefore, when one plots real duration against estimated duration for the click and no-click conditions, there is a difference in *slope* between the two functions. These slope effects are important pieces of evidence that are consistent with (although do not prove) that an internal clock-like process has been speeded up. Unfortunately, there is a raft of studies out there that show that stimuli in condition X are overestimated relative to condition Y and will argue that this is due to an increase in clock speed in condition X, but they fail to use a range of durations to show that the effect is indeed multiplicative or slope-like.

There are other well-documented slope effects which are more problematic to explain. For example, it has been well known for a long time that sounds are judged as longer than lights of the same actual duration (Goldstone & Lhamon, 1974). Wearden, Edwards, Fakhri, and Percival (1998) showed that this too is a slope effect and not an additive bias. Touch stimuli lie somewhere between sound and lights in their slope (Jones et al., 2009; Williams, Yüksel, Stewart, & Jones, 2019). Although

[1] There is also evidence from 'time-left' experiments, the methodology of which is somewhat complicated, see Gibbon and Church (1981) and Wearden (2002).

[2] Repetitive flashes of light have also been shown to have the same effect (e.g. Droit-Volet & Wearden, 2002; Ortega & Lopez, 2008; Wearden, Williams, & Jones, 2017).

the slope effect is consistent with a 'speeding up' effect, is this an explanation or just a description of this phenomenon? Is there a single clock that runs at different speeds for different modalities? Are there multiple modality-specific clocks that run at different speeds, and what would be the evolutionary advantage (or epiphenomenal cause) of this discrepancy in speed? As we go about our everyday life, we are not aware of this temporal distortion; in fact, the opposite is the case: visual and auditory events are temporally bound. By which I mean that when I hear and watch you talk to me, I see your lips move and I hear the sound of your voice in perfect synchronization; yet the speed of light and sound and the transduction and processing within each modality have different latency, while the brain binds the two events together.

A similar problem of explanation also exists for the phenomena that gives rise to the largest difference in slope, that between filled (a continuous tone or light) and unfilled (a period of silence or darkness); called the 'filled duration illusion' (Thomas & Brown, 1974; Wearden, Norton, Martin, & Montford-Bebb, 2007). The filled stimuli are grossly overestimated relative to the empty. Why would the internal clock run at different speeds for these two types of phenomena? Nobody knows, but if we accept the idea of a pacemaker-accumulator clock then it appears to do so.

Other evidence for this pacemaker-accumulator type internal clock is the fact that it can be used as a stopwatch, that is, one can start timing, stop timing during a gap and then continue timing again (Roberts & Church, 1978). Finally, people can easily make judgments of ordinality, that is, they can tell not just that a 15-second duration is different from a 40-second duration, but that it is shorter. This may seem trivial, but it causes some serious problems for alternative models to the internal clock.

I have attempted above to outline some of the compelling evidence consistent with the idea of an internal clock. However, there has always been a resistance to this internal clock idea. Some authors find the idea of a dedicated pacemaker pure speculation, or find the idea of unbounded accumulation biologically implausible. Nonetheless, these non-clock proponents would agree, that whatever you replace the clock with, it has to be able to do all of the tricks and have all of the operating characteristics outlined above, which can be more problematic than one might suppose. For example, if there is no accumulation or linear relation between objective and subjective, then it is hard to conceive of how such a system can perform ordinality judgments, perform like a stopwatch, or produce the characteristics of linear timing. Likewise, the multiplicative slope effects are hard to explain without invoking a mechanism that is operating at different speeds. Although, as discussed above, whether the internal clock offers an explanation or just a description of such effects is open to debate.

2 Retrospective Timing

The type of timing discussed so far is termed 'prospective' timing, in that people in the experiments are informed before being presented with the stimuli that they will subsequently be required to make a temporal judgment about it. However, it is possible to ask people to make timing judgments after they have received the stimuli, in circumstances where they were not informed that they would be required to make such a judgment. This is known as 'retrospective' timing. Up until the 1970s, this distinction between these two types of judgment was not made, and as such, there were a lot of seemingly conflicting studies, as these two types of timing have different characteristics. It is argued that, rather than relying on the internal clock, people make retrospective judgments based on inference, for example, the amount of information processed, events witnessed, distance traveled or more generally, the amount of information stored in their memory for the duration in question. It is thought that this type of timing is more akin to on-the-fly guesstimates that we make about duration in everyday life.

Initial studies of retrospective timing seemed to support the idea that they are based on the amount of information processed or 'work done', when these variables are increased then so are the length of the retrospective judgments. The clearest such example comes from the study that first distinguished between these types of judgment: Hicks, Miller, and Kinsbourne (1976).[3] Their participants had to sort cards into piles for 42 seconds with different levels of processing; one pile, two piles (red and

[3] These judgments had been distinguished from each other by William James (1890), but have been seemingly forgotten about by the timing field.

black) and lastly four piles (according to suit), the retrospective judgments increasing with the level of processing required. Well at least that is the received wisdom; if one actually examines their data, the mean duration judgments were 28.33 s (one pile), 38.33 s (two piles) and 33.17 s (four piles), so it was not a monotonic increase. However, they also had a prospective condition which showed a monotonic *decrease* in estimations as the information processing level increased: 52.92 s, 42.83 s, 31.15 s plus the mean of all three conditions was much closer to the actual duration (42.25 s compared to 32.28 s; the relative underestimation in retrospective timing is a common finding in other studies). This effect on prospective timing is often interpreted in terms of attention: the more non-temporal processing required, the less attention to temporal processing and the shorter the estimation.

Later studies have been far more mixed in their results. Professor John Wearden spent several years exploring retrospective timing and was eventually so exasperated by the inconsistent findings that he advised me 'whatever you do don't study retrospective timing, it will ruin your career!' A full discussion of Wearden's and other retrospective studies are beyond the scope of this discussion (see Block & Zakay, 1997 and chapter 6 of Wearden, 2016).

3 Passage of Time Judgments

In 2005, the term 'passage of time judgments' (POTJs) was suggested for a type of timing phenomenon that doesn't seem to correspond with either prospective or retrospective duration judgments (Wearden, 2005). It is instead a judgment of how quickly time seems to pass. There are many situations in which one feels time is either 'dragging' or 'flying' by.

But what determines this speed of time passing? An earlier model (Zakay, 1992) made some suggestions applied to situations in which one is waiting for something. Zakay argued that any situation could be analyzed in terms of how much attention is given to time, and that attention will determine how slowly time will seem to pass. This attention to time was proposed to be determined by two factors: temporal relevance (TR), which is how critical it is that the thing we are waiting for occurs on time and temporal uncertainty (TU), or how certain the timing of the event is. Increasing TR and/or TU will cause time to crawl by; reducing them will speed it up.

If we are waiting for a bus to take us to an important date, then TR will be high; if the bus service is unreliable, then TU will also be high. Thus, our waiting time at the bus stop will likely crawl by. If we reduce the TU in this situation, either by the bus service actually being reliable, or if we have a digital display at the bus stop that accurately displays waiting time, then time will pass by more quickly. If we are in a situation where both TR and TU are low, for example, reading a book on the beach on holiday, then time will seem to fly by.

4 Timing in Everyday Life

4.1 Prospective Timing

It seems clear that we must be doing some sort of continuous prospective timing all of the time. In general, we have a good feeling of how long we have been engaged in one particular activity or another; this is most clear when our expectations are violated. For example, when I push the call button for the elevator at the workplace car park, I have an expectation of how long it will take for it to arrive. Sometimes, it seems to take much longer than on other days. This simple example shows three things: firstly that I have a memory of how long the lift normally takes to arrive (on average); secondly that when I press the call button, some part of me starts to time the wait; and lastly, I perform a comparison between the average waiting time memory and the current wait time. If the difference is more than a certain amount, I will notice, and perhaps start fruitlessly jabbing at the call button.

It seems that there are lots of examples of this type of memory for duration, and the comparison with currently elapsed time that we do all the time: waiting for traffic lights to change, waiting for a waiter to come and take your order, waiting for the cash to be dispensed from an ATM. However, all of the clearest examples seem to involve *waiting*. In non-waiting situations, are we also constantly prospectively timing? The prospective/retrospective timing difference that we observe in the lab would not exist if we were constantly timing. On the other hand, we must be doing some prospective timing all of the time—else, why would we have evolved such a precise internal clock mechanism?

There may be a couple of solutions to this contradiction. Perhaps we are doing some low-level prospective timing all of the time, and this is enhanced or given extra cognitive resources when we are in a waiting situation, or when we are in a laboratory and are explicitly told to time a stimulus duration. Otherwise we rely on retrospective timing mechanisms. Alternatively, perhaps retrospective timing is actually prospective timing but with very low level of attention to time, and/or few cognitive resources allocated to it. This would go some way to explain why retrospective judgments of duration are typically underestimated relative to prospective judgments—as we saw earlier, if attention to timing is reduced even in an explicitly prospective timing situation, then estimates of duration tend to be reduced.

4.2 Retrospective Timing

I may say to a friend that 'It's been a long week/day!' Rather than pedantically pointing out that a week/day is always the same length, my friend will know what I mean by this, that I have had a hard or busy time. We have reached the end of the day/week and are looking back to the start. As we have seen, (some) studies of retrospective timing suggest the judgments are based on how much work we have done, and/or how many events have happened, so a retrospective judgment of a week/day 'feels' longer when it has been busy. This appears to have some prima facie validity, but we simply don't study such long durations in experiments. Alternatively/additionally, this could be hyperbole; we are saying 'it's been a long week' as a way of expressing how exhausting it was. Using time as a hedonistic metaphor in social communication is also something we see in everyday POTJs.

4.3 Passage of Time Judgments

Let's consider the following types of statements:

The time I spent with my friends in the pub just flew by, suddenly they were ringing the bell for last orders.

The two weeks on holiday just zoomed by, I couldn't believe it was time to pack up to go home already.

These are people remarking not on duration, but on how quickly time seemed to pass. They are examples of 'time flies when you're having fun'. An interesting characteristic about these situations is that people are not aware that time is 'flying' during the duration. It is only with some outside marker of duration—the ringing of the closing bell, the text from a friend—that one realizes that so much time has passed, and retrospectively concludes that time flew.

Contrast this with situations on which people report time crawling by:

I was so bored at work, the time just dragged by.

I thought the tedious film would never end.

In these situations, people may be aware of time passing differently *during the event* itself. Notice that these situations have high TR, and—with the tedious film—perhaps medium/high TU too, predicting the slow passage of time. These predictions of the Zakay (1992) model could also be explained by attention to time *keepers* rather than to time. In an absorbing, enjoyable situation, one's attention is drawn away from checking the time on a watch (indeed in a social situation, this could be interpreted as being rude). Whereas in an unenjoyable situation, one may keep checking the time in anticipation of the situation finally being over, so the clock on the wall will appear to crawl.

Other strange statements that people make, such as 'Christmas comes round quicker every year', or 'Time passes faster the older I get', also appear to be POTJs. The Christmas phenomenon seems particularly odd as it implies that one has a representation of how quickly Christmas should come around.

Why are we constantly taken by surprise? Why is this representation not updated? The same is true of time passing more quickly as we age, which is even more strange as it additionally implies that we have a memory of how quickly time seemed to pass when we were younger.

A common explanation of these phenomena is that, as we age, a year becomes a smaller percentage of our age. This is an old idea (e.g. Janet, 1877; James, 1890), but with no empirical evidence until the 1970s. In Lemlich's (1975) study, people were asked how quickly time passed now compared to when they were either half or a quarter of their age; results did not support the simple percentage rule. One had to be around four times as old before time went twice as fast. Wittman and Lehnhoff (2005), using a much wider age range and number of participants, found that this speeding-up effect actually peaks in middle age. Perhaps this is another example of people using time as a metaphor; expressing worries about mortality or being the frustration of where they are with their life/being stuck in routine. Baum, Boxley, and Sokolowski (1984) suggest this relationship with time is qualitatively different in the elderly. In people aged 62–94 years old, they found that quality of life was a better prediction of speed of time passing than age alone, that is, it passes more quickly for people who are happy.

Other POTJs can appear to be duration judgments. After a particularly dull, 20-minute conference speech, someone may remark to you, 'that talk felt like it was an hour long!', but if you were to ask the person for an accurate estimate of how long the speaker was talking, they would likely give an estimate of around 20 minutes, not an hour. So, again, this is an example of people using duration as hedonistic metaphor.

5 Relationships Between Types of Timing

5.1 Passage of Time Judgments and Prospective Timing

Before POTJs were separately defined, a lot of the phenomena of timing speeding up or slowing down were explained in terms of prospective timing (Zakay's model was originally called a 'prospective timing model'—POTJs at the time did not exist as a term). The idea being that either the internal clock runs faster in frustrating/unpleasant situations due to increased arousal, and so the world (including clocks) seems to run more slowly, or that attention to time (controlling the switch between pacemaker and accumulator) controls how much time is accumulated. In the example of being down the pub with friends, there is little attention to time, so less time is accumulated, so when the closing bell rings there is a discrepancy between accumulated time and actual time. This assumes that we are constantly prospectively timing, and that our sense of 'what time it is', or our temporal orientation, is determined by this internal clock.

There is very little experimental evidence to bring to bear on this relationship, but one intriguing study clearly suggests that POTJs are not determined by prospective timing. Droit-Volet and Wearden (2016) used smartphones to send the participants random alerts eight times a day for five days. When alerted, they made both a prospective time judgment (estimation and production) and made a POTJ. The results showed no relationship between the two types of judgment whatsoever.

5.2 POTJs and Retrospective Timing

One might suppose that a period of time that seems to drag as one passes through it would be judged as longer than a period of time that zips by. However, both experimentally and in everyday life, the opposite is the case. Understanding that these are two fundamentally different types of judgment allows us to understand this. A period of time which passes quickly is likely to be full of engaging activity; when one looks back over the period of time retrospectively, it seems longer, as there are more events/memories/units of information processed. Conversely, a tedious and boring period of time that drags will seem short retrospectively as there are fewer events and less information to remember. An example of this in the UK is the schools' six weeks holiday, which seems to zip by when in it, but when returning to school in September, July seems a very long time ago retrospectively.

5.3 Retrospective and Prospective Timing

As discussed earlier, it is an open question as to how much of everyday timing is retrospective or prospective. This division seems useful in experimental work, even if the results seem to be less clear-cut than was once first thought. One of the key questions in time perception, and in relating experimental work to everyday time experience, is the extent to which we are constantly prospectively timing, or not, and the degree to which retrospective timing is qualitatively different to prospective timing rather than being prospective timing with impoverished attentional resources.

6 Conclusion

Hopefully, this chapter has illuminated how complicated our relationship with time and duration is. Further, that some odd and often conflicting things that people report about time, or their timing behavior itself, can be far better understood when one categorizes them into different timing judgments. That one needs to be careful as to whether the phenomenon in question is a pure duration judgment or a hedonistic/feeling judgment. However, these categories appear not to be mutually exclusive, and until we have a better understanding of the blend of the different types of timing that people perform, and under what circumstances, our explanations will remain lacking. Like many other aspects of experimental psychology, the biggest challenge is reconciling what is now an impressive catalogue of lab-based data and models with real life behavior.

References

Baum, S. K., Boxley, R. L., & Sokolowski, M. (1984). Time perception and psychological well-being in the elderly. *Psychiatric Quarterly, 56*, 54–61.

Block, R. A., & Zakay, D. (1997). Prospective and retrospective duration judgments: A meta-analytic review. *Psychonomic Bulletin and Review, 4*, 184–197.

Droit-Volet, S., & Wearden, J. (2002). Speeding up an internal clock in children? Effects of visual flicker on subjective duration. *Quarterly Journal of Experimental Psychology, 55B*, 193–211.

Droit-Volet, S., & Wearden, J. (2016). Passage of time judgments are not duration judgments: Evidence from a study using experience sampling methodology. *Frontiers in Psychology, 7*, 176.

Gibbon, J. (1977). Scalar expectancy theory and Weber's law in animal timing. *Psychological Review, 84*, 279.

Gibbon, J., & Church, R. M. (1981). Time left: Linear versus logarithmic subjective time. *Journal of Experimental Psychology: Animal Behavior Processes, 7*, 87.

Gibbon, J., Church, R. M., & Meck, W. H. (1984). Scalar timing in memory. *Annals of the New York Academy of Sciences, 423*, 52–77.

Goldstone, S., & Lhamon, W. T. (1974). Studies of auditory-visual differences in human time judgment: 1. Sounds are judged longer than lights. *Perceptual and Motor Skills, 39*, 63–82.

Hicks, R. E., Miller, G. W., & Kinsbourne, M. (1976). Prospective and retrospective judgments of time as a function of amount of information processed. *American Journal of Psychology, 89*, 719–730.

James, W. (1890). *The principles of psychology*. New York: H. Holt.

Janet, P. (1877). Une Illusion D'Optique Interne. *Revue Philosophique de la France et de lÉtranger, 3*, 497–502.

Jones, L. A., Poliakoff, E., & Wells, J. (2009). Good vibrations: Human interval timing in the vibrotactile modality. *The Quarterly Journal of Experimental Psychology, 62*, 2171–2186.

Lemlich, R. (1975). Subjective acceleration of time with aging. *Perceptual and Motor Skills, 41*, 235–238.

Ortega, L., & Lopez, F. (2008). Effects of visual flicker on subjective time in a temporal bisection task. *Behavioural Processes, 78*, 380–386.

Penton-Voak, I. S., Edwards, H., Percival, A., & Wearden, J. H. (1996). Speeding up an internal clock in humans? Effects of click trains on subjective duration. *Journal of Experimental Psychology: Animal Behavior Processes, 22*, 307–320.

Roberts, S., & Church, R. M. (1978). Control of an internal clock. *Journal of Experimental Psychology: Animal Behavior Processes, 4*, 318.

Thomas, E. C., & Brown, I. (1974). Time perception and the filled-duration illusion. *Perception & Psychophysics, 16*, 449–458.

Treisman, M., Faulkner, A., Naish, P. L. N., & Brogan, D. (1990). The internal clock: Evidence for a temporal oscillator underlying time perception with some estimates of its characteristic frequency. *Perception, 19*, 705–748.

Wearden, J. (2016). *The psychology of time perception*. London: Springer.

Wearden, J. H. (2002). Traveling in time: A time-left analogue for humans. *Journal of Experimental Psychology: Animal Behavior Processes, 28*, 200–208.

Wearden, J. H. (2005). The wrong tree: Time perception and time experience in the elderly. In J. Duncan, L. Phillips, & P. McLeod (Eds.), *Measuring the mind: Speed, age, and control* (pp. 137–158). Oxford: Oxford University Press.

Wearden, J. H., Edwards, H., Fakhri, M., & Percival, A. (1998). Why "sounds are judged longer than lights": Application of a model of the internal clock in humans. *Quarterly Journal of Experimental Psychology, 51B*, 97–120.

Wearden, J. H., & Jones, L. A. (2007). Is the growth of subjective time in humans a linear or nonlinear function of real time? *The Quarterly Journal of Experimental Psychology, 60*, 1289–1302.

Wearden, J. H., Norton, R., Martin, S., & Montford-Bebb, O. (2007). Internal clock processes and the filled-duration illusion. *Journal of Experimental Psychology: Human Perception and Performance, 33*, 716.

Wearden, J. H., Williams, E. A., & Jones, L. A. (2017). What speeds up the internal clock? Effects of clicks and flicker on duration judgements and reaction time. *The Quarterly Journal of Experimental Psychology, 70*, 488–503.

Williams, E. A., Yüksel, E. M., Stewart, A. J., & Jones, L. A. (2019). Modality differences in timing and the filled-duration illusion: Testing the pacemaker rate explanation. *Attention. Perception and Psychophysics, 81*, 823–845.

Wittman, M., & Lehnhoff, S. (2005). Age effects in the perception of time. *Psychological Reports, 97*, 921–935.

Zakay, D. (1992). On prospective time estimation, temporal relevance and temporal uncertainty. In *Time, action and cognition* (pp. 109–117). Dordrecht: Springer.

Part II

Duration

5

The Temporal Oddball Effect and Related Phenomena: Cognitive Mechanisms and Experimental Approaches

Rolf Ulrich and Karin M. Bausenhart

Introspection suggests that time is an essential component of our cognition. Yet the concept of time itself is elusive because there is no adequate physical stimulus of time, and thus no sensory system is dedicated to the perception of time, unlike for vision or audition (see Grondin, 2001; Wearden, 2016). Hence, time might be a construction of our minds, allowing us to structure external and internal events in a coherent way. Given this distinct status of time, it is not astonishing that our perception of time is especially prone to illusions.

In particular, many studies have documented that the perception of time can be distorted by various non-temporal factors (Eagleman, 2008; Grondin, 2010). For instance, psychophysical studies have demonstrated that stimulus properties like stimulus intensity, stimulus magnitude, stimulus complexity, and the sensory modality of a stimulus modulate the perceived duration of a stimulus (for a review, see Matthews & Meck, 2016). For example, participants judge the duration of physically large stimuli longer than the duration of smaller ones (Casasanto & Boroditsky,

---

R. Ulrich (✉) • K. M. Bausenhart
University of Tübingen, Tübingen, Germany
e-mail: rolf.ulrich@uni-tuebingen.de; karin.bausenhart@uni-tuebingen.de

V. Arstila et al. (eds.), *The Illusions of Time*,
https://doi.org/10.1007/978-3-030-22048-8_5

2008; Mo & Michalski, 1972; Rammsayer & Verner, 2014). This size-time illusion occurs even when objects are imaged rather than visually perceived (Birngruber & Ulrich, 2018).

This chapter reviews an intriguing time illusion, the temporal *oddball effect* (OE). According to the OE, rare events are perceived as lasting longer than more frequent events of the same physical duration (Tse, Intriligator, Rivest, & Cavanagh, 2004). In Sect. 1, we describe the basic experimental paradigms that have been employed to induce the OE and also highlight some experimental caveats. Section 2 reviews studies that tested the generality of this effect, provided experimental variants, and proposed various accounts of the OE. Section 3 reviews evidence that perceived stimulus duration within oddball paradigms is driven by two independent mechanisms that exert opposing effects on perceived duration.

1 Basic Experimental Paradigms

The *serial-streaming paradigm* is the typical experimental paradigm to induce and study the OE (Fig. 5.1a). In this paradigm, participants see a stream of standard stimuli, for example, small black discs, in the center of a computer screen. All these standard stimuli have a constant duration (e.g., 800 msec). After a number of standards (e.g., 7 to 12), a deviant stimulus, the oddball, is randomly interspersed within the stream of homogenous standards. The oddball usually differs from the standards in one physical feature, for example, size, color, shape, or even tone pitch in case of auditory stimuli. Moreover, the oddball varies in duration—in some instances, the oddball's duration is somewhat shorter than the standard duration and in other instances, it is somewhat longer than the standard duration. Following each oddball presentation, the participant judges the oddball's duration in comparison to the duration of the standards. Specifically, the participant is usually asked to judge whether the oddball's duration was "shorter" or "longer" than the duration of the standards (i.e., comparative judgment).

A psychometric function is fitted to the proportion of "longer" judgments as a function of the oddball duration. From this function, the point of subjective equality (PSE) is estimated, that is, the duration of the

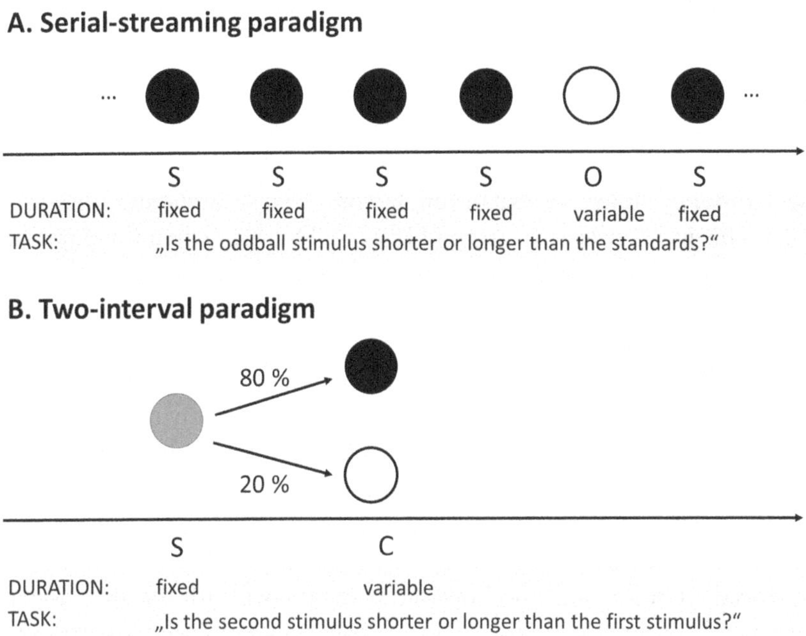

Fig. 5.1 Oddball effects can be observed in two general paradigms. In the serial-streaming paradigm (a), participants compare the duration of a rare oddball stimulus (O) to the duration of repeated standard stimuli (S) within a stimulus stream. In each trial of the two-interval paradigm (b), participants compare the duration of a comparison stimulus (C) to the duration of a single standard (S). The type of comparison stimulus varies from trial to trial, and some stimulus types are presented rather frequently (e.g., in 80% of all trials), while others are more rarely presented (e.g., in 20% of all trials)

oddball that appears equal to the duration of a standard. Using this psychophysical procedure, it has been repeatedly observed that participants tend to overestimate the duration of oddballs, as indicated by PSEs smaller than the standard duration. Less technically speaking, the duration of the oddball must be shorter than the duration of the standard in order to perceive both stimuli as equally long. For example, the size of OE would be equal to 50 msec for PSE = 750 msec and a standard duration of 800 msec.

In the pioneering OE study of Tse et al. (2004), the OE was investigated for various types of oddballs and standard durations ranging from

70 to 4200 msec. Oddball stimuli were, for example, an expanding disk among stationary disks, a stationary disk among expanding disks, a colored circle among black circles, or a square among circles. They also employed auditory oddballs and different psychophysical methods for assessing the OE. In order to assess the size of the OE, Tse et al. calculated the temporal subjective expansion factor (TSE = Standard duration/PSE). This index revealed a strong OE (i.e., TSE > 1) when the duration of the standards exceeded approximately 200 msec. In fact, the observed TSE was sometimes close to 1.4, suggesting that participants overestimated oddball duration by 40%.

Tse et al. (2004) attributed the OE to attention, assuming that unexpected events (like the oddball) receive more attention than expected ones (like the standard). More specifically, they argue that unexpected stimuli are assumed to "increase attention to the duration judgement," and thus "fewer temporal cues are missed" (p. 1184) during the presentation of an oddball than during the presentation of a standard. This enhanced attention account, however, is at variance with the attentional-gate model (Zakay & Block, 1997), accordingly enhanced attention to non-temporal information (i.e., the color of the oddball) should decrease rather than increase perceived stimulus duration (see Birngruber, Schröter, & Ulrich, 2014, p. 815; Macar, Grondin, & Casini, 1994; Thomas & Weaver, 1975; Ulrich, Nitschke, & Rammsayer, 2006). In brief, if attention is distracted from temporal information, this model holds that fewer rather than more pulses would be accumulated by an internal clock during the presentation of an oddball.

Thus, this view predicts that the duration of oddballs should be underestimated rather than overestimated, a prediction that is opposite to the observed OE. As noted by Seifried and Ulrich (2010), the large TSE observed in this original study was presumably partly due to placing the oddball durations not symmetrically around the standard duration but around the expected PSE. In fact, these authors observed a smaller TSE when a symmetrical arrangement of the oddball durations around the standard durations was used. With such a symmetrical arrangement, the TSE typically ranges from 5% to 20% (compare also Pariyadath & Eagleman, 2007). Nevertheless, the OE is rather robust, because it is not only observed with comparative judgments but also with equality judgments (Birngruber et al., 2014). Since equality judgments seem to be less prone to response biases than comparative judgments (Schneider & Komlos, 2008), this result supports the proposal that the origin of the OE might be perceptual.

The size of the OE seems also to vary between groups of different temporal experiences. For example, in a study by Nazari, Ebneabbasi, Jalalkamali, and Grondin (2018), musicians and non-musicians performed an auditory oddball task. Oddballs deviated from standards in pitch. Not only did musicians demonstrate a higher discrimination performance than non-musicians, they also revealed no oddball effect in contrast to non-musicians. Independent replications, however, would be required before one can safely conclude that the OE is modulated by musical experience.

A drawback of the serial-streaming paradigm is that the quantification of the OE can be problematic (Birngruber, Schröter, & Ulrich, 2015a). Within this paradigm, the OE has been quantified as the difference between the OE and the objective standard duration. Consequently, the OE may not only result from an overestimation of the oddballs but could also be due to an underestimation of the standards (see also Matthews, 2011). Moreover, the so-called negative time-order error could contaminate this difference measure, that is, the tendency of participants to underestimate the second of two successive stimuli (Allan, 1977).

In order to control for such nuisance, Birngruber et al. (2015a) presented four successive standards in the center of the screen. A marker indicated the serial position of each stimulus within the stream. The fifth stimulus could be either an oddball or a standard. Participants were asked to judge the duration of this last stimulus with respect to the duration of the preceding standards. The result revealed an overestimation of the last stimulus' duration even when this stimulus was a standard and thus identical to the four preceding stimuli. This indicates that the time-order error contributes to the OE in the classical serial-streaming paradigm.

Nonetheless, the overestimation was even larger when the last stimulus was an oddball instead of a standard, thus revealing a genuine temporal overestimation of oddball stimuli. The results of Birngruber et al. (2015a) also constrain the generality the OE. In two further experiments, the standards were not presented at the screen center, but arranged in a spatially predictable, circular order. In each trial, the first standard appeared at the 12 o'clock position, and the successive standards at 3, 6, and 9 o'clock positions. The final stimulus in this sequence, the target stimulus, again appeared at the 12 o'clock position. In 50% of all trials, this target was also a standard stimulus. In the remaining trials, one of four different oddballs appeared as target, such that the effective oddball probability was 12.5%. No reliable difference in PSE was observed between standard targets and oddball targets. In a further experiment, standard targets were presented in 70% of all trials, and one out of nine oddballs in 30% of all trials, yielding an effective oddball probability of 30%/9 = 3.3%. Even under this more extreme probability condition, no reliable PSE difference, and thus no OE, was observed. These results suggest that repeated stimuli in different locations are represented as distinct objects, and thereby make the oddball less unique.

The *two-interval paradigm* (Fig. 5.1b) is another psychophysical approach for assessing the OE (Ulrich et al., 2006). In each trial of this paradigm, a standard stimulus of constant duration is followed by a comparison stimulus with duration varying from trial to trial. Participants are asked to compare the durations of the two stimuli and to judge whether the duration of the second stimulus is shorter or longer than the duration of the first stimulus. As in the serial-streaming paradigm, the proportion of "longer judgments" at each comparison duration is used to generate a psychometric function from which the PSE and the difference limen (DL) can be estimated. In Ulrich et al., changes in expectancy were induced by presenting one type of comparison stimulus (e.g., a filled red circle) more frequently than another stimulus type (e.g., a blue square). More specifically, a frequent comparison of constant color and shape appeared in 70% of all trials, whereas in the remaining 30% of all trials, one out of nine infrequent comparisons differing in color and shape was presented. Thus, the frequency of an infrequent comparison was 30%/9 = 3.3%.

One advantage of this design is that it controls for the time-order error because PSEs are compared between frequent and infrequent comparisons, which are both equally prone to this error. Thus, any difference in perceived duration between these two conditions cannot be attributed to an artifact produced by the time-order error. Utilizing this paradigm,

Ulrich et al. (2006) observed smaller PSEs for frequent than infrequent comparisons, that is, the duration of infrequent comparisons was judged as longer than the duration of frequent comparisons. A reliable OE of about 15 msec, however, was only observed for longer standards (>400 msec), but not for short ones (100 msec). Moreover, steeper psychometric functions (i.e., a smaller difference limen) were observed for infrequent than for frequent comparisons, indicating that temporal discrimination sensitivity is higher for infrequent than for frequent comparisons.

Ulrich et al. favored an arousal account to explain this pattern of results. The basic idea is that novel events raise the level of arousal (Sokolov, 1963), thereby speeding up the pacemaker of the internal clock (e.g., Penton-Voak, Edwards, Percival, & Wearden, 1996). The higher the pacemaker's pulse rate, the more temporal pulses are emitted and accumulated within a given time interval. Since the number of pulses corresponds to perceived duration, this account can explain that infrequent stimuli (oddballs) appear to last longer than frequent ones. It also explains the counter-intuitive finding that the duration of infrequent stimuli can be resolved with higher temporal precision, because a higher pulse rate of the pacemaker would also promote temporal discrimination (e.g., Rammsayer & Ulrich, 2001). Moreover, this account can also accommodate the finding that the OE emerges only for longer durations, if one accepts the plausible auxiliary assumption that the increase in arousal takes some time to affect the timing mechanism. Electrophysiological support for this arousal account of the OE has been reported by Ernst et al. (2017).

In sum, then, the OE has been observed in two paradigms, the *serial-streaming paradigm* and the *two-interval paradigm*. Many follow-up studies on this phenomenon not only extended the empirical scope of the OE but also examined possible alternative explanations of the OE. In the following, we will review these studies in chronological order. To our mind, these studies demonstrated that the OE is a rather complex phenomenon of duration perception, which presumably cannot be broken down to a single explanation.

2 Repetition Suppression and Predictive Coding

As mentioned before, Tse et al. (2004) and Ulrich et al. (2006) assumed that the subjective duration of the oddball is expanded relative to the subjective duration of each single standard stimulus within the streaming paradigm. However, Pariyadath and Eagleman (2007, 2008) and Eagleman and Pariyadath (2009) have argued that the repetition of the standard stimuli diminishes the neuronal response to these stimuli, an adaptation effect that is also known as *repetition suppression*. Moreover, they assume that subjective stimulus duration reflects the size of this neuronal response. According to this alternative account, the OE results from a contraction of perceived standard duration rather than a dilation of perceived oddball duration. Others have favored, however, an alternative interpretation why the neuronal response should be larger for oddballs than for standards (Schindel, Rowlands, & Arnold, 2011). According to this alternative, referred to as *predictive coding*, an unpredicted stimulus like the oddball evokes more neuronal activation than a predicted stimulus like a standard stimulus. Thus, both repetition suppression and predictive coding have been suggested to contribute to the OE, and both accounts entail a relatively larger neuronal response associated with the oddball stimulus than with the standard stimuli.

Thus, both views might be regarded as two sides of the same coin because they both imply that oddballs evoke a larger neuronal response than standards.[1] Nonetheless, it is conceivable that repetition suppression is linked to bottom-up processes because stimulus adaptation may reside at low processing levels, whereas prediction and expectation are typically associated with processes operating at higher cognitive levels. According to this consideration, then, the two views would not merely be two sides of the same coin but would reflect mechanisms operating at different

[1] Note that the predictive coding view is akin to the OE accounts suggested by Tse et al. (2004) and Ulrich et al. (2006), who have argued that the OE reflects a higher pacemaker rate due to increased attention and immediate arousal, respectively. These accounts assume a higher activation of the pacemaker for oddballs than for standards, and hence interpret the OE within the traditional internal clock framework.

processing stages. In the following, we will review several findings that are pertinent to these two accounts.

First, these neuronal accounts are consistent with the observation that oddballs presented later in the stimulus stream are judged to be longer than oddballs presented earlier in the stimulus stream (Birngruber et al., 2014; Kim & McAuley, 2013; Nazari et al., 2018; Pariyadath & Eagleman, 2012). According to the repetition suppression account, the neuronal response to standard stimuli weakens with each additional presentation of a standard. Consequently, the duration of an oddball at a late position within the stimulus stream should appear relatively long compared to the subjective duration of the preceding standard, and also compared to an oddball stimulus presented at earlier positions within the stimulus stream. Likewise, the predictive coding approach assumes that repetition of a standard stimulus enhances its predictive coding. Thus, a larger violation of this prediction would result when the oddball occurs at a rather late stimulus position, which in turn would evoke an especially large neuronal response, and thus prolong perceived oddball duration.

However, Kim and McAuley (2013) have suggested a different account for this position effect. They argue that oddballs in later positions are perceived more quickly than at earlier positions because subjects develop a *temporal* expectation for the occurrence of the oddball during the presentation of the stimulus stream. The longer the stimulus stream goes on without an oddball presentation, the higher the probability that the oddball will appear at one of the remaining positions. This increase in probability leads to higher temporal expectations at later positions within the stimulus stream. There is converging evidence that such temporal expectations enhance the speed of stimulus detection (Seifried, Ulrich, Bausenhart, Rolke, & Osman, 2010), and as a result may increase perceived stimulus duration (Mo & George, 1977; Grondin & Rammsayer, 2003). Nevertheless, this account cannot be the sole cause of the OE because the OE also emerges in the two-interval paradigm (Ulrich et al., 2006) in which temporal expectation does not differ between frequent and infrequent stimuli. Thus, although this temporal expectation account might contribute to the OE, it seems not sufficient to fully explain it (see also Matthews, 2015, Experiments 8 and 9).

Second, an OE cannot be elicited when the standards appear at different spatial locations but only when they appear at the same location (Birngruber et al., 2015a). Presumably, repeated standards at different spatial locations suffer less from bottom-up adaptation than standards appearing at the same spatial location as in the standard serial-streaming paradigm. Conversely, stimuli presented at different locations may not produce a stable predictive code, and consequently an OE would not emerge. However, New and Scholl (2009) have reported conflicting results. In their experiments, an odd visual stimulus appeared at a randomized peripheral position together with a central target that prompted participants to judge its presentation duration. The odd peripheral stimuli did produce an OE. In a similar vein, the oddball effect is enhanced when a visual oddball is accompanied by an auditory stimulus, but not when an auditory oddball is accompanied by a visual stimulus (Chen & Yeh, 2009). More research would be required to constrain the exact stimulus properties that are capable of eliciting an OE when stimulus position varies from trial to trial.

Third, both repetition suppression and predictive coding suggest that the size of the OE should depend on the similarity between the oddball and the standards: The more the oddball resembles the standards, the smaller should be the neuronal response elicited by the oddball, and accordingly, the magnitude of the OE. Such a gradual effect was reported by Schindel et al. (2011). In one of their experiments, a white bar in the middle of the computer screen served as test stimulus. The oddball was a gray bar and had the same orientation as the standard, or was rotated from the standard by a small amount (±15°) or a larger amount (±45°). In fact, the size of the OE increased with the amount of oddball rotation, a finding that is in line with the repetition suppression account.

In an additional experiment, however, Schindel et al. (2011) provided evidence against the view that the OE emerges from low-level visual adaptation. For example, a dim oddball among bright standards still elicited an OE. This particular result is at variance with a simple link between the magnitude of the neuronal response and perceived duration because the neuronal response of bright stimuli is stronger than that of weak stimuli. Based on the results of their study, Schindel et al. concluded that a reduction of the neuronal response cannot exclusively account for the

OE. Instead, they favor predictive coding as an account for the observed OE in their study.

A gradual modulation of the OE can not only be observed for visual stimuli but also for auditory ones (Kim & McAuley, 2013). In their first experiment, they presented a stream of sine tones as standards and oddballs, and the frequency of the oddballs could be either close to or far from the frequency of the standards. For instance, among a stream of 400 Hz standards, there were oddballs of either 500 Hz (near condition) or 700 Hz (far condition). In general, the PSE in the far condition was smaller than the PSE in the near condition, indicating that the OE is gradually modulated by pitch distance. However, Fromboluti and McAuley (2018) provide novel results suggesting that not the similarity between the pitch of the oddball and the standard is crucial for eliciting an OE, but whether or not an oddball's pitch is outside a pitch window that is established by context pitch stimuli (pitch-window hypothesis). It has to be seen whether their results generalize to the visual domain. If so, this would neither support predictive coding nor repetition suppression.

Fourth, predictive coding as well as repetition suppression can also account for effects of immediate stimulus repetition on time perception in the two-interval paradigm (Birngruber, Schröter, & Ulrich, 2015b; Matthews, 2011). For example, in the study by Matthews, participants saw two successively presented images and judged whether the duration of the second image was shorter or longer than that of the first stimulus. On repeat trials, the two images (e.g., house—house) were identical, whereas on novel trials, the two images were different (e.g., car—telephone pole). Larger PSE values were observed for repeated than for novel images, demonstrating that the duration of repeated stimuli is shortened or that of novel ones is increased. Because repeat and novel trials were equally likely, the observed repetition effect cannot be attributed to stimulus expectation induced by stimulus probability (i.e., Ulrich et al., 2006). Additional experiments demonstrated that the repetition effect (a) only emerges for immediate but not for delayed repetitions and (b) can also be established with alternative judgment tasks. The immediate repetition effect was conceptually replicated by Birngruber et al. (2015b). These authors used pseudowords and also letter strings instead of images of natural scenes and confirmed that the immediate repetition effect also emerges for semantically meaningless stimuli. To our mind, the immediate repetition effect is the strongest evidence so far for the repetition suppression/predictive coding accounts.

Fifth, it has also been examined whether a violation of high-level stimulus expectations elicits an OE (Cai, Eagleman, & Ma, 2015). These authors examined whether violations of expectations that are based on overlearned sequences would produce an OE. On each trial, participants watched such a sequence, as for example, four digits in ascending order (1-2-3-4), which was followed by a target digit in the fifth stimulus position. Participants were to judge the presentation duration of the target digit, which could be a 5 (expected), a 4 (repeated), or a 6 (unexpected). The authors conjectured that if repetition would reduce perceived duration, a 4 after 1-2-3-4 should be judged as shorter than a 6. Moreover, if high-level expectation is involved in the perception of the target digit, a 5 after 1-2-3-4 should be judged as shorter than a 6. Consistent with their prediction, the duration of a repeated digit appeared shorter than that of an unexpected digit. However, their data did not suggest a difference between the perceived duration of expected and unexpected digits. These and further results of their study led the authors to the conclusion that perceived duration is reduced by stimulus repetition but not by high-level expectation such as predictive coding.

Thus, the results of their study support the repetition suppression account of the OE, and point to lower processing levels as the source of this effect. However, this cannot explain why the duration of infrequent comparisons is judged as longer than the one of frequent comparisons in the two-interval paradigm (Ulrich et al., 2006). Therefore it seems possible that the OE is a sign of different mechanisms operating at different processing levels, a point that we will consider in the remainder of this article.

3 Can a Unitary Account Explain the OE?

Experimental evidence supporting the notion that more than one mechanism operates on the perceived duration of odd stimuli was reported by Matthews (2015). He adapted an experimental approach that was used by neuroscientists to examine whether repetition suppression effects can be attributed to low-level adaptation processes or to high-level predictive coding. In a two-interval paradigm, participants saw pairs of faces and judged whether the second face appeared longer than the first one. As in Birngruber et al. (2015b) and in Matthews (2011), there were novel and repeat trials. One block of trials contained less repeats than novel trials (low repetition rate), whereas another block contained more repeats than novel trials (high repetition rate). As in the two previous studies, participants judged the duration of novel stimuli as longer as the duration of repeated stimuli. However, in contrast to a pure low-level adaptation process, this effect was modulated by repetition rate. More specifically, the effect was less pronounced when the repetition rate was high than when it was low.

Theoretically most important, this interaction effect is at variance with the predictive coding account, which implies that the effect should increase (rather than decrease) when a block contains more repeated than novel stimuli. This effect could be observed with other types of stimuli and judgment tasks as well. For example, in one experiment, the effect actually reversed, that is, repeated stimuli were actually judged to appear longer than novel ones when repeats were frequent, and thus especially expected.

In conclusion, these results of Matthew's study can neither be attributed to an account in which stimulus expectation (i.e., predictive coding) compresses perceived duration, nor can they solely be attributed to repetition suppression. Matthews has concluded that the complete pattern of his results argues against a unitary explanation of the OE. Instead, he suggested that stimulus repetition (low-level adaptation) and repetition expectation (top-down expectation) exert opposing effects on perceived duration: low-level adaptation decreases the perceived duration of a repeated stimulus, whereas top-down expectation increases the perceived duration of an expected stimulus (see also Matthews & Gheorghiu, 2016; Matthews & Meck, 2016).

The latter prediction was rather directly tested by Birngruber, Schröter, Schütt, and Ulrich (2017). These authors introduced a novel paradigm to isolate the effect of top-down stimulus expectation from repetition and probability effects. On each trial, one of two visual stimuli was presented and subjects classified this stimulus as short or long in a temporal bisection task. Immediately before stimulus presentation, participants indicated which of the two stimuli they expected (self-generated expectations; Bernstein & Reese, 1965). In line with the prediction of the top-down expectation account, the duration of expected stimuli was judged longer than the duration of unexpected ones. This outcome contrasts with the interpretation of previous studies according to which stimulus expectation should shorten perceived duration (e.g., Matthews, 2011; Pariyadath & Eagleman, 2007; Schindel et al., 2011; Ulrich et al., 2006). Makwana and Srinivasan (2017) reached basically the same conclusion as Birngruber et al. (2017) did with a somewhat different experimental setup, namely that an intended outcome is perceived longer than an unintended one even if both outcomes are of identical physical duration.

The idea that high-level expectations increase rather than decrease perceived duration receives further support from a study conducted by Wehrman, Wearden, and Sowman (2018) employing the serial-streaming paradigm rather than the two-interval paradigm. They sought to distinguish the effects of repetition suppression on expectation by adapting established expectancy manipulations from the variable foreperiod paradigm in reaction time research.

For example, in one of their experiments (positional orienting), the oddball could either appear in the fourth or sixth serial position. Before each stimulus stream, a cue provided advance information about the serial position of the oddball. This cue was valid in 70% of the trials. The authors reasoned that if an invalid cue indicated position 6 (i.e., the oddball was actually presented at position 4), the oddball would be unexpected compared to when the oddball would be presented at position 4 after a valid cue. In contrast, if an invalid cue indicated position 4, then, after the fourth stimulus position passes without presentation of the oddball, it must occur with probability one at position 6. Thus, the oddball would be expected at this later position, despite the invalid cue. The results of this and further experiments are consistent with the idea that expected oddballs are perceived as being longer than unexpected ones and thus generalize the aforementioned effects of higher-level expectancy to the serial-streaming paradigm.

4 Conclusions

In this article, we reviewed several studies on the temporal oddball effect. This effect has been studied within the serial-streaming paradigm and the two-interval design. There is no doubt that the oddball effect is a pronounced and robust cognitive illusion. Consequently, the oddball paradigm has also been used as an experimental tool to address other research questions, such as whether monetary reward alters the perception of time (Failing & Theeuwes, 2016).

However, explaining the cognitive basis of the oddball effect has turned out to be a challenge. The currently most promising explanation assumes that two mechanisms exert opposing effects (see Skylark & Gheorghiu, 2017): (a) repetition suppression or stimulus adaptation operating at low-level perceptual stages decreases the perceived duration of repeated stimuli, and thus the duration of a novel stimulus, the oddball, is perceived longer compared to the preceding repeated stimuli; (b) high-level stimulus expectation, however, involves that the duration of an expected stimulus appears longer than that of unexpected stimuli. Future research should be directed toward generalizing these results to other sense modalities and to how these findings connect to theories of time perception and the neurosciences.

References

Allan, L. G. (1977). The time-order error in judgments of duration. *Canadian Journal of Psychology/Revue Canadienne de Psychologie, 31*, 24–31. https://doi.org/10.1037/h0081647

Bernstein, I. H., & Reese, C. (1965). Behavioral hypotheses and choice reaction time. *Psychonomic Science, 3*, 259–260.

Birngruber, T., Schröter, H., Schütt, E., & Ulrich, R. (2017). Stimulus expectation prolongs rather than shortens perceived duration: Evidence from self-generated expectations. *Journal of Experimental Psychology: Human Perception and Performance, 44*, 117–127. https://doi.org/10.1037/xhp0000433

Birngruber, T., Schröter, H., & Ulrich, R. (2014). Duration perception of visual and auditory oddball stimuli: Does judgment task modulate the temporal oddball effect? *Attention, Perception & Psychophysics, 76*, 814–828. https://doi.org/10.3758/s13414-013-0602-2

Birngruber, T., Schröter, H., & Ulrich, R. (2015a). Introducing a control condition in the classic oddball paradigm: Oddballs are overestimated in duration not only because of their oddness. *Attention, Perception, and Psychophysics, 77*, 1737–1749. https://doi.org/10.3758/s13414-015-0868-7

Birngruber, T., Schröter, H., & Ulrich, R. (2015b). The influence of stimulus repetition on duration judgments with simple stimuli. *Frontiers in Psychology, 6*, 1213. https://doi.org/10.3389/fpsyg.2015.01213

Birngruber, T., & Ulrich, R. (2018). Perceived duration increases not only with physical, but also with implicit size. *Journal of Experimental Psychology: Learning, Memory, and Cognition, 45*, 969–979. https://doi.org/10.1037/xlm0000628

Cai, M. B., Eagleman, D. M., & Ma, W. J. (2015). Perceived duration is reduced by repetition but not by high-level expectation. *Journal of Vision, 15*, 1–17. https://doi.org/10.1167/15.13.19

Casasanto, D., & Boroditsky, L. (2008). Time in the mind: Using space to think about time. *Cognition, 106*, 579–593. https://doi.org/10.1016/j.cognition.2007.03.004

Chen, K.-M., & Yeh, S.-L. (2009). Asymmetric cross-modal effects in time perception. *Acta Psychologica, 130*, 225–234. https://doi.org/10.1016/j.actpsy.2008.12.008

Eagleman, D. M. (2008). Human time perception and its illusions. *Current Opinion in Neurobiology, 18*, 131–136. https://doi.org/10.1016/j.conb.2008.06.002

Eagleman, D. M., & Pariyadath, V. (2009). Is subjective duration a signature of coding efficiency? *Philosophical Transactions of the Royal Society of London. Series B, Biological Sciences, 364*, 1841–1851. https://doi.org/10.1098/rstb.2009.0026

Ernst, B., Reichard, S. M., Riepl, R. F., Steinhauser, R., Zimmermann, S. F., & Steinhauser, M. (2017). The P3 and the subjective experience of time. *Neuropsychologia, 103*, 12–19. https://doi.org/10.1016/j.neuropsychologia.2017.06.033

Failing, M., & Theeuwes, J. (2016). Reward alters the perception of time. *Cognition, 148*, 19–26. https://doi.org/10.1016/j.cognition.2015.12.005

Fromboluti, E. K., & McAuley, J. D. (2018). Perceived duration of auditory oddballs: Test of a novel pitch-window hypothesis. *Psychological Research*. https://doi.org/10.1007/s00426-018-1124-2

Grondin, S. (2001). From physical time to the first and second moments of psychological time. *Psychological Bulletin, 127*, 22–44.

Grondin, S. (2010). Timing and time perception: A review of recent behavioral and neuroscience findings and theoretical directions. *Attention, Perception and Psychophysics, 72*, 561–582.

Grondin, S., & Rammsayer, T. (2003). Variable foreperiods and temporal discrimination. *Quarterly Journal of Experimental Psychology Section A: Human Experimental Psychology, 56*, 731–765. https://doi.org/10.1080/02724980244000611

Kim, E., & McAuley, J. D. (2013). Effects of pitch distance and likelihood on the perceived duration of deviant auditory events. *Attention, Perception & Psychophysics, 75*, 1547–1558. https://doi.org/10.3758/s13414-013-0490-5

Macar, F., Grondin, S., & Casini, L. (1994). Controlled attention sharing influences time estimation. *Memory & Cognition, 22*, 673–686. https://doi.org/10.3758/BF03209252

Makwana, M., & Srinivasan, N. (2017). Intended outcome expands in time. *Scientific Reports, 7*, 6305. https://doi.org/10.1038/s41598-017-05803-1

Matthews, W. J. (2011). Stimulus repetition and the perception of time: The effects of prior exposure on temporal discrimination, judgment, and production. *PLoS One, 6*, e19815. https://doi.org/10.1371/journal.pone.0019815

Matthews, W. J. (2015). Time perception: The surprising effects of surprising stimuli. *Journal of Experimental Psychology: General, 144*, 172–197. https://doi.org/10.1037/xge0000041

Matthews, W. J., & Gheorghiu, A. I. (2016). Repetition, expectation, and the perception of time. *Current Opinion in Behavioral Sciences, 8*, 110–116. https://doi.org/10.1016/j.cobeha.2016.02.019

Matthews, W. J., & Meck, W. H. (2016). Temporal cognition: Connecting subjective time to perception, attention, and memory. *Psychological Bulletin, 142*, 865–907. https://doi.org/10.1037/bul0000045

Mo, S. S., & George, E. J. (1977). Foreperiod effect on time estimation and simple reaction time. *Acta Psychologica, 41*, 47–59. https://doi.org/10.1016/0001-6918(77)90010-5

Mo, S. S., & Michalski, V. A. (1972). Judgment of temporal duration of area as a function of stimulus configuration. *Psychonomic Science, 27*, 97–98. https://doi.org/10.3758/BF03328900

Nazari, M. A., Ebneabbasi, A., Jalalkamali, H., & Grondin, S. (2018). Time dilation caused by oddball serial position and pitch deviancy: A comparison of musicians and nonmusicians. *Music Perception, 35*, 425–436.

New, J. J., & Scholl, B. J. (2009). Subjective time dilation: Spatially local, object-based, or a global visual experience? *Journal of Vision, 9*, 4. https://doi.org/10.1167/9.2.4

Pariyadath, V., & Eagleman, D. (2007). The effect of predictability on subjective duration. *PLoS One, 2*, e1264.

Pariyadath, V., & Eagleman, D. M. (2008). Brief subjective durations contract with repetition. *Journal of Vision, 8*, 1–6. https://doi.org/10.1167/8.16.11

Pariyadath, V., & Eagleman, D. M. (2012). Subjective duration distortions mirror neural repetition suppression. *PLoS One, 7*, e49362. https://doi.org/10.1371/journal.pone.0049362

Penton-Voak, I. S., Edwards, H., Percival, A., & Wearden, J. H. (1996). Speeding up an internal clock in humans? Effects of click trains on subjective duration. *Journal of Experimental Psychology: Animal Behavior Processes, 22*, 307–320.

Rammsayer, T., & Ulrich, R. (2001). Counting models of temporal discrimination. *Psychonomic Bulletin & Review, 8*, 270–277.

Rammsayer, T. H., & Verner, M. (2014). The effect of nontemporal stimulus size on perceived duration as assessed by the method of reproduction. *Journal of Vision, 14*, 17. https://doi.org/10.1167/14.5.17

Schindel, R., Rowlands, J., & Arnold, D. (2011). The oddball effect: Perceived duration and predictive coding. *Journal of Vision, 11*, 17. https://doi.org/10.1167/11.2.17

Schneider, K. A., & Komlos, M. (2008). Attention biases decisions but does not alter appearance. *Journal of Vision, 8*, 3. https://doi.org/10.1167/8.15.3

Seifried, T., & Ulrich, R. (2010). Does the asymmetry effect inflate the temporal expansion of odd stimuli? *Psychological Research, 74*, 90–98. https://doi.org/10.1007/s00426-008-0187-x

Seifried, T., Ulrich, R., Bausenhart, K. M., Rolke, B., & Osman, A. (2010). Temporal preparation decreases perceptual latency: Evidence from a clock paradigm. *The Quarterly Journal of Experimental Psychology, 63*, 2432–2451. https://doi.org/10.1080/17470218.2010.485354

Skylark, W. J., & Gheorghiu, A. I. (2017). Further evidence that the effects of repetition on subjective time depend on repetition probability. *Frontiers in Psychology, 8*, 1915. https://doi.org/10.3389/fpsyg.2017.01915

Sokolov, E. N. (1963). Higher nervous functions: The orienting reflex. *Annual Review of Physiology, 25*, 545–580.

Thomas, E. A. C., & Weaver, W. B. (1975). Cognitive processing and time perception. *Perception & Psychophysics, 17*, 363–367. https://doi.org/10.3758/BF03199347

Tse, P. U., Intriligator, J., Rivest, J., & Cavanagh, P. (2004). Attention and the subjective expansion of time. *Perception & Psychophysics, 66*, 1171–1189.

Ulrich, R., Nitschke, J., & Rammsayer, T. (2006). Perceived duration of expected and unexpected stimuli. *Psychological Research, 70*, 77–87.

Wearden, J. H. (2016). *The psychology of time perception*. London: Palgrave Macmillan.

Wehrman, J. J., Wearden, J., & Sowman, P. (2018). The expected oddball: Effects of implicit and explicit positional expectation on duration perception. *Psychological Research*, 1–15. https://doi.org/10.1007/s00426-018-1093-5

Zakay, D., & Block, R. A. (1997). Temporal cognition. *Current Directions in Psychological Science, 6*, 12–16. https://doi.org/10.1111/1467-8721.ep11512604

6

Why the Intrinsic Value of Hedonic Sensations Is Not Quantifiable

Ingmar Persson

1 A Discrepancy Between the Duration and Intensity of Sensations

While the hedonist view that the *only* things that are intrinsically good and bad for us are pleasure and pain is dubious, it is virtually incontestable that they are intrinsically good and bad for us. I shall here focus on the simplest kinds of pleasures and pains, sensations of *physical* pleasure and pain, *hedonic sensations*. Examples are the pleasures of an orgasm and of smelling freshly baked bread when hungry, and the pains of a toothache and of something very hot or cold. Hedonic sensations stand in contrast to such *emotional* pleasures and pains as the pleasure of success and revenge, and the pains of bereavement and humiliation. The value of emotional pleasures and pains is a more complex matter, since they have objects whose value might add or detract from the value of their feeling

I. Persson (✉)
University of Gothenburg, Gothenburg, Sweden

The Oxford Uehiro Centre for Practical Ethics, Oxford, UK
e-mail: ingmar.persson@filosofi.gu.se

V. Arstila et al. (eds.), *The Illusions of Time*,
https://doi.org/10.1007/978-3-030-22048-8_6

component, for instance, the fact that it is bad to take revenge might detract from the goodness of the pleasure we get from it. This complication does not arise as regards physical or sensory pleasure and pain.

A standard view is that the intrinsic value hedonic sensations have for the subjects feeling them is a function of their *duration* and *intensity*. Prima facie, it might appear that both of these variables are quantifiable and, thus, that the intrinsic value these sensations have for their subjects is quantifiable. The principal aim of this chapter is to explain that this impression is false because sensory intensity is not quantifiable, that is, there is not even a rough numerical scale along which it can be measured.

It is plausible to think that it is intrinsically good for us to feel sensations of pleasure because we like feeling them for their own sake, and intrinsically bad for us to feel sensations of pain because we dislike feeling them for their own sake. It is a matter of philosophical controversy how this 'because' is to be understood, whether or not the goodness/badness is *definable* in terms of the liking/dislike of them (because of their strength and duration), but this issue need not be settled here. It is, however, important that the value of these sensations is not confused with the variables determining it, for example, the intrinsic badness of a pain with its intensity. With these points in mind, let us now turn to the question of whether the value of hedonic sensations is quantifiable.

Suppose that on two occasions you are feeling a rather mild pain which lasts for one minute. On one occasion this pain is prolonged to ten minutes, without any change of its intensity. On the other occasion the intensity of the physical stimulation causing this pain becomes ten times stronger, making your pain agonizing, while its duration is constant. At first blush, it might seem that these two procedures would result in pains that are equally bad, that the nine extra minutes of pain of the same intensity would equal increasing the intensity of the original stimulation nine times. The intrinsic badness of a pain for its subject, determined by its duration and intensity, would then be straightforwardly quantifiable. But if after having undergone these two procedures, you are asked about their relative badness, it is quite likely that you would judge that feeling the much more intense pain for one minute was clearly worse for you. The purpose of this chapter is to bring out the difference between the duration and intensity of hedonic sensations that accounts for the validity of such judgments.

In the case of the duration of a hedonic sensation, there is a distinction between *how long it feels like being* (or *is felt to be*) and *for how long it is felt*. Thus, suppose that a rather intense pain and pleasure are both felt for ten minutes. Then, typically, the pain feels like lasting longer than pleasure. Time flies when we have fun, as we say. Now imagine you first experienced only brief instances of the pain and pleasure—lasting 20 seconds, say—and that you found that experiencing this instance of pleasure is as intrinsically good for you as experiencing this instance is intrinsically bad for you, so that it is just about worth undergoing the pain to enjoy the pleasure. If you then were to experience instances of this pain and pleasure which unbeknownst to you are both ten minutes long, you might judge that feeling this longer pain was worse for you than feeling the equally long pleasure was good for you, since the pain feels like lasting longer than the pleasure. However, if you are afterward informed that the periods during which you felt the pain and pleasure were in fact equally long, you will likely correct your original judgment and concede that the value of the longer experiences is equal just like the value of the shorter experiences. Looking back, you would not think that you made your life worse by having both the longer pain and pleasure instead of having neither. Your knowledge of how long the sensations were felt for trumps your impression of how long they felt like being.

This is not quite accurately put, however. Strictly speaking, the pain does not *feel* like being longer than the pleasure. Although what we feel or, more generally, perceive has *some* temporal extension, since we perceive duration and change, it is very short, at most a second. For if we listen to a clock that ticks every second, we do not *hear* the tick preceding it when we hear a tick; we *remember* having heard it. So, it rather *seemed* to you that the period during which you felt the pain was longer than the period during which you felt the pleasure, but in fact it was not. You were subject to a sort of *temporal illusion*. Both the pleasure and the pain started and ended roughly when the physical stimuli causing them started and ended, and this marks a ten-minute period in both cases.

It could be objected that the fact that the ten-minute pain seems to have gone on for longer than the pleasure adds to its badness. To eliminate this factor, we should stipulate that our memory does not extend for more than one minute; perhaps, we have taken a pill having this effect. This ensures that the intrinsic badness of this pain is as much determined by its actual duration as the one-minute pain. Its intrinsic badness is equal to the intrinsic badness of having a series of ten discrete instances of a one-minute pain of the same intensity under the same memory restriction, and this is plausibly ten times as bad as having one instance of this pain.

With respect to the intensities of hedonic sensations, we can also distinguish between how intense a sensation seems to be and how intense it actually is. For instance, a pain can seem more intense if it is believed to be a symptom of a life-threatening condition. Accordingly, we must stipulate that you are assured that your pain is completely harmless. Then, if on the basis of the most scrupulous attention to your sensation, you make a judgment about its intensity, you cannot be subject to any illusion about it. Its intensity is something present in the moment, so it is something that, strictly speaking, you can feel, and on the basis of what you are feeling, you can make the most reliable judgments about what you feel. There is no ground by reference to which your most careful experientially based judgments about the intensities you are feeling could be corrected. By contrast, no attention to what you are feeling could enable you to judge accurately the duration of your sensation.

Sensations of pain and pleasure have locations in time, which coincide roughly with the temporal location of the physical stimulation causing them. But it is usually possible to determine more precisely when the physical stimulation begins and ends than when the sensation caused begins and ends. Therefore, the temporal boundaries of sensations are rougher or vaguer, but they do exist, and they could inform us that though, on the basis of experiencing them, one sensation seemed to be longer or shorter than another, their duration was roughly the same. But there is no such common scale in terms of which both the intensities of sensations and the intensities of their physical causes can be measured.

Or rather, given the dualist assumption that having hedonic sensations (alongside other sensory or perceptual states) are distinct from any physical states, there is no dimension or scale along which both the intensity of hedonic sensations and the intensity of their physical causes can be measured. I regard this dualist assumption as eminently plausible, but shall not attempt to defend it here.

Suppose you are feeling a rather acute pain, and the physical stimulation causing it is increased four times. Then, on the basis of careful attention to the pain this increase causes, you might well judge it to be *more* than four times as strong, and that its intensity makes this pain intrinsically *worse* for you than a four times as long pain of the original intensity. If you are afterward told that the physical cause was only made four times as strong, this provides you with no reason to revise your judgment about the intensity of the pain you felt and concede that it, too, was only (roughly) four times as strong. For your judgment was about the (felt) intensity of *the pain*, and this is not an intensity on the same scale as the intensity of the physical stimulation; these intensities do not share any scale or dimension as the duration of a sensation of pain and its physical cause share one and the same temporal dimension. Your judgment stands firm that the new sensation was more than four times as intense as the original sensation and, thus, that it would be worse for you to have this pain than the pain that would result were the duration of the original pain instead increased four times.

2 Non-transitivity of Sameness with Respect to Sensory Intensity

I have elsewhere (2013, chapter 9, 2014, 2017, chapter 4.2) argued that the relation of sameness or perfect similarity is non-transitive with respect to the intensity and value of hedonic sensations. This argument turns on these features being *supervenient*. They are supervenient on intensity of the physical stimuli causing these sensations.

According to an informal understanding of the notion of supervenience, to say that S is a supervenient property of X means that there have to be other properties of X, basic or subvenient properties, B, *in virtue* or *because of* which X has S, properties that *determine* or *explain* X's having S. I shall say no more about this explanatory relation than that I take it to imply that supervenient properties are not *logically entailed* by or *identical* to their subvenient properties. To give a few illustrations of what I take this to exclude, X's having some less specific or more determinable property, such as being colored, is not supervenient on X's being yellow because the former property is obviously entailed by the latter. Similarly, the weight and spatial properties of a thing do not supervene on the weight and spatial properties of its proper parts, since if you know the weight and spatial properties of all parts, you could calculate the weight and spatial properties of the thing they constitute. Nor is a dispositional property like X's being brittle supervenient on X's having a certain molecular structure if being brittle *is* having this structure (identified in terms of how it responds to certain causes). But, as remarked, I am proceeding on the assumption that the painfulness and pleasantness of hedonic sensations and their intensity are not identical to anything physical, such as their stimuli and their intensity. Consequently, room is left for supervenience to kick in here.

It follows from the informal notion of the supervenience of a property that if there is a difference between X and Y with respect to S—if X has S but Y lacks it, or if X has S to a greater or smaller degree than Y—there must be a difference in respect of B between X and Y. Otherwise, the difference in respect of S cannot be explained in terms of B. By contrast, supervenience does not imply that, if there is no difference, or perfect similarity, between X and Y with respect to S, there is no difference between them with respect to B. Differences in respect of B which are insufficient to generate differences in respect of S may well exist.

For the purposes of my argument, I need not plunge any deeper into the notion of supervenience, since I have already secured the simple implication of supervenience that forms the first premise of my argument:

> *Simp*: If S is a property of objects that supervenes on their having B, then, for all objects X, Y, and Z, even if both X and Y, and Y and Z are perfectly similar or the same with respect to S, it is logically possible that there are differences with respect to B between both X and Y, and Y and Z.

To illustrate, even if the physical stimulations X and Y are felt to be equally painful by the subject, and the same is true of Y and Z; it may be that there are differences between X and Y and Y and Z as regards B that are too minute to be registered and transmitted by the subject's nervous system to the brain.

The next step in the argument is a claim about the possible differences as regards B between X and Y, and Y and Z of which *Simp* speaks:

> *Add*: Even if there are differences in respect of B between X and Y, and between Y and Z, neither of which are sufficient for differences in respect of S between X and Y, or between Y and Z, but X is perfectly similar to Y, and Y to Z with respect to S, it is *logically possible* that there are differences in respect of B between X and Z that are sufficient for a difference with respect to S between X and Z.

The fact that the differences with respect to B neither between X and Y nor between Y and Z are sufficient for there to be a difference with respect to S between X and Y, or between Y and Z, must be compatible with there being *another* difference in respect of B between X and Z which *is* sufficient to manifest itself in a difference with respect to S between X and Z. Since B and S are distinct properties, the sufficiency in question is contingent, for instance, causal. But, evidently, the fact that neither the difference in respect to B between X and Y nor between Y and Z is contingently sufficient for a difference in respect to S between them cannot *logically entail* that the difference in respect to B between X and Z—which may be twice as big as either of the two other differences—is not contingently sufficient for a difference as regards S between X and Z. For instance, the following is clearly logically possible: the difference between X and Y and between Y and Z with respect to B (e.g. pain-producing properties) is one unit each, but the difference in this respect between X and Z is two units, and a difference of two units is minimally sufficient to give rise to a difference as regards S (e.g. intensity of pain) for the subject.

This additive possibility is one reason for the name of the second step of the argument. Another reason is that it is an additional premise, supplying the link between supervenience and transitivity. This link comes out in the third step:

> *Trans*: If *Add* is true, it must be false that the relation of perfect similarity or sameness with respect to S is transitive, that is to say, it must be false that it is a *logically necessary* truth that if X and Y, and Y and Z, are perfectly similar or the same with respect to S, then X and Z are perfectly similar or the same with respect to S.

If *Add* is true, it must be logically possible that there be a difference with respect to S between X and Z, even though there is no such a difference between X and Y, or Y and Z, for, as observed, the latter perfect similarities are compatible with there being a difference with respect to B between X and Z which *suffices* to manifest itself in a difference as regards S. If it is logically possible that there is a difference which suffices for the manifestation of another difference, it must be logically possible that the second difference obtains.

Now, from *Add* and *Trans*, we may infer by means of *modus ponens*:

> *Conclusion*: The relation of perfect similarity or sameness with respect to a supervenient property S is not transitive.

According to *Simp*, this is true of S because of something that follows from the fact that S is supervenient, namely that there might be differences in the subvenient properties, though there is no difference in the supervenient ones. So, that there is no transitivity of perfect similarity or sameness as regards these properties follows from their supervenience.

It is, then, logically possible that even though the difference with respect to B between X and Y, Bxy, is insufficient to create a difference in respect of S between X and Y, and another difference with respect to B between Y and Z, Byz, is insufficient to create a difference with respect to S between Y and Z, a third difference with respect to B between X and Z, Bxz, *is* sufficient to create a difference in respect of S between X and Z. The heart of my argument is that, since this possibility statement is incompatible with the necessity statement that expresses transitivity as regards sameness with respect to S—that it is necessary that if X and Y, and Y and Z, are the same with respect to S, so are X and Z—the latter statement cannot be true if the former is.

It might be objected that it is impossible that one and the same thing, Y's S-ness, can be perfectly similar to both X's and Z's S-ness, when X and Z are different with respect to S. But this is in fact not impossible if S is a supervenient property, and how something intrinsically is with respect to S can remain the same, though the stimulation which is its cause varies within a certain range. For then the stimulation Y may lie within the same range as the stimulation X and within the same range as stimulation

Z—at a point at which these ranges overlap—though X and Z do not lie in the same range.

Imagine, again, that it takes a difference of two units of physical stimulation for there to be any difference in respect of S (intensity of pain, say) and that X consists in one unit of stimulation, Y in two units, and Z in three units. X is then S-in-virtue-of-one-unit, Y is S-in-virtue-of-two-units, and Z is S-in-virtue-of-three-units. By hypothesis (since a one-unit difference in stimulation is insufficient to give rise to a difference as regards S), X and Y are qualitatively identical as regards S-ness, and so are Y and Z. Hence, being S-in-virtue-of-one-unit-*or*-two-units expresses one kind of S-ness, and so does being S-in-virtue-of-two-units-*or*-three-units. But being S-in-virtue-of-one-unit-or-two-units and being S-in-virtue-of-two-units-or-three-units do not express one kind of S-ness, since X which is S-in-virtue-of-one-unit and Z which is S-in-virtue-of-three-units differ in respect of S. Nevertheless, since Y is S-in-virtue-of-two-units, and this is a common element of the two disjunctive properties, Y is both S-in-virtue-of-one-unit-or-two-units and S-in-virtue-of-two-units-or-three-units. So, Y can after all exhibit just one kind of intrinsic S-ness and still be perfectly similar in respect of intrinsic S-ness to both X and Z, though X and Z are distinct from each other as regards intrinsic S-ness.

This brings out that the non-transitivity of perfect similarity as regards supervenient properties implies that these properties are less precise than their subvenient bases. It is the possibility of there being differences in respect of the subvenient bases—while there is perfect similarity or no difference in respect of the supervenient properties—which implies that perfect similarity in the latter respect cannot be transitive. Of course, this can be so only if supervenient properties are not identical or reducible to their subvenient bases.

In the publications mentioned earlier, I have tried to reply to various objections to this argument, and shall now only reply to an objection that readily presents itself if the differences between the intensity and duration of sensations noted in Sect. 1 are overlooked (see Temkin, 2014). It might then be objected that just as I have argued that it is possible that the pain of Y is equally intense as the pain of X, and the pain of Z is equally intense as the pain of Y, while the pain of X is more (or less) intense than the pain Z, it could be argued that it is possible that the duration of Y is the same as the duration of X, and the duration of Z is the same as the duration of Y, whilst the duration of X is longer (or shorter) than the duration of Z. What this argument shows, it might plausibly be claimed, is that the durations of X and Y and of Y and Z are only *roughly* the same, and it is well known that the relation of being roughly the same is not transitive. Likewise, it might be concluded, what my argument brings out is merely that the intensities of X and Y, and Y and Z are only roughly the same.

However, this objection overlooks the difference between the duration and intensity of sensations earlier highlighted. Both sensations and their stimuli have location in time, though the location of sensations is less precise. So, the durations of the sensations of X and Y, and Y and Z are indeed only roughly the same but, according to my conception, they do not supervene on the more precise durations of the stimuli, for the former are simply less determinate than the latter. By contrast, the intensities of these sensations are not only roughly the same. Their sameness might be as precise as sameness in respect of intensity of sensations can be. Certainly, there are subvenient differences in the magnitudes of their physical stimuli, but these magnitudes are distinct from the intensity of sensations. There is no common scale along which sensory intensities have a less precise place than the underlying physical magnitudes, as there is in the case of duration.

In the case of the intensity of sensations, their *esse* is *percipi*: if no difference in intensity is felt, there *is* no difference. If after the most scrupulous attention to a pair of sensations, we cannot notice any difference in their intensity, we have the best warrant for claiming that there *is* no difference. There is no need to postulate any such difference with respect to the sensations of X and Y, and Y and Z in order to explain the (noticeable) difference as regards the felt intensity between X and Z because this difference is explicable in terms of other differences on the subvenient level.

By contrast, as regards properties that do not supervene on other properties, perfect similarity or identity must be transitive. Imagine that we observe a difference in duration (length, weight, etc.) between A and C, though according to our most accurate measurements of A and B, and of B and C, the members of these pairs are the same. Then, provided that the comparison of A and C is correct, we are forced to conclude that the duration of A and B, or B and C, is only roughly the same. Since their possession of duration does not supervene on any other kind of property of theirs, a difference with respect to duration between A and C cannot be explained in terms of differences as regards any other property.

At the outset, we imagined two rather mild pains each lasting one minute, and prolonged the physical stimulus of one of them to ten minutes and intensified the physical stimulus of the other ten times. Given that the subject's memory lasts only for a single minute, the intrinsic badness of the ten-minute pain will be ten times as great (all things considered), since it is roughly ten times as long, and the intensity is the same (these being the only relevant dimensions).

By contrast, there is no reason to expect that the pain caused by the ten times stronger stimulus will only be ten times as intense as the original sensation and, thus, will make the pain ten times as bad in itself, for this intensity does not have even a rough position on the same scale as any magnitude of the physical stimulus has a more exact place. Therefore, there is no reason to expect that sensory intensities march even roughly in step with changes of the intensities of their physical cause, as sensory durations march roughly in step with changes of the durations of their physical cause. They *might* do so: the intrinsic badness of a pain caused by a stimulus two or three times as strong as the stimulus of the original pain might be roughly the same as a two- or three-minute pain. But this is not *necessarily* so; the pain caused by a ten times stronger stimulus might be considerably worse than the ten-minute pain.

I think it is only because we can make such comparisons of how much worse (or better) a boost of intensity would make a sensation to how much worse (or better) a prolongation of its duration would make it that we can give some sense to talk about the intensity of one sensation being roughly a certain number of times stronger or weaker than the intensity of another sensation. For there is no numerical scale on which sensory intensity can be even roughly plotted.[1] The non-transitivity of sameness

[1] Contrast Mayerfeld who, though he concedes 'the inescapable roughness and uncertainty of intensity estimates', claims: 'The less intense suffering is preferable to the more intense suffering even though the total quantity involved (intensity multiplied by duration) is greater' (1999, pp. 134–135). There is no such 'total quantity'.

with respect to their intensity excludes such a scale, for example, on a numerical scale, the intensity of Y would have the same number as the intensity of X and Z, though they do not have the same number as each other. Their number is not only rough, for nothing could have a more precise place on the relevant scale.

In not being quantifiable, hedonic sensations are similar to emotional pleasures and pains, such as the pleasure of reading a poem or getting revenge, since these are surely not quantifiable due to the contribution of the value of their objects.

References

Mayerfeld, J. (1999). *Suffering and moral responsibility.* New York: Oxford University Press.

Persson, I. (2013). *From morality to the end of reason: An essay on rights, reasons and responsibility.* Oxford: Oxford University Press.

Persson, I. (2014). Internal and external grounds for the nontransitivity of "Better/Worse than". *Law, Ethics and Philosophy, 2*, 125–145.

Persson, I. (2017). *Inclusive ethics: Extending beneficence and egalitarian justice.* Oxford: Oxford University Press.

Temkin, L. S. (2014). Intransitivity and the internal aspects view. *Law, Ethics and Philosophy, 2*, 145–158.

7

The Temporal Dynamic of Emotion Effects on Judgment of Durations

Sylvie Droit-Volet

1 Introduction

Time is a sort of psychological illusion because it varies under the influence of our emotions. As early as 1890, the philosopher Jean-Marie Guyau stated that the "estimation of durations is only a phenomenon of internal optic… related to emotion of pleasure and pain" (pp. 85–86). Since then, psychologists have studied this phenomenon but only very sporadically (Angrilli, Cherubini, Pavese, & Manfredini, 1997; Falk & Bindra, 1954; Hare, 1963; Langer, Wapner, & Werner, 1961; Thayer & Schiff, 1975). It was only after the experiment conducted by Droit-Volet, Bruno, and Niedenthal (2004) that a real explosion of research into emotion and time judgment occurred. The question is why? The reason probably lies in the task used in this experiment (temporal bisection), which was initially developed for use in animals and has yielded data for the internal clock theories which suggest the existence of a physiological

S. Droit-Volet (✉)
CNRS, UMR 6024, Université Clermont Auvergne,
Clermont-Ferrand, France
e-mail: sylvie.droit-volet@uca.fr

V. Arstila et al. (eds.), *The Illusions of Time*,
https://doi.org/10.1007/978-3-030-22048-8_7

mechanism dedicated to time processing (Gibbon, 1977; Gibbon, Church, & Meck, 1984; Treisman, 1963). The results obtained with this task have thus allowed researchers to imagine that a simple physiological internal clock mechanism may also account for time illusions under the influence of emotion. This has allowed them to reintroduce, what Janet (1928) called, "rational, reasonable and universal" in subjective time distortion. The need for the rational, for simplification, is found in every area of science. We need to simply recall the virulent attack made by the American behaviorist John Watson (1924) against introspective psychology and its study of sensations and affects, which he considered illogical. Researchers therefore considered that they were able to explain emotional time distortions through a simple physiological mechanism (e.g., acceleration of an internal clock), without having to have recourse to complex models or mysterious processes of the time consciousness, specific to the human being.

For the internal clock models, the same clock mechanism is involved in the processing of all durations ranging from a few hundred milliseconds to several seconds or minutes. However, the processes underlying the judgment of durations necessarily change as a function of the time scale considered. For example, it is likely that the brain accumulates temporal units (pulses, neural oscillations), emitted by a clock-like system, for the representation of short durations. However, it is unlikely that we are able to continuously accumulate these pulses over a duration of several minutes, without interruption and without storage or temporal reconstruction in memory (Droit-Volet, Monceau, Berthon, Trahanias, & Maniadakis, 2018). For long durations, one can also become aware of the passage of time and experience it as being stretched or contracted. However, it is unlikely that one becomes aware of the passage of time for shorter durations of a few hundred milliseconds. Obviously, we can produce a verbal judgment of very short durations as we can for longer ones, that is, we can note that a given duration is brief, but we do not really experience the feeling of lasting ("sentiment de durer"), that is to say the subjective feeling of an extension of internal time. Through these examples, we can understand that the influence of emotions on time judgment differs depending on the temporal scale examined, because temporal mechanisms but also emotional state change over the course of time.

2 Methodology for Studying Emotion and Time Judgment

Studies on emotion and time have used a wide variety of emotional stimuli (e.g., pictures from the International Affective Picture System (IAPS), facial expressions), that have been validated and standardized in the domain of the affective sciences (for reviews, see Droit-Volet, Fayolle, Lamotte, & Gil, 2013; Lake, 2016; Thönes, Arnau, & Wascher, 2018). However, one initial methodological problem in these studies lies in the fact that physical characteristics of the stimuli used, rather than their emotional dimension, could influence temporal judgment. It is well known that the size, the complexity or even the color of visual stimuli influence the perception of time (Thönes, Von Castell, Iflinger, & Oberfeld, 2018). Shibasaki and Masataka (2014) showed that the duration of a red screen was judged longer than that of a blue screen presented for the same period. And the color red is frequently used in high-arousal (HA) pictures. However, using an evaluative conditioning method, Kliegl, Watrin, and Huckauf (2015) succeeded in dissociating the picture from the emotion that it arouses. In this conditioning situation, the participants initially learned the association between a ringing sound (neutral stimulus) and emotional pictures (unconditional stimulus). They then had to estimate the duration of this sound in the absence of these emotional pictures (conditional stimulus). The duration was judged longer when the ringing sound had been previously associated with the high-arousal unconditional stimulus. As argued by Kliegl et al. (2015, p. 1350), this demonstrates that "the overestimation of emotional stimuli cannot be attributed to perceptual differences between neutral and emotional stimuli".

Another methodological problem, and probably the most critical, is that the emotion produced in response to an emotional stimulus is usually short-lived. It does not persist over time. Using the event-related potential method, Gan, Wang, Zhang, Li, and Luo (2009) have shown that temporal processing is modulated by the emotional dimension of pictures (facial expression) in a very short temporal window, after which attentional bias attenuates emotional effects. The instability of the emotional state may therefore explain the inconsistency of results in emotional studies. For instance, Angrilli et al. (1997) showed that the emotional arousal effect on temporal reproduction decreased between 2 and 6 s, resulting in underestimations for the 6-s duration, whereas overestimations were observed for shorter durations of 2 and 4 s. Bar-Haim, Kerem, Lamy, and Zakay (2010) also observed temporal overestimations with threat stimuli (fearful face) compared to neutral stimuli (neutral face) for 2 and 4 s, but not for 8 s. They concluded that participants become accustomed to emotional stimuli when their exposure is prolonged and that this therefore decreases their effects on timing.

To test the effect of emotion on longer durations (<2 s), some researchers therefore decided to use aversive stimuli (electric shock, aversive sound) that are thought to induce a more intense and less transitory arousal level (Fayolle, Gil, & Droit-Volet, 2015; Droit-Volet, Mermillod, Cocenas-Silva, & Gil, 2010). The results obtained with this type of stimuli were clear and consistent, as reported below. However, even if these stimuli were randomly delivered during a time interval, it is difficult to know whether the time distortions were due to the expectation of aversive stimuli that would in itself produce an emotion or to a post-stimulus emotional reaction. In addition, the durations tested in these studies did not exceed 8 s. More recently, in our laboratory, we thus performed an experiment on the temporal judgment of a long sequence of emotional sounds (Droit-Volet, submitted). The results showed a temporal overestimation of highly arousing sounds for durations lasting up to 60 s. However, beyond this duration, the emotional effect decreased again. This could simply be due to the fact that emotional stimuli become less arousing the longer they are presented for. Indeed, the results indicated that the duration of the stimulus itself affected their emotional rating. The same emotional stimuli were indeed perceived as being less arousing and more pleasant when their durations were longer (Fig. 7.1). However, when the habituation effect was reduced by limiting the number of trials (three trials for each emotional stimulus), time distortions appeared for long durations of several minutes in the same way as for shorter durations. This is consistent with the studies describing temporal distortions for a singular event (e.g., skydivers' jumps, Campbell & Bryant, 2007; three-min video clip of the September 11th terrorist attack, Anderson, Reis-Costa, & Misanin, 2007). However, it considerably limits the

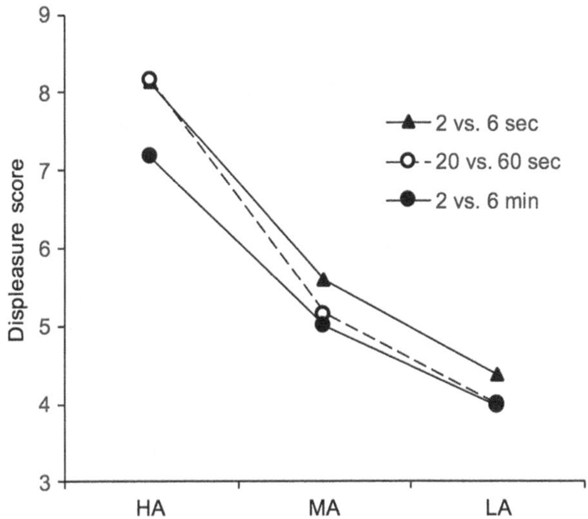

Fig. 7.1 Displeasure score plotted against the high-arousal (HA), the middle-arousal (MA) and the low-arousal (LA) emotional stimuli for the 2 vs. 6-s, the 20 vs. 60-s and the 2 vs. 6-min duration condition (from Droit-Volet, submitted)

experimental investigation of temporal processes that makes it necessary to present several durations across several trials (Wearden, 2016). In sum, investigating temporal illusions on different time scales is difficult because of the temporal dynamic of emotion.

One solution may be to examine the evolution over time of the physiological reaction to emotional stimuli. However, this greatly complicates the experimental protocols. This is why most temporal studies have not checked whether the emotional stimuli used really induce the emotions they are thought to induce in their participants. However, this is not really problematic as these experiments have generally used very short durations (<1–2 s) and obtained significant results. Some studies, nevertheless, have verified whether their participants do indeed feel the intended emotion by means of verbal scales, such as the Self-Assessment Manikin (Bradley & Lang, 1994). The verbal self-reports are considered as validated measures of emotion. As stated by Ledoux and Hofmann (2018, p. 69), "behavioral, physiological and subjective responses are all products of the same system". However, a part of the emotional system can operate non-consciously. In fact, the verbal self-reports take account of the subjective perception of emotion rather than emotion per se. It nevertheless remains the "gold standard for assessing the content of immediate emotional experience, or is at least the only easily accessible measure".

Some rare, although increasingly frequent, temporal studies have recorded physiological reaction to emotional stimuli by measuring heart rate, skin conductance level (Angrilli et al., 1997; Mella, Conty, & Pouthas, 2010; Droit-Volet et al., 2010; Fayolle et al., 2015; van Hedger, Necka, Barakzai, & Norman, 2017) or, more recently, pupillary diameter (Wöllner, Hammerschmidt, & Albrecht, 2018). The activity of the autonomic nervous system (ANS) is indeed a major component of emotional arousal. However, most authors have simply verified that their stimuli produce the expected physiological reaction (increase of arousal), but have not analyzed the correlation between the physiological reaction and the time judgment. This might be due to the lack of data demonstrating a clear linear relationship between physiological activities and time judgment. Meissner and Wittmann (2011) found a significant correlation between accuracy in time reproduction and individual abilities to feel a bodily signal (one's own heartbeat). However, their results have not been replicated in other studies, or at least not for all stimulus durations (Schötz et al., 2015; Pollatos, Laubrock, & Wittmann, 2014). Cellini, Mioni, Levorato, Grondin, and Stablum (2015) have nevertheless shown that participants with a higher heart rate variability are less variable in their time estimates (higher time sensitivity). This is reminiscent of the results on meditation showing an improvement in time sensitivity when the participants were trained to focus their attention on their breathing rate or the different parts of their body (Droit-Volet & Heros, 2017; Kramer, Weger, & Sharma, 2013). We can thus assume that the results on bodily signals accounts for attention abilities (individual or trained) that affect temporal performance, rather than time-specific processes. To our knowledge, only one recent study, conducted by van Hedger et al. (2017), found a significant correlation between a measure of sympathetic nervous activity and increased temporal reproductions for negative pictures. However, this correlation was not found for all the stimulus durations tested. Therefore, on the one hand, it is difficult to record and analyze physiological measures, while, on the other, the links between changes in ANS activity and temporal distortions are complex. The situation becomes even more complex when one examines emotions other than those involved in the threat conditions. As stated by Feldman-Barrett (2006, p. 41), "it is not possible to confidently claim that there are kinds of emotion with unique and invariant autonomic signatures". This explains the reluctance of researchers working on time to engage in the long analyses of physiological responses to emotion.

3 The Effect of Emotion on the Perception of Short Durations

Recent reviews conclude that emotional effects on timing are "robust", at least for short durations, from 300 ms to 1–2 s, and for emotions of negative valence and high-arousal level (e.g., Droit-Volet et al., 2013; Lake, 2016). Indeed, whatever the emotional stimuli and the procedure used, the results systematically reveal time distortions for negative high-arousal emotion compared to low-arousal or neutral emotion, with the stimulus durations being judged to last longer. For example, in the seminal bisection experiment performed by Droit-Volet et al. (2004), the participants were initially presented with a short (400 ms) and a long (1600 ms) standard duration in the form of a pink oval. They were then presented with comparison durations of either the same duration as the standard durations or with intermediate durations (600, 800, 1000, 1200 and 1400 ms). Their task was to judge whether the comparison durations were more similar to the short or the long standard duration. However, in this testing phase, the pink oval was replaced by facial expressions of anger, happiness, sadness or neutral expressions. In this task, the results are presented in the form of a psychophysical function, with the proportion of "long" responses plotted against the comparison durations. As illustrated in Fig. 7.2, the psychophysical function remained orderly in this emotional condition, with the proportion of long responses increasing with the length of the comparison durations. This demonstrates that emotion does not disrupt time judgment. By using several ranges of

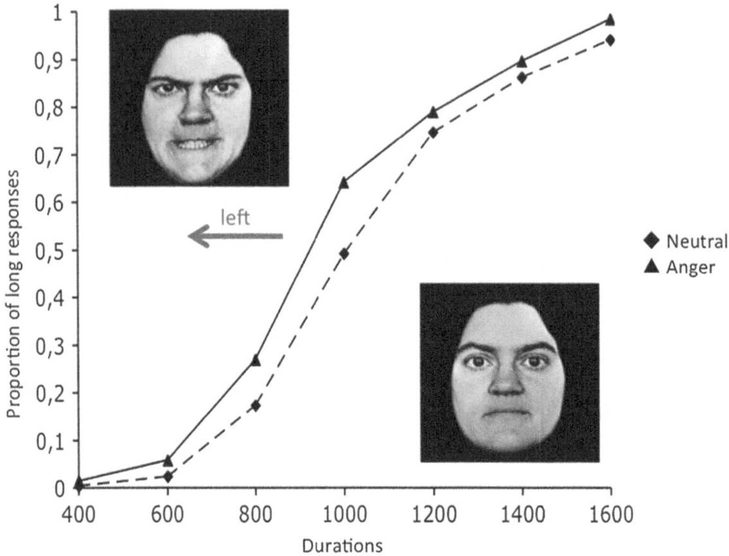

Fig. 7.2 Proportion of long responses plotted against stimulus duration for the neutral and the angry facial expression (adapted from Droit-Volet et al., 2004)

durations, recent studies have confirmed that the scalar properties of variance are preserved in emotional contexts, with the stimulus durations always being discriminated and the variance (standard deviation) of the temporal estimates always increasing with the length of the duration (Fayolle et al., 2015). There is therefore no violation of the fundamental scalar properties of time under the influence of emotion. In other words, our fundamental sense of time is not lost. This can be explained by the fact that internal time distortions serve for the preparation of different adaptive behaviors (e.g., acceleration of time when action is urgent in a threatening situation). We cannot, however, exclude the possibility that time judgment may be disrupted in the case of an intense emotional shock, a trauma that seizes one body and soul. Indeed, the intensity of emotion generated in the laboratory remains fairly low compared to the emotional events encountered in everyday life. That is one of the limits of laboratory studies.

Whatever the case may be, the psychophysical function in the bisection task described above was significantly shifted toward the left for the angry faces compared to the other faces, with a lower point of subjective equality (PSE). The PSE is the stimulus duration giving rise to 50% of "long" responses. Consequently, the participants responded "long" more often for the negative high-arousal stimulus (angry faces) than for the neutral or the low-arousal stimulus (sad faces) (lengthening effect), although their durations were similar.

Some researchers have then suggested that the emotional effect on timing is greater in threatening conditions than in other emotional conditions because the function of the emotion of fear is to enable the subject to act as quickly as possible (see Droit-Volet et al., 2013). When one is confronted with danger, the heart rate increases and the muscles contract, that is, every part of the body is prepared to act. Consequently, our brain with its internal clock system is overrun as the body functions accelerate in readiness for action. Indeed, when threat is imminent, it is urgent to be ready for action in order to survive (fight, flight or freeze when faced by a predator). Consequently, the mediating factor of the acceleration of internal time could be a motivation for action (action tendencies), in the form of avoidance in the case of fear. To my knowledge, the role of the motivation for action has not been demonstrated in the emotional studies on timing. In their exploratory experiment, Gable and Poole (2012) suggested "that approach motivation, rather than heightened arousal or attentional capture, hastens the perceived passing of time" (p. 6). This could explain why, in our own different studies, the low-arousal emotional stimuli (sadness) did not produce any temporal bias (shortening effect) compared to neutral stimuli, even though they are theoretically associated with the slowing down of an internal clock. Indeed, there is no urgency to "not act". The internal time distortion would thus be located at the level of motor functions, that is, movement. In Latin, emotion is referred to as "animi motus", which means movement. However, we are still faced with the methodological problem that the physiological slowing down could be not sufficiently great to be able to produce significant changes in timing behavior. Recently, Wöllner et al. (2018) observed a shortening effect in temporal estimates of video excerpts (slow motion), which was associated with a decrease in physiological arousal. Whatever the case may be, the role of the motivational component of emotion (action tendencies) in emotional effects on timing has not been clearly established because Gable and Poole have mainly worked on positive emotion without systematically comparing different types of movements (approach vs. avoidance) and emotional valence (negative vs. positive). Therefore, although different components of emotion influence time judgment (see Fig. 7.3 for the description of dynamic effects of different components of emotion on time judgment), it is, as we will discuss later, physiological arousal and its regulation that appear to be the major mechanism underlying the lengthening effects observed with high-arousal emotions in the short duration judgment tasks (Mella et al., 2010; Gil & Droit-Volet, 2012; Cheng, Tipples, Narayanan, & Meck, 2016). Moreover, emotional effects would be at their greatest in threat conditions simply because they produce the strongest arousal level.

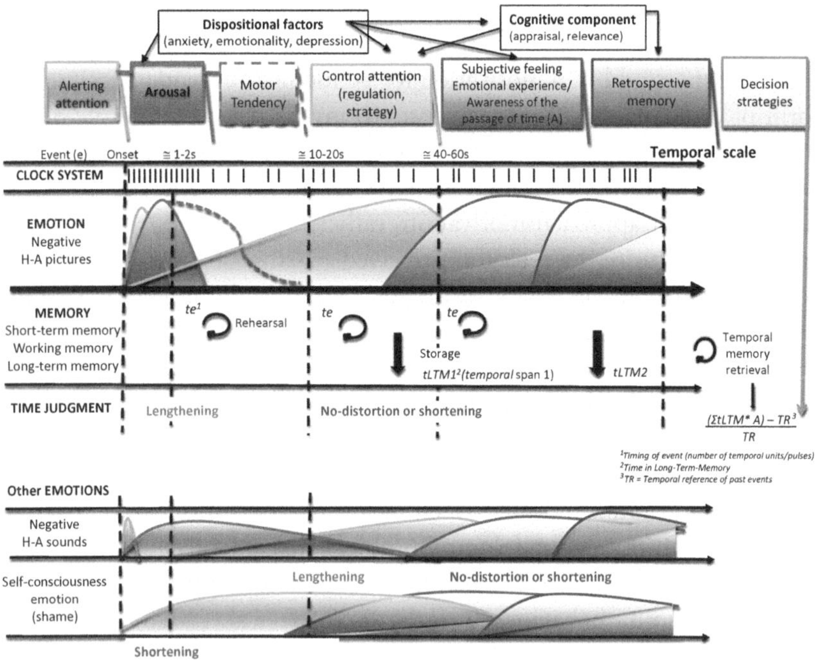

Fig. 7.3 Dynamic effects of different components of emotions on the judgment of durations for the negative high-arousal emotion and other emotions (negative high-arousal sounds and self-consciousness emotion)

Obviously, the importance of emotional arousal for timing depends on the emotional situation. Each emotional stimulus, each emotion category serves a particular function, and the underlying mechanisms differ in consequence (Frijda, 1986; Izard, 1977). For example, the self-consciousness emotion of shame does not produce a temporal lengthening that is characteristic of an arousal effect, but a shortening that is characteristic of an attentional effect (Gil & Droit-Volet, 2011). To take another example, although disgust is a negative high-arousal emotion, the lengthening effect is higher for the disgusting pictures from the IAPS (i.e., body mutilation) than for the disgusted faces (Grondin, Laflamme, & Gontier, 2014). But the emotional meaning differs between faces indicating a source of disgust and a disgusting visual scene. Therefore, as already reported, some cognitive components of emotion (appraisal and relevance) also contribute to the emotional effects on time judgment (Fig. 7.3).

4 The Unconscious and Automatic Effect of Emotional Arousal on the Judgment of Short Durations

The lengthening of perceived time under the effect of highly arousing emotion has mainly been discussed within the framework of pioneering models suggesting that an internal clock system is at the basis of time processing (Gibbon, 1977; Gibbon et al., 1984; Treisman, 1963). This does not mean that, in the field of time, researchers are particularly conservative and attached to old models. They are perfectly aware that the pioneering models of the internal clock have been updated (Buonomano, 2017; Kononowicz, van Rijn, & Meck, 2017; Merchant, Harrington, & Meck, 2013). They also agree that they are not physiologically plausible and that time judgment is not determined by a mere mechanism (a clock that count pulses emitted by a pacemaker) located in a dedicated area, but rather by a distributed neural circuit in which some cerebral areas are shared by the processing of other magnitudes (space, length). In this section, we do not want to present the different models of timing that have already been well described in numerous reviews. We simply want to argue that the term "internal clock is still useful shorthand for describing any mechanism that measures the passing of time" (Coull & Droit-Volet, 2018). This simply means that the representation of stimulus duration depends on the amount of something (ticks, pulses, neural oscillations) accumulated online in the brain by an accumulator-like system (timer) during its presentation. The lengthening of perceived duration with high-arousal emotions has thus been logically explained in terms of the level of induced arousal, which speeds up the "internal clock" underlying the representation of time (more pulses emitted per unit of time).

There is nevertheless a discussion in the literature about the real mechanisms underlying the lengthening effect produced by high-arousal emotion. Indeed, a lengthening effect can be obtained when the attention devoted to the processing of time increases as well when arousal increases, thereby speeding up the clock system. In fact, it is difficult, even impossible, to dissociate these mechanisms because both arousal and attention increase together in threatening conditions. Individuals watch carefully to detect any sign of danger, for example, the slightest movement of a predator. For example, in Lui, Penney, and Schirmer's (2011) experiment, the participants were presented with two successive durations: a reference duration and a comparison duration shorter or longer than this reference duration. The latter duration was preceded by an emotional picture. In this condition, the participants responded shorter more often than longer with the negative high-arousal picture than the neutral one. According to Lui et al. (2011), this demonstrates the attentional role (as distractors) of emotional stimuli. Some attentional mechanisms (alerting attention) are thus involved in the orientation of attention toward emotional stimuli to be timed (Fig. 7.3). However, this attentional effect is brief and probably insufficient to significantly mask the physiological effect of arousal on timing.

There is indeed ample evidence of the role of arousal in timing when high-arousal emotional stimuli are used. As previously reported, by measuring physiological responses to emotion, it has been proven that time dilations for short durations are associated with actual changes in physiological arousal level. In addition, one behavioral marker of the arousal effect on clock speed is the increase in the magnitude of temporal bias as the length of durations increases (multiplicative effect). An attentional effect (early switch closure allowing the accumulation of supplementary pulses) results not in a multiplicative but an additive effect, with a constant temporal bias regardless of duration length. This multiplicative effect is difficult to observe because researchers generally work with a limited range of short durations. However, Fayolle et al. (2015) recently conducted a bisection task testing different ranges of durations in a threatening condition. As illustrated in Fig. 7.4, the extension of temporal bias (lengthening) with the high-arousal emotion increased linearly with the length of the duration range. This is consistent with the multiplicative effect that is characteristic of the clock speed effect. The defensive system therefore activates a series of physiological and behavioral phenomena, including the acceleration of the system that forms the basis for the representation of time.

According to most specialists of emotion (e.g., Darwin, 1872; Ekman & Davidson, 1994; Ledoux, 2007), the functioning of fear mechanisms, which ensures the survival of individuals, is innate and automatic. In line with this idea, we observe the same time distortion when both children

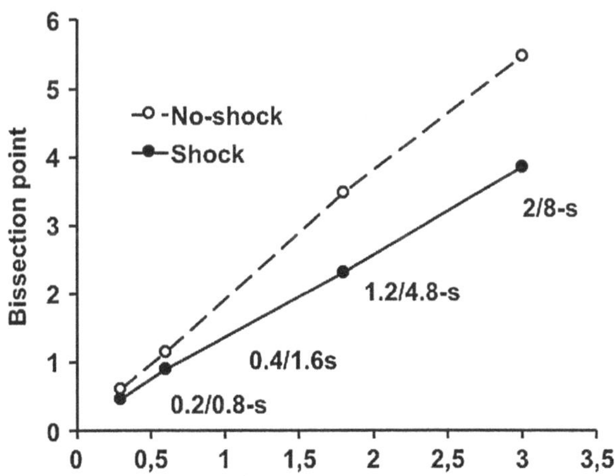

Fig. 7.4 Difference in the bisection point between the trials with and without an electric shock in the 0.2–0.8-s, 0.4–1.6-s, 1.2–4.8-s and 2.0–8.0-s duration ranges (from Fayolle et al., 2015)

and adults are confronted with angry faces, and this at different levels of the ontogenetic scale (Gil, Niedenthal, & Droit-Volet, 2007). This time distortion occurs independently of the inter-individual differences in cognitive abilities assessed by neuropsychological tests (Droit-Volet, Fayolle, & Gil, 2016). A similar time distortion is also observed with automatic time processing. We recently used an implicit time judgment task in which the participants did not receive any temporal instruction (Droit-Volet, 2016; Droit-Volet & Berthon, 2017). They were simply presented with two signals separated by a reference interval and had to press as fast as possible after the second signal. After a learning phase with the reference interval, the participants were presented with variable probe intervals, but their task remained the same and took the form of a simple reaction time (RT) task. In this condition, the curve plotting the RT against the probe interval formed a U shape with a peak (faster RT) close to the reference interval, indicating an implicit accurate representation of time. However, in the testing phase, we introduced an emotional picture during the probe interval. The results then showed a shifting of the U-shaped curve toward the left, with a peak at a lower temporal value for the high-arousal pictures than for the low-arousal pictures or for the no-picture condition (Fig. 7.5). This shows that time distortion due to an acceleration of the internal clock is observed even when time processing is unconscious.

In line with these automatic and unconscious effects of emotional arousal on timing, when adults are asked about the different factors which they know to affect the flow of time, they never evoke the factors related to basic high-arousal emotion (Lamotte, Chakroun, Droit-Volet, & Izaute, 2014). They are not conscious of the impact of their anger on the timing of short events. In conclusion, the studies described above report automatic and unconscious time distortions resulting from the rapid effects on timing of the physiological component of emotion (arousal) and/or its motivational component (action tendencies) (Fig. 7.3). Then, subsequent higher order cognitive processes (top-down processes) related to both the cognitive component of emotion (appraisal) and the conscious feelings (subjective experience of emotion and subjective experience of the passage of time) can influence time judgment, and especially that of durations longer than 1–2 s (see the component process model of emotion, Scherer, 1982; Sander, Grandjean, & Scherer, 2005).

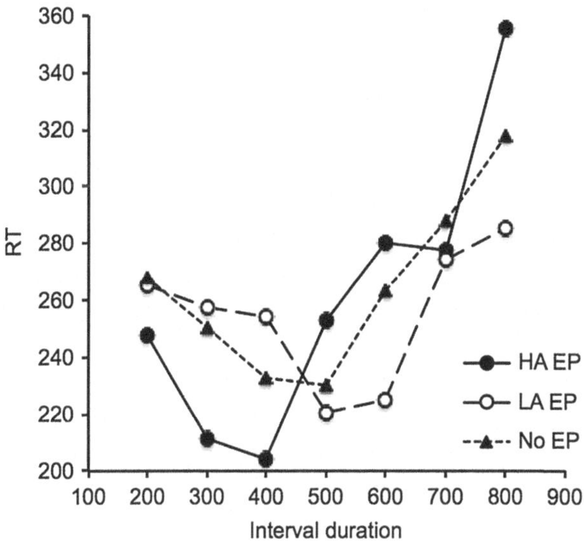

Fig. 7.5 Reaction time plotted against interval duration for the high-arousal, the low-arousal and the no emotional picture in the implicit timing task (adapted from Droit-Volet & Berthon, 2017)

5 The Effect of the Conscious Component of Emotion and Time on the Perception of Long Durations

A small number of time studies have investigated the cognitive components of emotion. It has nevertheless been demonstrated that higher order control processes modify the automatic effects of emotion on time judgment. Droit-Volet, Lamotte, and Izaute (2015) have shown how explicit knowledge of (belief in) a phenomenon increases or decreases the temporal overestimation induced by the perception of high-arousal emotional stimuli. Mereu and Lleras (2013) have also shown that the feeling of control over the emotional events alters the way we judge time. In the same way, Lamotte, Izaute, and Droit-Volet (2012) have shown that the participants who were more aware of being subject to time distortions were more accurate in their temporal estimates. The exact nature of the influence of cognitive components of emotion on timing is nevertheless unknown. It necessarily depends on the stage at which they occur during temporal information processing (Fig. 7.3). In the case of automatic emotional effects on the timing of short durations, it is very likely that these cognitive processes occur "later", that is, regulating/inhibiting initial arousal effects. However, they could also block the transfer of pulses into the accumulator by keeping the attention-switch open, as a function of their relevance/meaning for individual (see Zakay & Block's attentional model, 1996). Indeed, several studies have shown that the magnitude of time distortions changes as a function of dispositional factors (anxiety, emotionality, fearfulness, phobias) (Bar-Haim et al., 2010; Buetti & Lleras, 2012; Tipples, 2008, 2011; Lake, Meck, & LaBar, 2016). The cognitive components of emotion might therefore play a key role at every moment of the dynamic of temporal information processing, in other terms, at different levels of hierarchical processing of temporal information.

Recent studies have also experimentally investigated the subjective experience of the pace of time (passage-of-time judgment) and showed that it also influences the verbal judgment of durations but mainly that of long durations. Assessing the subjective experience of the passage of time in everyday life using the Experience Sampling Methodology, Droit-Volet and Wearden (2016) did not find any significant relationship between the passage-of-time judgment and the judgment of durations shorter than 2 s. Indeed, we may have the impression of waiting for a very long time in a queue at the baker's, but also know that little time has passed. This absence of a link has been observed for longer durations, lasting up to 33 s (Droit-Volet, Trahanias, & Maniadakis, 2017). However, for very long durations of more than 60 s (from 2 to 32 min), a significant relationship appears between these two forms of temporal judgment. The awareness of the passage of time even becomes a good predictor of the judgment of durations in the range of minutes. This suggests that the passage-of-time judgments and the judgments of very long durations share certain mechanisms, likely based on retrospective memory processes, which are themselves based on the content of the temporal interval, namely our subjective emotional experience (Droit-Volet & Dambrun, 2019; Droit-Volet et al., 2018).

The awareness of changes in the flow of time is indeed significantly linked to the subjective experience of our emotion. In our different studies, participants have always reported that time passes faster when they feel happy or more excited, and drags when they feel sad or sleepy (Droit-Volet & Wearden, 2016; Droit-Volet et al., 2017). However, we do not know exactly what they verbally reported regarding their subjective experience of time (Wearden, 2015). Indeed, this phenomenon has received little experimental attention beyond the philosophical works on phenomenological time (Husserl, 1964; Merleau-Ponty, 1945; Minkowski, 1968). Our recent studies conducted with Michaël Dambrun on altered awareness states produced by meditation revealed that the awareness of time during meditation is accompanied by the awareness of a modification of the experience of our body in space (body boundaries) with the feeling of a loss of sense of self, in other words a dissolution of the "Ego" (see also Berkovich-Ohana, Dor-Ziderman, Glicksohn, & Goldstein, 2013; Wittmann, 2015). This dissolution of our body boundaries is significantly correlated with a modification of the awareness of the passage of time. This supports the theory of the embodiment of time (Effron, Niedenthal, Gil, & Droit-Volet, 2006; Droit-Volet, 2014; Wittmann, 2014) in which bodily signals are at the core of our subjective experience of time. Finally, when the duration extends (duration of the activity/event), individuals engage in an introspective analysis of their own internal states: emotion, bodily signal, cognitive and physical effort. They can then feel a change in their internal state, that is, in their own "internal time". However, to be able to declare that time passes faster or slower, they use a reference scale, that is, their conception that time flows at a continuous and homogeneous rate, as our watches indicate. The comparison between their internal time—the self in time (self-duration)—and their representation of external time (world duration) leads them to say that time goes faster or slower (Droit-Volet, 2018).

The awareness of changes in the flow of time would thus emerge through experience of the extension of internal time, that is, the extension of the self in time. It has indeed been demonstrated that the best predictor of the passage-of-time judgment is subjective emotional experience. In fact, the psychological time is "elastic" with automatic distortions in the judgment of short durations related to the state of our internal time measurement tool, but also the awareness of changes in the passage of time that emerges with the extension of "internal time". The effect of emotion on duration judgment therefore changes with the range of durations tested, with self-awareness and memory processes playing a more prominent role in the case of long durations. It is therefore important to further examine the temporal dynamic of emotional effects on the judgment of durations of different lengths if we are to propose a general model of timing that goes beyond the simple model of an internal clock and takes into account our consciousness of the passage of time.

Acknowledgments This work was supported by a grant (TIMESTORM) from European Commission, Horizon 2020 research and innovation action (H2020-FETPROACT-2014).

References

Anderson, M. J., Reis-Costa, K., & Misanin, J. R. (2007). Effects of September 11th terrorism stress on estimated duration. *Perceptual and Motor Skills, 104*, 799–802.

Angrilli, A., Cherubini, P., Pavese, A., & Manfredini, S. (1997). The influence of affective factors on time perception. *Perception & Psychophysics, 59*, 972–982.

Bar-Haim, Y., Kerem, A., Lamy, D., & Zakay, D. (2010). When time slows down: The influence of threat on time perception in anxiety. *Cognition and Emotion, 24*, 255–263.

Berkovich-Ohana, A., Dor-Ziderman, Y., Glicksohn, J., & Goldstein, A. (2013). Alterations in the sense of time, space, and body in the mindfulness-trained brain: A neurophenomenologically-guided MEG study. *Frontiers in Psychology, 4*, 912. https://doi.org/10.3389/fpsyg.2013.00912

Bradley, M. M., & Lang, P. J. (1994). Measuring emotion: The self-assessment manikin and the semantic differential. *Journal of Behavior Therapy and Experimental Psychiatry, 25*(1), 49–59.

Buetti, S., & Lleras, A. (2012). Perceiving control over aversive and fearful events can alter how we experience those events: An investigation of time perception in spider-fearful individuals. *Frontiers in Psychology, 3*, 337. https://doi.org/10.3389/fpsyg.2012.00337

Buonomano, D. (2017). *Your brain is a time machine*. New York: W.W. Norton & Company Ltd.

Campbell, L. A., & Bryant, R. A. (2007). How time flies: A study of novice skydivers. *Behaviour Research and Therapy, 45*, 1389–1392.

Cellini, N., Mioni, G., Levorato, I., Grondin, S., & Stablum, F. (2015). Heart rate variability helps tracking time more accurately. *Brain and Cognition, 101*, 57–63.

Cheng, R. K., Tipples, J., Narayanan, N. S., & Meck, W. H. (2016). Commentary—Clock speed as a window into dopaminergic control of emotion and time perception. *Timing & Time Perception, 4*(1), 99–122.

Coull, J., & Droit-Volet, S. (2018). Explicit understanding of duration develops implicitly through action. *Trends in Cognitive Sciences, 22*(10), 923–937.

Darwin, C. (1872/1998). *The expression of the emotions in man and animals*. Oxford: Oxford University Press.

Droit-Volet, S. (2014). What emotions tell us about time. In D. Lloyd & V. Arstila (Eds.), *Subjective time: The philosophy, psychology, and neuroscience of temporality*. Cambridge, MA: MIT Press.

Droit-Volet, S. (2016). Emotion and implicit timing. *PLoS One, 11*(7), e0158474. https://doi.org/10.1371/journal.pone.0158474

Droit-Volet, S. (2018). Intertwined facets of subjective time. *Current Directions in Psychological Science, 27*(6), 422–428.

Droit-Volet, S., & Berthon, M. (2017). Emotion and implicit timing: The arousal Effect. *Frontiers in Psychology, 8*, 176. https://doi.org/10.3389/fpsyg.2017.00176

Droit-Volet, S., Brunot, S., & Niedenthal, P. (2004). Perception of the duration of emotional events. *Cognition & Emotion, 18*(6), 849–858.

Droit-Volet, S., & Dambrun, M. (2019). Awareness of the passage of time and self-consciousness: what do meditators report? *PsyCH Journal*. https://doi.org/10.1002/pchj.270

Droit-Volet, S., Fayolle, S., & Gil, S. (2016). Emotion and time perception in children and adults: The effect of task-difficulty. *Timing and Time Perception, 4*(1), 7–29.

Droit-Volet, S., Fayolle, S., Lamotte, M., & Gil, S. (2013). Time, emotion and the embodiment of timing. *Timing and Time Perception, 1*(1), 99–126.

Droit-Volet, S., & Heros, J. (2017). Time judgments as a function of mindfulness meditation, anxiety and mindfulness awareness. *Mindfulness, 8(2)*, 266–275.

Droit-Volet, S., Lamotte, M., & Izaute, M. (2015). The conscious awareness of time distortions regulates the effect of emotion on the perception of time. *Consciousness and Cognition, 38, 155–164.* https://doi.org/10.1016/j.concog.2015.02.021

Droit-Volet, S., Mermillod, M., Cocenas-Silva, R., & Gil, S. (2010). The effect of expectancy of a threatening event on time perception in human adults. *Emotion, 10*(6), 908–914.

Droit-Volet, S., Monceau, S., Berthon, M., Trahanias, P., & Maniadakis, M. (2018). The explicit judgment of long durations of several minutes in everyday life: Conscious retrospective memory judgment and the role of affects? *PLoS One, 13*(4), e0195397.

Droit-Volet, S., Trahanias, P., & Maniadakis, M. (2017). Passage of time judgments in everyday life are not related to duration judgments except for long durations of several minutes. *Acta Psychologica, 73*, 116–121.

Droit-Volet, S., & Wearden, J. (2016). Passage of time judgments are not duration judgments: Evidence from a study using Experience Sampling Methodology. *Frontiers in Psychology, Section Cognition, 7*, 176. https://doi.org/10.3389/fpsyg.2016.00176

Effron, D., Niedenthal, P., Gil, S., & Droit-Volet, S. (2006). Embodied temporal perception of emotion. *Emotion, 6*(1), 1–9.

Ekman, P., & Davidson, R. (1994). *The nature of emotion: Fundamental questions*. New York: Oxford University Press.

Falk, J. L., & Bindra, D. (1954). Judgment of time as a function of serial position and stress. *Journal of Experimental Psychology, 39*, 327–331.

Fayolle, S., Gil, S., & Droit-Volet, S. (2015). Fear and time: Fear speeds up the internal clock. *Behavioural Processes, 120*, 135–140.

Feldman-Barrett, L. (2006). Are emotions natural kinds? Perspectives on psychology. *Association for Psychological Science, 1*(1), *1*–58.

Frijda, N. H. (1986). *The emotions*. Cambridge and New York: Cambridge University Press.

Gable, P. A., & Poole, B. D. (2012). Time flies when you're having approach-motivated fun: Effects of motivational intensity on time perception. *Psychological Science, 23*, 879–886. https://doi.org/10.1177/095679761143581

Gan, T., Wang, N., Zhang, Z., Li, H., & Luo, Y.-J. (2009). Emotional influences on time perception: Evidence from event-related potentials. *Cognitive Neuroscience and Neuropsychology, 20*, 839–843.

Gibbon, J. (1977). Scalar expectancy theory and Weber's law in animal timing. *Psychological Review, 84*, 279–325.

Gibbon, J., Church, R. M., & Meck, W. (1984). Scalar timing in memory. In J. Gibbon & L. Allan (Eds.), *Annals of the New York Academy of Sciences, 423: Timing and time perception* (pp. 52–77). New York: New York Academy of Sciences.

Gil, S., & Droit-Volet, S. (2011). Time perception in response to ashamed faces in children and adults. *Scandinavian Journal of Psychology, 52*, 138–145.

Gil, S., & Droit-Volet, S. (2012). Emotional time distortions: The fundamental role of arousal. *Cognition & Emotion, 26*(5), 847–862.

Gil, S., Niedenthal, P. M., & Droit-Volet, S. (2007). Anger and temporal perception in children. *Emotion, 7*, 219–225.

Grondin, S., Laflamme, V., & Gontier, E. (2014). Effect on perceived duration and sensitivity to time when observing disgusted faces and disgusting mutilation pictures. *Attention, Perception, & Psychophysics, 76*, 1522–1534.

Guyau, M. (1890). *La genèse de l'idée de temps*. Paris: Felix Alcan.

Hare, R. D. (1963). The estimation of short temporal intervals terminated by shock. *Journal of Clinical Psychology, 19*(3), 378–380.

Husserl, E. (1964). *Leçons Pour une Phenomenology de la Conscience Intime du Temps*. Paris: Presse Universitaire de France.

Izard, C. E. (1977). *Human emotions*. New York: Plenum.

Janet, P. (1928). *L'évolution de la mémoire et de la notion de temps*. Paris: L'Harmattan.

Kliegl, K. M., Watrin, L., & Huckauf, A. (2015). Duration perception of emotional stimuli: Using evaluative conditioning to avoid sensory confounds. *Cognition & Emotion, 29*(8), 1350–1367.

Kononowicz, T. W., Van Rijn, H., & Meck, W. H. (2017). A critical review of neural timing signatures: Before, during and after the to-be-time interval. *Stevens' Handbook of Experimental Psychology and Cognitive Neuroscience, 1*, 1–38.

Kramer, R. S. S., Weger, U. W., & Sharma, D. (2013). The effect of mindfulness on time perception. *Consciousness and Cognition, 22*, 846–852.

Lake, J. L. (2016). Recent advances in understanding emotion-driven temporal distortion. *Current Opinion in Behavioral Sciences, 8*, 214–219.

Lake, J. L., Meck, W. H., & LaBar, K. S. (2016). Discriminative fear learners are resilient to temporal distortions during threat anticipation. *Timing & Time Perception, 4*(1), 63–78.

Lamotte, M., Chakroun, N., Droit-Volet, S., & Izaute, M. (2014). Metacognitive Questionnaire on Time: Feeling of Passage of Time. *Timing & Time perception, 2*(3), 339–359.

Lamotte, M., Izaute, M., & Droit-Volet, S. (2012). The consciousness of time distortions and their effect on time judgment: A metacognitive approach. *Consciousness and Cognition, 21*, 835–842.

Langer, J., Wapner, S., & Werner, H. (1961). The effect of danger upon the experience of time. *American Journal of Psychology, 74*, 94–97.

Ledoux, J. E. (2007). The amygdala. *Current Biology, 23*, 868–874.

Ledoux, J. E., & Hofmann, S. G. (2018). The subjective experience of emotion: A fearful view. *Current Opinion in Behavioral Science, 19*, 67–72.

Lui, M. A., Penney, T. B., & Schirmer, A. (2011). Emotion effects on timing: Attention versus Pacemaker accounts. *PLoS One, 6*(7), e21829.

Meissner, K., & Wittmann, M. (2011). Body signals, cardiac awareness, and the perception of time. *Biological Psychology, 86*, 289–297.

Mella, N., Conty, L., & Pouthas, V. (2010). The role of physiological arousal in time perception: Psychophysiological evidence from an emotion regulation paradigm. *Brain Cognition, 75*, 182–187.

Merchant, H., Harrington, D. L., & Meck, W. H. (2013). Neural basis of the perception and estimation of time. *Annual Reviews of Neuroscience, 36*, 313–336.

Mereu, S., & Lleras, A. (2013). Feelings of control restore distorted time perception of emotionally charged events. *Consciousness & Cognition, 22*, 306–314.

Merleau-Ponty, M. (1945). *Phénoménologie de la perception*. Domont: Gallimard.

Minkowski, E. (1968). *Le temps vécu*. Brionne: Imago Mundi.

Pollatos, O., Laubrock, J., & Wittmann, M. (2014). *Interoceptive focus shapes the experience of time. PLoS One, 9*(1), e86934. https://doi.org/10.1371/journal.pone.0086934

Sander, D., Grandjean, D., & Scherer, K. R. (2005). A systems approach to appraisal mechanisms in emotion. *Neural Networks: The Official Journal of the International Neural Network Society, 18*, 317–352.

Scherer, K. R. (1982). Emotion as a process: Function, origin, and regulation. *Social Science Information, 21*, 555–570.

Schötz, E., Otten, S., Wittmann, M., Schmidt, S., Kohls, N., & Meissner, K. (2015). Time perception, mindfulness and attentional capacities in transcendental meditators and matched controls. *Frontiers in Psychology, 6*, 1215.

Shibasaki, M., & Masataka, N. (2014). The color red distorts time perception for men, but not for women. *Scientific Reports, 4*, 5899.

Thayer, S., & Schiff, W. (1975). Eye-contacts, facial expression, and the experience of time. *The Journal of Social Psychology, 95*, 117–124.

Thönes, S., Arnau, S., & Wascher, E. (2018). Cognitions about time affect perception, behavior, and physiology—A review on effects of external clock-speed manipulations. *Consciousness & Cognition, 63*, 99–109.

Thönes, S., Von Castell, C. J., Iflinger, J. I., & Oberfeld, D. (2018). Color and time perception: Evidence for temporal overestimation of blue stimuli. *Scientific Reports, 8*, 1688.

Tipples, J. (2008). Negative emotionality influences the effects of emotion on time perception. *Emotion, 8*, 127–131.

Tipples, J. (2011). When time stands still: Fear-specific modulation of temporal bias due to threat. *Emotion, 11*, 74–80.

Treisman, M. (1963). Temporal discrimination and the indifference interval: Implications for a model of the "internal clock". *Psychological Monographs, 77*, 1–13.

van Hedger, K., Necka, E. A., Barakzai, A. K., & Norman, G. J. (2017). The influence of social stress on time perception and psychophysiological reactivity. *Psychophysiology, 54*(5), 706–712.

Watson, J. B. (1924). *Behaviorism*. New York: People's Institute.

Wearden, J. H. (2015). Passage of time judgements. *Consciousness and Cognition, 38*, 165–175.

Wearden, J. H. (2016). *The psychology of time perception*. London: Palgrave Macmillan.

Wittmann, M. (2014). Embodied time: The experience of time, the body, and the self. In V. Arstila & D. Lloyd (Eds.), *Subjective time: The philosophy, psychology, and neuroscience of temporality* (pp. 507–523). Cambridge, MA: MIT Press.

Wittmann, M. (2015). Modulations of the experience of self and time. *Consciousness and Cognition, 38*, 172–181.

Wöllner, C., Hammerschmidt, D., & Albrecht, H. (2018). Slow motion in films and video clips: Music influences perceived duration and emotion, autonomic physiological activation and pupillary responses. *PLoS One, 13*(6), e0199161.

Zakay, D., & Block, R. A. (1996). The role of attention in time estimation processes. *Advances in Psychology, 115*, 143–164.

8

Hidden Duration: Time-Lag in the World and Mind

Kristoffer Sundberg

1 Introduction

When we look up at the sun, it seems to be immediately available to us. As children, we are surprised to learn that we actually see it as it was some eight minutes in the past. This claim is often followed by the observation that if the sun were to go out of existence, then we wouldn't experience the event until eight minutes later. My own reaction as a child was one of uncertainty. The sun suddenly didn't seem as immediately available to me through my senses. "Is it still out there?" I thought to myself. Still, there was *something* present to my perception. If the sun had ceased to be, what was that thing I was experiencing? Was what I was seeing something akin to an echo, rather than the sun itself? The time-lag argument (or "the time-gap argument" as it is sometimes referred to) is an argument that takes as its basis the empirical fact that there is a time lag, or time gap, between the perceiver and the perceived and uses this observation to undermine the idea that we perceive everyday objects (and other everyday

K. Sundberg (✉)
University of Oslo, Oslo, Norway
e-mail: kristoffer.sundberg@ifikk.uio.no

V. Arstila et al. (eds.), *The Illusions of Time*,
https://doi.org/10.1007/978-3-030-22048-8_8

entities like processes, events, and instances of properties). A very early example is provided by Leibniz who argues:

> Since rays of light need time—however little—to reach us, it is possible that the object should be destroyed during the interval and no longer exist when the light reaches the eye; and something which no longer exists cannot be the present object of our sight. (Leibniz, 1996, p. 135)

The difference between seeing distant stars and seeing nearby objects is a difference in degree, not in kind. The time-lag argument risks generalizing to all perception since in all perception there will be, however short, a time gap between the perceiver and the perceived.

Philosophical theories about perception are partly differentiated in terms of, and motivated by, how they account for perceptual failures. As Fiona Macpherson and Clare Batty put it: "Identifying new instances of illusion and hallucination provides much needed, important data for testing theories of experience and perception—theories that are frequently motivated, and should be judged, by their ability to account for cases of illusion and hallucination" (Macpherson & Batty, 2016, p. 264). As modern philosophers of perception typically use "illusion" and "hallucination", illusions are cases where we perceive something, but where what we perceive appears other than it actually is, and hallucinations are cases where we have an experience as of perceiving an object (and/or other entity) but where that object (and/or other entity) does not exist (see, e.g., Smith, 2002; Brewer, 2011; Crane & French, 2017).

Both types of perceptual failure have been suggested to threaten the idea that we perceive objective external reality, an idea we can refer to as "Perceptual Realism", by, supposedly, forcing us to accept that we are aware of something radically different than the everyday objects (and other entities) that we typically take ourselves to perceive. When philosophers talk of *the* problem of perception, they are referring to such challenges to the idea that we perceive everyday objects (Smith, 2002; Crane & French, 2017 provide excellent overviews). The modern debate is less concerned with answering *if* these challenges can be met, than with answering how these challenges are *best* met. Each solution offers certain advantages and drawbacks. Rather than focusing on the details of these challenges or on the details of the suggested solutions, I here want to highlight the alternative challenge to Perceptual Realism provided by "the time-lag argument".

The temporal distance between the perceiver and the perceived provides interesting challenges for Perceptual Realism. While illusions and hallucinations have received much attention, the time-lag argument, though not completely ignored, has not figured prominently in the debate.[1] I will argue that, just as the discovery of new forms of hallucinations and/or illusions provide data against which we can test and evaluate the various theories of perception, so too does the time-lag argument.

Possibly, the importance of the time-lag argument has been underestimated. It is, I think, symptomatic that A. D. Smith, in his book-length defense of Direct Realism, spends nine chapters on arguments from illusion and hallucination, but only a few sentences on the time-lag argument (Smith, 2002). If the time-lag argument has been underestimated, I believe there are two potential explanations for this. First, while the classic literature talks, as have I initially, of *the* time-lag argument, I will argue that there are several distinct time-lag arguments. I will especially focus on two of them: a hallucination-like time-lag argument and an illusion-like time-lag argument. I will argue that neither the hallucination-like time-lag argument nor the illusion-like time-lag argument is identical to the time-lag argument as it has typically been stated. If we deny that the past exists, then all perception is, like hallucination, perception of something which does not exist. If objects and events in the world are presented to us *as* simultaneous with the experience, then the experience is illusory, given that we perceive objects and events in the past. These arguments are independent so that answering one does not answer the other. One way of underestimating the importance of the time lag, then, is to acknowledge and answer only one of the time-lag arguments.

Second, an assumption that accounting for illusions and hallucinations provides theoretical tools that makes answering the time-lag argument trivial might explain why the time-lag argument has received so little attention compared to the arguments based on illusion and

[1] Philosophers who have stressed the importance of the time-lag argument include Suchting (1969), Carrier (1969), Houts (1980) Power (2013), Le Poidevin (2015), and Johansson (2017).

hallucination. While I have described the two versions of the time-lag argument as "illusion-like" and "hallucination-like", respectively, the different versions of the time-lag argument bring unique challenges for Perceptual Realism compared to that of traditional illusions and hallucinations. Unlike the arguments from illusion and hallucination, the time-lag argument(s) does (do) not rely on perception malfunctioning. The challenge(s) of the time lag remains even in ideal circumstances.

I will discuss how these challenges relate to recent philosophical theories about perception. I will argue that both challenges can be overcome given certain assumptions about perception and about the nature of time. However, only those specific assumptions allow Perceptual Realism to be maintained. As such, the time-lag argument, like the argument from illusion and the argument from hallucination, impose certain limits and challenges that any feasible theory of perception has to be able to account for.

2 The Classic Time-Lag Argument

Before moving on to the time-lag argument, we need to make some distinctions. We can distinguish *distance perception* from *contact perception* (cf. Coates, 2007). Distance perception relies on a *medium* for transmitting information of the object perceived. (Vision depends on the transfer of light, auditory perception on the transfer of sound waves, and olfactory perception on the transfer of odorous molecules.) Contact perception (tactile perception and gustatory perception) *does not* rely on a medium for transmitting information of the object. Though I will begin by stating the time-lag argument in terms of distance perception, I will later argue that the argument goes through for contact perception as well.

I consider "see", as well as "hear", "taste", and so on, and generally "perceive", to be success verbs in the sense that seeing x implies that there exists an x that is seen by me. I consider "experience", "perceptual experience", and "visual experience" to be a report on the subjective experience of the perceiver that does not necessarily imply that there is anything that is perceived (cf. Macpherson, 2013; Lewis, 1980; Siegel, 2010). On this use of terminology, a subject who suffers a hallucination as of a pink elephant has an experience as of a pink elephant, but the subject does not see (or perceive) a pink elephant. I will later discuss disjunctivism, a view which denies this distinction, but it will make the presentation easier if we assume it for now. Also, as I use the terminology, the time lag is a time lag between the subject and what the subject perceives (sees, hears, etc.), not between the subject and the subject's experience. With these distinctions in mind, let us turn now to the time-lag argument.

Does it matter whether the object still exists when the subject has the experience? W. A. Suchting (1969) provides both an excellent statement of the time-lag argument (in its classic form) and a great response to it. Suchting assumes that we perceive objects in virtue of perceiving the state they are in, an assumption I share. So, for example, we see an object in virtue of it having a certain shape, color, size, and so on. This fixes the temporal location of the perceived object. Even in a case where an object is unchanging, the state of the object being property F at t_1 is different from the state of the object being F at t_2. The crucial assumption behind the time-lag argument, according to Suchting, is that for a person P, an object O, and a time t, "[i]f at t, P sees a state of O, then what P sees (at t) is a state of O at t" (Suchting, 1969, p. 47). While Suchting does not give this assumption a name, we can refer to this as "the Simultaneity Principle". Because of the time it takes for the relevant energy (light, sound waves, etc.), a physical object could only be perceived in an earlier state than the time of the subjective experience. If we accept the Simultaneity Principle, the immediate object of awareness is perceived in the state it is at t, but we cannot, at t, due to the time lag, perceive a state of any physical object at t. Thus, we do not perceive an ordinary physical object at t. On this version of the argument, which I take to be the classic version of the argument, the non-existence of the object plays no part and the argument generalizes to all perception.[2]

---

[2] Suchting points out that on the version of the time-lag argument that he develops and eventually provides an argument against:

> no use has been made of the consideration that in some cases the object transmitting the light may actually have disappeared by the time the light reaches earth. Such a point may serve to make the argument more dramatic and persuasive, but it is far from clear what it adds to its systematic force. The essential problem would seem to be independent of the

Suchting points out that those endorsing the principle typically seem to consider it self-evident and in no need for argument, or, alternatively, insisting that it follows from reflections on ordinary language (Suchting, 1969, p. 49f.).[3] I, like Suchting, find these arguments unconvincing, but rather than going through these arguments in detail, I will here focus on what follows if the principle is denied. That is, I will primarily be concerned with what follows if the classic time-lag argument fails. As far as I know, Suchting was the first to make the implicit assumption of the Simultaneity Principle explicit and to argue that denying the principle was the key to disarming the threat from the (classic) time-lag argument. If we deny the Simultaneity Principle, then there can be a temporal distance between the perceiver and the perceived. Suchting concludes that we perceive objects in past states.[4]

question of the existence or non-existence of the object at the time when it is apparently seen. (Suchting, 1969, p. 48)

[3] An exception is L. S. Carrier (1969) who argues for "the Simultaneity Principle" by claiming that we can only see what exists. Carrier obviously assumes that the past does not exist. A similar idea is suggested by Ronald Houts (1980). However, the view that the past does not exist is but one possible view of the ontology of time. Contra Carrier and Houts, the existence or non-existence of the past is still very much an open question. Due to a perceived conflict with special relativity, the view that the past does not exist, or generally, that there is a metaphysically privileged present, is somewhat controversial (see Bourne, 2006 for an excellent overview, as well as a defense of Presentism, the view that only the present exists). An assumption that the past does not exist underlies what I will refer to as the hallucination-like time-lag argument. Thus, there *can* be an overlap between the classic time-lag argument and the hallucination-like time-lag argument. I will insist that they are distinct because one could accept "the Simultaneity Principle" without denying that the past exists, and one can accept the hallucination-like time-lag argument without accepting "the Simultaneity Principle".

[4] Why should we think that we only perceive the past, even if we admit that the process enabling the perception takes time? What decides where and when the entity perceived (seen, heard, etc.) is located? It can be helpful to compare perception with shooting. Like shooting, perception has (typically) a target. A shot is fired from a specific position in spacetime and hits something at a different position in spacetime. What decides the aim for the perception? Where and when should we look to find the actual or presumed target of the perception? There are two natural suggestions: based on the relevant causal process involved in linking the experience and a physical object, or based on where the experience presents the object as being located. We could refer to the latter as the object's "*apparent* location".

One could insist that the apparent location decides where any relevant target for the perceiving has to be. To see the problem with that suggestion, consider a type of thought experiment suggested by Grice (1961) and Soteriou (2000). Let us assume that there are two qualitatively identical red balls, where one is positioned straight in front of the subject, "ball 1", and one is off to the subject's

However, this still leaves two challenges: (1) All perception of physical objects (and their features) is of the past, but we may doubt that the past exists. If, for example, only the present exists, then all perception is perception of what does not exist. (2) At t, do we perceive the object *as* being in state F *at t* (or as "now", or as in the present)? If so, our experience is illusory with respect to the temporal location of what is perceived. The two challenges are, what I will refer to as, the hallucination-like time-lag argument and the illusion-like time-lag argument. Let us consider them in turn, beginning with the latter.

3 The Illusion-like Time-Lag Problem

While the spatial distance between the perceiver and the perceived is typically obvious, the temporal distance usually remains hidden from the perceiver. As young, we are *surprised* to learn that we see the sun as it was eight minutes ago. The duration of the process leading up to the experience is (typically) hidden.[5] Since any exceptions are arguably rare and controversial, I will generally assume that the time gap between the

right, "ball 2". Let us also assume that there is a mirror between the subject and ball 1. Further, the mirror is angled so as to reflect light from ball 2 to the subject. As a result of light from ball 2 being reflected by the mirror, the subject has an experience as of a red ball in front of him. In this scenario, does the subject see ball 1 or ball 2? If the target of the perception (seeing in this case) is based on the apparent position, then the subject would see ball 1. After all, the subject has an experience as of a red ball in front of him right where ball 1 in fact is. However, intuitively, the subject in this scenario sees ball 2. Beyond intuition, a sign that the subject sees ball 2, rather than ball 1, is that the subject's experience is counterfactually dependent on ball 2, and not on ball 1. Similar considerations hold for the temporal case. The experience will counterfactually depend on the state the object is in when the relevant energy (light, sound, odorous molecules) was emitted or reflected off of the object. I conclude that it is the relevant causal connection that decides the target of the perception, rather than the apparent location.

[5] I add the qualification "typically" because there are, *possibly*, cases where the time gap becomes apparent and is experienced as such. Because of the great difference of the speed of light and sound waves, audio-visual cross-modal perception offers a type of example. Such cases would include perceiving thunder and lightning, perceiving distant explosions, and seeing and hearing a drummer hit a drum from a distance. In those cases, the same event is first seen and then heard. Possibly, the auditory experience represents the event as past. A suggestion for a purely auditory case might be found in the case of direct hearing and hearing echoes. Possibly, hearing the echoes *as* echoes implies hearing the event that the echoes present as past. Finding purely visual examples is even more difficult. One option would be, if one accepts the controversial thesis that we can see *through* photographs and TV screens (as argued by Kendall Walton, 1984), then one could set up a TV screen to be slightly delayed compared to other TV

perceiver and the perceived is hidden. If that is so, experience does not (typically) present an event or a state of an object as past. Does perceptual experience present an event or a state of an object *as* present (as simultaneous with the experience, or as "now")?

There are at least three alternatives. Alternative 1: experience is noncommittal about whether a seen state of an object or an event is present or past (or future?) (Prosser, 2007; Le Poidevin, 2015). Alternative 2: experience presents the state of the object, or the event, as existing now/in the present/simultaneously with the experience (Peacocke, 1999; Almäng, 2014; Searle, 2015). Alternative 3: experience includes a demonstrative component that presents the event or the state of the object as being located at, *that* time, where "that" refers to a specific location in time, but not necessarily the time when the subject undergoes the experience. Alternative 2 is clearly controversial and the question is too complicated to settle here. Instead, I will examine the conditional question: *if* Alternative 2 is accepted, what are the consequences for theories of perception?

If Alternative 2 is the right account, then the time lag implies that perceptual experience is illusory (inaccurate) with respect to the time of the state of the object or of the event. This suggests an illusion-like time-lag argument:

1. (Perceptual Realism.) We perceive the states of everyday physical objects and events.
2. Perceptual experience presents the states of object or events as present (or as simultaneous with the experience).
3. The target of our perception is determined by a special causal relation between the potential target of perception and potential perceiver and this causal relation involves a transfer of energy or matter (such as light, sound waves, or odorous particles).
4. Because of the finite speed of the perceptual medium (light, sound waves, or odorous particles) and the finite speed of nerve signals, if something is perceived, it is perceived in a past state.

---

screens. In such a case, the subject might experience the screen with the delay as showing something akin to an echo and the experience might present the event as past.

5. Perceptual experience is illusory (inaccurate) with respect to the temporal location (or tense) of the state of the object or of the event perceived.

If Alternative 2 is accepted (here appearing as the second premise of the argument), then it threatens to make *all* perceptual experience more or less illusory. Jan Almäng, a defender of Alternative 2, agrees that "there is always a time lag between an event and the perception of the event. Events located very close to us occur slightly before we perceive them, even though they are perceived as present. So if we perceive tensed properties, then all perceptions must be illusory" (Almäng, 2014, p. 374).[6]

The degree to which this consequence is problematic will vary between different theories of perception. According to representationalism, perceptual experience has a content which can be more or less accurate.[7] For example, an experience may be partly accurate in virtue of accurately representing an object's shape but misrepresenting its color. On such an account, there is no obvious reason for why an illusion with respect to the temporal location (or of tense) would be more problematic than other kinds of illusion.

However, this consequence will be especially bad for certain forms of disjunctivism.[8] Disjunctivism is a strategy for defending Direct Realism against the argument from illusion and the argument from hallucination by denying that the account given of perceptual experience in cases of illusion and hallucination generalizes to veridical perception. Thus, a disjunctivist can accept that in cases of illusion and hallucination, we are aware of something else than everyday objects, for example, mind-dependent sense data, but deny that this implies that we are not aware of everyday objects in veridical perception. While the traditional view, "conjunctivism", holds that both hallucinatory experience and

[6] Almäng further points out that the conclusion, that all perception is illusory, does not require experience to present events as tensed because "if we assume that the experience of having a perception is given as simultaneous with the perceived event, all perceptions would be illusory whether or not we assume that events are perceived as tensed" (Almäng, 2014, p. 374).

[7] Defenders of a representational account of perception (broadly construed) include Searle (1983), Smith (1989), Dretske (1995), Tye (2009), Chalmers (2010), Burge (2010), and Siegel (2010).

[8] Disjunctivism is defended by Campbell (2002), Fish (2009), and Brewer (2011), and in various papers by Martin (1997, 2004, 2006).

veridical experience are perceptual experience, the disjunctivist claims that while hallucinations can be mistaken for perceptual experience by the subject, hallucinations are a different kind of mental state (and this is true even if hallucination and perception involve the same brain states).

Disjunctivists disagree on whether to group illusions with hallucinations or with veridical perception.[9] If illusions are grouped with hallucinations, then it follows from the conclusion of the illusion-like time-lag argument that all presumed perception is illusory with respect to the temporal location and that there is no perceptual experience. Even worse, according to "negative disjunctivism" (as defended by Mike Martin, 1997, 2004, 2006), illusions and hallucinations have no positive features beyond being indiscriminable from veridical perception. This is typically referred to as a "negative epistemic" view of non-veridical perception. However, if the illusion-like time-lag argument is accepted, there is no veridical perception. This is a problem because, on the negative epistemic account, veridical perception is used to define non-veridical perception. Perhaps, the negative epistemic view can rely on the metaphysical possibility of veridical perception. (It is, for example metaphysically, but not physically, possible that light and nerve signals are instantaneous.) Alternatively, one might, as Suchting (1969, p. 54) does, insist that if the time gap is smaller than what we can differentiate, then the experience is not illusory.[10] This would imply a sharp limit beyond which we cannot perceive (where the distance would be defined by the speed of the energy involved and by our powers of discrimination). Such a limit to perception is revisionary and counterintuitive. It might, for example, imply that we cannot see the sun, the moon, distant airplanes, and/or lightning).

However, the key point is not to argue against disjunctivism. Instead, I have wanted to highlight that the illusion-like time-lag argument is a different challenge for various forms of Perceptual Realism than the classic time-lag argument, and that the illusion-like time-lag argument can be used to test our theories of perception. For example, one might

[9] Byrne and Logue (2009) offer a good overview of various forms of disjunctivism.

[10] Strictly speaking, Suchting's argument concerns the accuracy or inaccuracy of judgments made on the basis of perception, rather than the accuracy of the perceptual experience itself, but reading him as if talking directly about perceptual experience yields an interesting argument.

prefer representationalism to disjunctivism based on the ease with which representationalism can cope with the illusion-like time-lag argument.[11]

4 The Hallucination-like Time-Lag Problem

The right response to the classic time-lag argument, I suggested following Suchting (1969), is that we perceive the past. However, this potentially presents a new problem for Perceptual Realism. A reasonable assumption is that we can only perceive what exists. A hallucination, as we saw above, is standardly understood as a case of having an experience as of an object (or other entity) in a case where no such object exists. According to some theories about the ontology of time, the past does not exist. An example of such a view is Presentism. According to Presentism, only what is present exists (Bourne, 2006). The time lag implies a conflict between

[11] As an example, John Searle seems to have something like the illusion-like time-lag argument in mind when he argues that his theory is superior to disjunctivism:

> A good test case for anybody's theory of perception is, Can it account for the veridical perception of objects that ceased to exist millions of years ago? I now see a star through a telescope that I know ceased to exist twenty-seven million years ago. In one respect, the experience is not veridical because, again, all experiences are of the here and now, and it seems to me that the star is existing here and now when I know in fact it does not. All the same, I know that I am seeing that particular star. Now I can draw a picture of that, indeed I have drawn such a picture in this chapter. The star causes in me a visual experience. I would like to see the picture drawn by the Disjunctivists. I do not think they can draw a coherent picture. In what sense exactly for them is the star a constituent part of the experience of it? It is pretty tough to describe that for an object that ceased to exist twenty-seven million years ago. (Searle, 2015, p. 198)

When Searle says that all experiences are of the here and now, he could be read as endorsing the Simultaneity Principle, that is, the principle says that we only perceive what is simultaneous with the subject undergoing the experience. On the other hand, in seeming contradiction, Searle says that the subject is seeing that particular star that ceased to exist 27 million years ago. If the star is seen, it can only be seen in the past when it existed, so what the subject sees is not simultaneous with the subject undergoing the experience. As I have argued, the right lesson to draw from the classic time-lag argument is that perception is always of the then and there, rather than the here and now. The confusion is, I think, a result of Searle not distinguishing the classic time-lag argument from the illusion-like time-lag argument. What Searle *ought* to say is that there is a case to be made for his theory having an advantage over disjunctivism with respect to the illusion-like time-lag argument because his theory, but not (certain forms of) disjunctivism, can account for the fact, if it is a fact, that we perceive past objects as present.

Perceptual Realism and Presentism (and any other theory about the nature of time according to which the past does not exist).

The conflict could be stated as follows:

1. (Perceptual Realism.) We perceive the states of everyday physical objects (and other entities).
2. We can only perceive what exists.
3. The target of our perception is determined by a special causal relation between the potential target of perception and potential perceiver, and this causal relation involves a transfer of energy or matter (such as light, sound waves, or odorous particles).
4. Because of the finite speed of the perceptual medium (light, sound waves, or odorous particles) and the finite speed of nerve signals, if something is perceived, it is perceived in a past state.
5. (Presentism.) The past does not exist.
6. We do not, in contradiction with (1), perceive states of everyday physical objects (or other entities).

Since 1 and 6 imply a contradiction we have to give up *either* 1 (Perceptual Realism) *or* 5 (Presentism). Previous literature has suggested that, due to the time lag, Presentism is not compatible with Direct Realism (Power, 2013; Sundberg, 2013). As I have suggested here, Presentism (and other theories about the ontology of time which denies the existence of the past) is in conflict with the broader category of Perceptual Realism. There is no reason to restrict Perceptual Realism to Direct Realism. If we can make sense of Indirect Realism as a mediated awareness of everyday physical objects, then Presentism and the time lag is a problem for Indirect Realism too. After all, if we are aware of x (a physical object) in virtue of being aware of y (e.g., a sense datum), this requires the existence of an x, otherwise there is nothing to be indirectly aware of.

Is this an argument against Presentism or is it an argument against Perceptual Realism? The answer depends on one's previous commitments. If one has a strong commitment to Presentism, then this is an argument against Perceptual Realism, but if one has a strong commitment to

Perceptual Realism, then this is an argument against Presentism. If we assume Presentism, then all perception is hallucination-like in that it is an experience of what does not exist. Let us therefore refer to this as the hallucination-like time-lag argument.

One may be tempted to respond by denying that we see stars, hear cars, and smell flowers. Rather, the objection would go, we see the *light* from stars, we hear the *sound* of cars and we smell the *scent* (smell?) of flowers. While the sources of the light, sound, and scent are located in the perceiver's past, the light, sound, and scent are now stimulating the sensory organs of the perceiver. We could refer to this as "the perceptual medium response". Possibly, the idea is most tempting for olfactory perception. Macpherson and Batty point out that:

> you might smell a floral scent and your experience might thereby represent that floral scent. But it is plausible to think that your experience does not also represent the object that has that scent—a bunch of flowers, or a jar of perfume, or any other number of objects. After all, scents can be around when their sources are not, or are no longer, present. (Macpherson & Batty, p. 270)[12]

However, as we have seen based on the time lag, this is true for auditory and visual perception too. If Presentism is true, then *all* distance perception involves a perceptual medium (light, sound waves, or scents) which is around when their sources are not. (Though the object might still be around, the state it was in when it figured as the causal *relatum* is in the past.) Just based on this argument, there is no reason to think that olfactory perception is special.

Nevertheless, independently of whether such a response is revisionary or in line with how we think of perception in the various sensory modalities, there is a problem with this response. To see the problem, we have to ask ourselves whether we perceive an object when our sensory organs come into contact with the thing in question, or when we have an experience of the object (or, if we want to leave room for unconscious perception, when the brain has processed the information). An unconscious person

[12] See Alvin Goldman (1977, p. 269f) for a similar point.

may rest a hand upon an object, but could hardly be said to perceive the object. The correct answer must, therefore, be that we perceive an object when we have an experience of it (or when the brain has processed the information from it). This requires information to be sent from the sensory organ to the brain. Because of the finite (and, compared to light, slow) speed of the nerve signals, the time-lag argument returns. This shows that the time-lag argument applies to contact perception (tactile and gustatory perception) as well as to distance perception (visual, auditory, and olfactory perception). Contact perception is also, it turns out, perception of the past.

The hallucination-like time-lag argument provides a challenge for Perceptual Realism, which is independent of the classic time-lag argument and the illusion-like time-lag argument. Providing an answer to the latter two does not defuse the challenge from the former. As such, it provides a unique challenge for Perceptual Realism. This challenge, too, can help us evaluate the relative merit of various theories of perception. I will give two examples.

First, as for the illusion-like time-lag argument, the hallucination-like time-lag argument provides a more severe challenge for disjunctivism. While a conjunctivist accepting Presentism would be forced to deny that we perceive the states of everyday objects, he or she could still maintain that there is perceptual experience. A disjunctivist would deny this since, for them, only veridical perception counts as perceptual experience. The problem of a negative epistemic account of hallucination, that was mentioned in the previous section, re-appear. Again, if veridical perception is impossible, then defining non-veridical perception solely in terms of them being indiscriminable from veridical perception is problematic. There is, then, a strong tension between Presentism and disjunctivism.

As a second example, the Presentist scenario and the hallucination-like time-lag scenario is relevant for the debate on whether the content of perception is singular (particular) or general (existential). On the former account, the object enters into the content such that two experiences as of qualitatively identical objects, nevertheless, have different content. Such an account is defended by Soteriou (2000). On the latter account, experiences as of qualitatively identical objects have the same content.

On that account, the content of perceptual experience is completely general (Lewis, 1980; McGinn, 1982; Davies, 1992).

The relevance in the present context is that, the Presentist scenario implies that, due to the time lag, the experience is hallucination-like. On the assumption that the singular content targets the place and time determined by the relevant causal chain, there exists, at the targeted time, no object that can enter into the content. The experience will then be non-veridical. Even so, it is tempting to say that some perceptual experiences would be accurate or veridical in the Presentist scenario. After all, judgments formed on the basis of perceptual experience would, in many cases, be very accurate, perceptual experience would be systematically very reliable and perceptual experience would be a great evolutionary advantage. A general perceptual content, unlike a singular content, could be accurate in the type of scenario we are considering. Of course, if the objects of perception are represented as simultaneous with the experience, as per the illusion-like time-lag argument, then this aspect of the content will be inaccurate. Michael Tye, who defends a singular content view, refers to general content views as views endorsing "the Existential Thesis" and claims that "[o]ne immediate objection to the Existential Thesis, then, is that, as yet, it lacks a clear motivation" (Tye, 2009, p. 79). Similarly, he points out, a view which accepts that perceptual experience has both singular and general content, the "the Multiple-Contents Thesis" as Tye refers to it, "is in somewhat better shape, but it is unnecessarily complicated: there is no need to suppose that in cases of veridical visual experience there is an additional existential content over and above the singular content" (Tye, 2009, p. 94). The hallucination-like time-lag argument could supply the motivation for the Existential Thesis, and it could motivate the added complication of the Multiple-Contents Thesis.

5 Conclusion

I have argued that there are three distinct time-lag arguments. They each pose a different challenge for Perceptual Realism. The point is not that they offer an insurmountable challenge for Perceptual Realism. Just as the modern debate about illusions and hallucinations is less concerned with answering *if* these challenges can be met, than with answering how these challenges are *best* met, the different versions of the time-lag argument offer an opportunity to test and evaluate the various theories of perception. To ignore the time-lag arguments in favor of exclusively considering illusions and hallucinations, then, is to not utilize all the data available for evaluating our theories of perception.

Acknowledgements Thanks are due to Jan Almäng, Jola Feix, Anders Nes, Simon Skau, and Sebastian Watzl for helpful discussions of many of the ideas in this chapter.

References

Almäng, J. (2014). Tense as a feature of perceptual content. *Journal of Philosophy, 111*(7), 361–378.

Bourne, C. (2006). *A future for presentism*. Oxford: Clarendon Press.

Brewer, B. (2011). *Perception and its objects*. Oxford: Oxford University Press.

Burge, T. (2010). *Origins of objectivity*. Oxford: Oxford University Press.

Byrne, A., & Logue, H. (2009). Introduction. In *Disjunctivism: Contemporary readings*. Cambridge, MA: MIT Press.

Campbell, J. (2002). *Reference and consciousness*. Oxford: Clarendon.

Carrier, L. S. (1969). The time-gap argument. *Australasian Journal of Philosophy, 47*(3), 263–272.

Chalmers, D. (2010). *The character of consciousness*. Oxford University Press.

Coates, P. (2007). *The metaphysics of perception: Wilfrid Sellars, perceptual consciousness and critical realism*. London: Routledge.

Crane, Tim, & French, Craig. (2017). The problem of perception. In E. N. Zalta, *The Stanford encyclopedia of philosophy* (Spring 2017 ed.). https://plato.stanford.edu/archives/spr2017/entries/perception-problem/

Davies, M. (1992). Perceptual content and local supervenience. *Proceedings of the Aristotelian Society, 66*, 21–45.

Dretske, F. I. (1995). *Naturalizing the mind*. Cambridge, MA: MIT Press.

Fish, W. (2009). *Perception, hallucination, and illusion* (1st ed.). Oxford: Oxford University Press.

Goldman, Alvin (1977). Perceptual objects. In *Synthese*, Vol. 35, No. 3, Perception II, Part I (July, 1977), pp. 257–284.

Grice, H. P. (1961). The causal theory of perception. *Proceedings of the Aristotelian Society Supplementary Volumes, 35*, 121–168.
Houts, R. W. (1980). Some implications of the time-lag argument. *Philosophy and Phenomenological Research, 41*(1/2), 150–157.
Johansson, I. (2017). John Searle on perception—Review. *Metaphysica, 18*(2), 259–275.
Le Poidevin, Robin. (2015). The experience and perception of time. *The Stanford encyclopedia of philosophy* (Summer 2015 ed.), Edward N. Zalta (Ed.), URL https://plato.stanford.edu/archives/sum2015/entries/time-experience
Leibniz, G. (1996). In P. Remnant & J. Bennett (Eds.), *Leibniz: New essays on human understanding* (Cambridge Texts in the History of Philosophy) (pp. 52–184). Cambridge: Cambridge University Press.
Lewis, D. (1980). Veridical hallucination and prosthetic vision. *Australasian Journal of Philosophy, 58*(3), 239–249.
Macpherson, F. (2013). The philosophy and psychology of hallucination: An introduction. In F. Macpherson & D. Platchias (Eds.), *Hallucination: Philosophy and psychology*. Cambridge, MA: MIT Press.
Macpherson, F., & Batty, C. (2016). Redefining illusion and hallucination in light of new cases. *Philosophical Issues, 26*(1), 263–296.
Martin, M. G. F. (1997). The reality of appearances. In A. Byrne & H. Logue (Eds.), *Disjunctivism: Contemporary readings*. Cambridge, MA: MIT Press.
Martin, M. G. F. (2004). The limits of self-awareness. *Philosophical Studies, 120*, 37–89.
Martin, M. G. F. (2006). On being alienated. In T. Gendler & J. Hawthorne (Eds.), *Perceptual experience* (pp. 354–410). Oxford: Oxford University Press.
McGinn, C. (1982). *The character of mind*. Oxford: Oxford University Press.
Peacocke, C. (1999). *Being known*. Oxford: Oxford University Press.
Power, S. E. (2013). Perceiving external things and the time-lag argument. *European Journal of Philosophy, 21*(1), 94–117.
Prosser, S. (2007). Could we experience the passage of time? *Ratio, 20*(1), 75–90.
Searle, J. R. (1983). *Intentionality: An essay in the philosophy of mind*. Cambridge: Cambridge University Press.
Searle, J. R. (2015). *Seeing things as they are: A theory of perception*. Oxford: Oxford University Press.
Siegel, S. (2010). *The contents of visual experience*. New York: Oxford University Press.
Smith, A. D. (2002). *The problem of perception*. Cambridge, MA: Harvard University Press.

Smith, D. W. (1989). *The circle of acquaintance: Perception, consciousness, and empathy*. Dordrecht: Kluwer Academic Publishers.

Soteriou, M. (2000). The Particularity of Visual Perception. *European Journal of Philosophy, 8*(2), 173–189.

Suchting, W. A. (1969). Perception and the time-gap argument. *Philosophical Quarterly, 19*(January), 46–56.

Sundberg, K. (2013). Ingvar Johansson and the bridging problem. In C. Svennerlind, J. Almäng, & R. D. Ingthorsson (Eds.), *Johanssonian investigations: Essays in honour of Ingvar Johansson on his seventieth birthday*. Heusenstamm: Ontos.

Tye, M. (2009). *Consciousness revisited: Materialism without phenomenal concepts*. Cambridge, MA: MIT Press.

Walton, K. (1984). Transparent pictures: On the nature of photographic realism. *Critical Inquiry, 11*(2), 246–277.

9

Modulations in the Experience of Duration

Marc Wittmann, Tijana Jokic, and Eric Pfeifer

1 Body Timing in Pavlov's Dogs and in Humans

The functioning of the human body and mind are only understandable with reference to time. Bodily processes like the breathing cycle and the heart rate typically have rhythmic patterns with specific frequencies. These autonomic body movements have been shown to influence

M. Wittmann (✉)
Institute for Frontier Areas of Psychology and Mental Health, Freiburg, Germany
e-mail: wittmann@igpp.de

T. Jokic
School of Philosophy, Psychology and Language Sciences, University of Edinburgh, Edinburgh, UK

E. Pfeifer
Catholic University of Applied Sciences Freiburg, Freiburg, Germany
e-mail: eric.pfeifer@kh-freiburg.de

V. Arstila et al. (eds.), *The Illusions of Time*,
https://doi.org/10.1007/978-3-030-22048-8_9

perception, cognition, and affective processing, including the perception of time (Garfinkel et al., 2014; Matthias, Schandry, Duschek, & Pollatos, 2009; Wittmann & Meissner, 2018). Two further rhythmic time domains can be identified on a high- and a low-frequency scale. On the high-frequency scale, global brain activation on multiple temporal and spatial scales is recorded with the electro- and magneto-encephalogram (Buzsáki, 2006). This oscillatory neural activity in the time domain of milliseconds to seconds is organized and woven into a spatio-temporal continuum to enable conscious awareness (Northoff, 2014). On a low-frequency scale, biological functions are controlled by endogenous clocks with a periodicity of around 24 hours, clearly experienced as a daily recurrence of wakeful activity and sleep (Golombek, Bussi, & Agostino, 2014; Roenneberg, Daan, & Merrow, 2003). The body's circadian clocks are intrinsically (genetically) regulated, and the regular external entrainment by the sun (the main *zeitgeber*) keeps the clocks in synchrony. This enables an organism as a functional unit to synchronize and coordinate internal activities and anticipate regular changes in the environment.

Between these two levels of biological fluctuations, further body cycles function to synchronize the organism with the environment. Ivan Pavlov's research on the classic conditioning of body reactions to external stimuli clearly disclosed impressively accurate and precise timing abilities of the organism. For example, the dogs' visceral systems learned the duration of a two-minute interval. A dog was repeatedly fed exactly two minutes after a buzzer had been sounded. After the learning phase, saliva (the conditioned reaction, CR) was elicited two minutes after the conditioned stimulus (CS, the buzzer) had been presented although no food was provided (as described in Todes, 2014, p. 292). The organism had learned the temporal contingencies of the CS-CR relationship. A learned body-timing behavior of an astonishingly accurate nature was also shown for 30-minute intervals. When a dog was repeatedly fed exactly every 30 minutes together with the sound of a metronome, saliva was secreted only when the metronome sounded at exactly 30 minutes and not after 29 minutes (Pavlov, 1928; see Table, p. 42). Pavlov concluded from these series of experiments that "changes in practically all the component tissues and parts of the organism are capable of influencing the cerebral hemispheres. This continuous cycle of direct and indirect influences upon nervous activity constitutes the physiological basis for the estimation of duration of time." (Pavlov, 1928, p. 42).

In search of the underlying mechanisms of the human awareness of time, Hugo Münsterberg (1889) had already concluded that—based on his psychophysiological studies performed in Freiburg—the sense of duration relied on the sensation of tension in different organs which is caused by muscle contractions of the skeletal muscle tissue, as well as of the internal organs. For example, in experiments where he was his own subject, Münsterberg found out that the accuracy of reproducing temporal intervals in the range between 6 and 60 seconds was higher when the onset and the offset of the to-be-judged temporal interval coincided with the onset of his breathing cycle. These early empirical findings of a link between the perception of duration and physical processes are complemented by recent conceptualizations. Craig (2009, 2015) proposed that the experience of time is generated by emotional and visceral processes of the interoceptive neural system. Time is perceived by the constantly ascending somatic signals culminating in the insular cortex, which create a series of conscious emotional moments over time. Duration is thereafter created by these successive moments of self-realization, which are essentially formed by changing body states. Our empirical studies support this idea. Functional-neuroimaging (fMRI) technology shows how recorded brain activity increases with increasing interval length in the left and right dorsal posterior insula when people have to judge the duration of multiple-second intervals (Wittmann, Simmons, Aron, & Paulus, 2010; Wittmann et al., 2011). In our psychophysiological studies, heartbeat intervals increased (the heart rate slowed down), and skin-conductance levels decreased steadily when subjects were exposed to auditory or visual intervals whose durations had to be estimated (Meissner & Wittmann, 2011; Otten et al., 2015). Together, these studies show that the judgment of duration in the multiple-second range, which is associated with the feeling of time passing, is at least partly based on the processing of physical states and associated neural networks.

Timing behavior in the range of milliseconds to a few seconds is governed by processes in sensorimotor areas of the brain (Merchant, Harrington, & Meck, 2013; Pedrosa et al., 2016). Indices of body arousal, such as the cardiac vagal tone (high frequency band (HF)), contribute to the accuracy of timing in the milliseconds (Cellini et al., 2015) and seconds range (Pollatos, Yeldesbay, Pikovsky, & Rosenblum, 2014). It is likely that time is represented by different networks interacting in a distributed manner in the brain (Buhusi, Oprisan, & Buhusi, 2018). Future research will have to delineate how the various functional and neuroanatomic components coordinate physical sensations and how the related neural substrates contribute to the perceived passage of time (Wittmann, 2013, 2016; Wittmann & Meissner, 2018). Oscillatory models of subjective time will help answer the question of how the brain processes duration. Neural oscillations of various frequencies underlie all cognitive functions, including the judgment of time intervals, such as when subjects have to estimate the duration or have to synchronize their behavior with regularly occurring events (Kononowicz & van Rijn, 2015; van Wassenhove, 2016).

2 Inter-individual Differences in the Perception of Time

Regarding any measurable biological or psychological function, humans differ considerably in how they experience and behave. Individual differences naturally also occur in the timing of body functions, as well as in the conscious perception of time. For example, differences in the circadian sleep-wake cycle, which are generated by genetic variations can be seen in the individually preferred timing of sleep onset and offset, where different chronotypes range from extreme early types ("larks") to extreme late types ("owls") with the majority of people in between (Wittmann, Dinich, Merrow, & Roenneberg, 2006). While all chronotypes require an average sleep of eight hours, larks find it difficult to stay up late in the evening, but spontaneously wake up at an early hour, and owls go to bed later (or even in the early morning hours) and sleep later into the day. Stable inter-individual differences also exist in the range of milliseconds. The detection of temporal order seems to be a unique individual bias and may result from structural constraints of the brain. In longitudinal studies, the intra-individual stability of the temporal-order detection of two short events is typically shown within and across the visual and auditory modality (Fink, Ulbrich, Churan, & Wittmann, 2006; Grabot & van Wassenhove, 2017).

Personality traits contribute to the perception of longer durations. When performing a tedious task in the multiple-second range, more boredom-prone individuals feel that time passes more slowly (Watt, 1991), and such individuals relatively overestimate duration (Danckert & Allman, 2005). Impulsivity as a personality trait is defined as a propensity to react to the immediate situation without thinking about future consequences. According to this definition, impulsivity is an extreme form of temporal shortsightedness (Retz Lucci, 2013; Wittmann & Paulus, 2016), and it complies with the concept of hedonistic present orientation (Zimbardo & Boyd, 1999). Impulsive persons feel more easily trapped in time when they cannot act on their desires and then tend to overestimate duration as a symptom of their frustration (Moreira, Pinto, Almeida, & Barbosa, 2016; Wittmann & Paulus, 2008). Within the embodiment framework of subjective time, impulsivity is the strong urge for immediate gratification which is generated through the interoceptive (body) system (Turel & Bechara, 2016). Empty waiting time without distraction accordingly leads to relatively higher bodily arousal levels and to a more focused attention to time. Both factors contribute to the feeling of a slower passage of time and an overestimation of duration (Zakay, 2014).

Waiting without distraction can be described as a situation in which one becomes aware of oneself and of time. Although the sketched relationship between waiting, subjective time, and impulsivity is not really astonishing—it does comply with our everyday observations, that is, when watching bored teenagers—this conceptual complex has not been assessed systematically. We therefore aimed at empirically confirming whether an impulsive present orientation of a person confined to a waiting situation would result in stronger negative affect and accordingly lead to the feeling of an expansion of duration (Jokic, Zakay, & Wittmann, 2018). Each student participant spent exactly 7:30 minutes in a small closed room at their University. Prior to the actual waiting situation, subjects handed their watches and electronic devices to the instructor. Bags which could contain distracting material were locked away. After filling out the administrative paperwork and two self-report questionnaires on impulsivity and time perspective, the participants entered the waiting room and were told to sit down and simply wait for the researcher to come back. The instructor reentered the room after exactly 7:30 minutes and asked the subjects to report their subjective impressions while waiting alone. The question concerned subjective time (see below) and the emotional state; the latter was assessed by the self-assessment-manikin (SAM) scales of valence and arousal (Bradley & Lang, 1994) and a 100-mm visual analogue scale (VAS) asking about the degree of relaxation.

All 82 subjects' answers to the questions of "How often did you think about time?" (Mean: 45.4; S.D.: 28.3; range: 5–97) and "How fast did time pass for you?" (Mean: 59.7; S.D.: 21.4; range: 3–100) on the VAS scales were normally distributed with an empirical mean close to the spatial mean of the scale (50 mm). The reported waiting time for the 7.5 minutes was on average 8.4 minutes (S.D.: 3.0 min.) with a considerable range of answers between 2.5 and 19 minutes. As can be seen in Fig. 9.1,

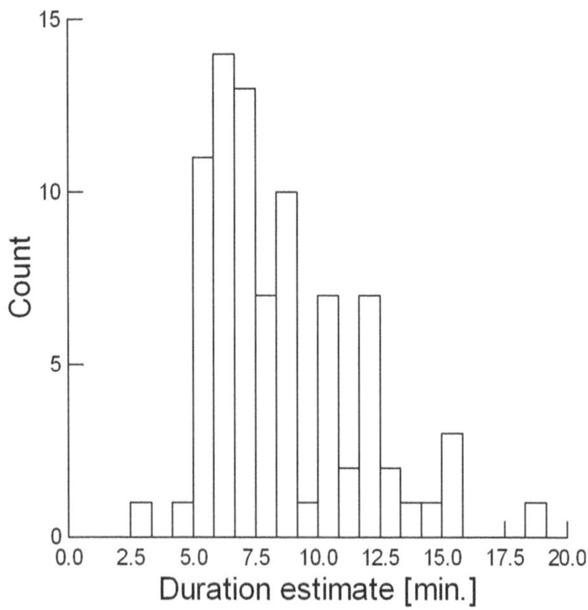

Fig. 9.1 The distribution of felt duration in 82 participants as assessed after a 7.5-minute waiting period

the modus and the two second-highest bars in the histogram lie just below 7.5 minutes, but a skew of the distribution to the right shows many estimates beyond the mean. The median that splits the distribution into the 50% shortest and 50% longest estimates lies with 7.75 minutes close to the 7.5-minute waiting period.

How does this inter-individual variation in duration estimates depicted in Fig. 9.1 relate to the emotional state during the waiting period? The longer the duration estimate for the waiting time, the more negative the emotion (SAM valence; $r = -0.303$, $p = 0.006$) and the less relaxed participants were ($r = -0.278$, $p = 0.011$). Similar correlations occur for the other two variables of subjective time; the more participants thought about the time during the waiting period, the more arousal they reported (SAM arousal; $r = 0.301$, $p = 0.006$). The faster subjects felt time to pass, the more positive emotions were reported (SAM valence; $r = 0.304$, $p = 0.005$) and the lower the emotional arousal level was (SAM arousal; $r = -0.361$, $p = 0.001$). These correlations confirm the experience that an overestimation of time and a felt slower passage of time coincides with an increased awareness of time and is related to lower affective well-being, more arousal, and less relaxation.

How do self-reported personality traits (inter-individual differences) associate with duration estimates? Correlations show that the longer the duration estimate, the higher the self-rated impulsivity ($r = 0.500$, $p < 0.001$), the stronger the present hedonistic orientation ($r = 0.279$, $p = 0.011$), and the stronger the fatalistic present orientation ($r = 0.276$, $p = 0.012$). Regarding the passage of time, the faster the time seems to pass while waiting, the lower the self-reported impulsivity ($r = -0.512$, $p < 0.001$) and the lower the present hedonistic orientation ($r = -0.306$, $p = 0.005$). Impulsivity and the two present orientation dimensions of hedonism and fatalism, all three of which correlate positively with each other (Baumann & Odum, 2012; Mueller, Berger, Tucha, & Falter, 2014), are temperamental indicators for overestimating duration. The related concept of these three personality-trait indicators is a strong present orientation, typically at the expense of a more pronounced future orientation. A multiple-path mediation analysis integrates the three domains of personality, time perception, and affective well-being. The analysis shows that the more impulsive individuals are, the less relaxed they are while waiting, and the more they overestimate duration and perceive a slower passage of time (Jokic et al., 2018).

This may have been the first systematic study to investigate inter-individual differences in the relationship between subjective time and affect in a real waiting situation spanning several minutes. Studies previously investigated judgments in the range of seconds or only a few minutes combined with a tedious task in a somewhat artificial laboratory setting (for articles of reference, see Jokic et al., 2018). An exception is the study by Droit-Volet, Monceau, Berthon, Trahanias, and Maniadakis (2018) who had individuals estimate intervals of daily life with duration between 2 and 32 minutes using cell phone technology. We similarly wanted to create a natural situation, a typical waiting situation as it occurs in daily life. Physical time is not isomorphic to the subjective time. Our research, which we consider to be paradigmatic for further investigations, shows that individual differences lead to specific emotional reactions in a waiting situation which, in turn, influence how we perceive time. Waiting is a ubiquitous situation in daily life. Long waiting times are actually considered the most common daily hassle in industrialized nations. For example, a representative study by the German Society for Consumer Research (Gesellschaft für Konsumforschung, GfK) in 2016 found out that the situation which annoys Germans most (55.6% of participants responded to this item) is "avoidable long waiting times." Strategies for developing architectural surroundings and other means to help individuals have a pleasant time waiting could emerge from applied research.

3 Intra-individual Differences in the Perception of Time

During normal waking, our states of consciousness vary considerably as fluctuations between drowsiness and focused attention and between relaxation and extreme excitement. We may be in the passive mood of boredom or highly engaged in a pleasant activity, and these states can change quickly. Mainstream psychology and artificial intelligence, in its attempt to mimic human cognition have long neglected the obvious fact that the conscious awareness of an observer succumbs to these "tides of mind" (Gelernter, 2016). These fluctuations in states of consciousness can be described by variations in the sense of self and time. In an extended conceptualization of self and time, the "narrative self" is related to the sensed continuity across longer time intervals relying on mental time-traveling in the past and an imagined future (Berkovich-Ohana & Glicksohn, 2014). Here we focus on the immediate sense of the "bodily self" and the associated passage of time in the range of seconds. This focus is justified, as we want to assess the sense of time as it is presently experienced in a given situation lasting seconds to minutes. Since subjective duration is related to bodily self-awareness across time on this basic level, the awareness of the self and the awareness of time are mutually up- and down-regulated. In states of boredom we feel ourselves intensively, and time drags. In contrast, in states of flow one is less aware of oneself and may underestimate duration. This happens because in flow states, one is fully immersed in challenging, but joyful activities, such as performing sports, doing fulfilling work, or simply playing (Csikszentmihalyi & Csikszentmihalyi, 1988). Attention is fully focused on the object of activity and away from oneself.

Altered states of consciousness are states that clearly deviate from everyday waking consciousness. These states can be induced through psychological techniques, such as meditation, rhythm-induced trance, or sensory deprivation (Block, 1979; Vaitl et al., 2005). They more frequently appear when one is absorbed in a musical experience or when playing a musical instrument (Schäfer, Fachner, & Smukalla, 2013), or they can be elicited through psychotropic substances (Shanon, 2001; Studerus, Gamma, Kometer, & Vollenweider, 2012). During these different altered states, a joint distortion of the feelings of self and time can occur, sometimes culminating in reports that self and time were hardly or not experienced at all during that period of time (Berkovich-Ohana, Dor-Ziderman, Glicksohn, & Goldstein, 2013; Wittmann, 2015, 2018).

According to personal reports, a variety of body-focused meditation and movement techniques exist that change the conscious experience of the bodily self and subjective time. In a systematic theoretical study, at least four functional aspects were identified in body-induced meditation, that is, the state of mindfulness, namely attention regulation, body awareness, emotion regulation, and changes in the perspective of the self (Hölzel et al., 2011). The topic of subjective time in meditation research has only recently begun to attract more attention (Wittmann & Schmidt, 2014). Since body-focused meditation and movement techniques, such as Yoga, Tai-Chi, and various forms of meditation, including classic sitting meditation, are widely practiced today, experienced individuals can be tested to assess changes in personality trait and temporary states of experience during or after meditation. Experimental studies assessing time perception after body-oriented (mindfulness) meditation point to subjective time dilation of presented temporal intervals in the milliseconds-and-seconds range (Droit-Volet, Fanget, & Dambrun, 2015; Kramer, Weger, & Sharma, 2013). These outcomes are believed to result from an increased (more mindful) interoceptive awareness after meditation (Wittmann, 2015).

The meditation-induced subjective prolongation of duration complements many studies showing how time intervals are relatively overestimated when individuals are presented with static or dynamic stimulus material (photos, movie clips, sounds) which elicit body-arousing emotions (Droit-Volet & Gil, 2009; Dirnberger et al., 2012; Wackermann, Meissner, Tankersley, & Wittmann, 2014). These results add to the corpus of experimental work that supports the theory that the sense of time is created through the constantly ongoing accumulation of bodily signals (Craig, 2009, 2015; Wittmann, 2013, 2016). Most experimental studies focus on more easily controllable, shorter durations in the range of milliseconds or seconds. One study, however, investigated time judgments of intervals lasting several seconds (Pollatos, Laubrock, & Wittmann, 2014). An overestimation of felt duration was seen in retrospective time judgments of emotional video clips lasting 45 seconds. A fear-inducing video (from a horror movie) elicited relative subjective time dilation in participants when compared to a documentary video and a funny cartoon of equal duration. The funny cartoon was actually judged to have lasted the shortest of the three video clips, an effect that coincides with the saying that "time flies when you are having fun." The overestimation effect of the horror video was enhanced in a separate trial arm in which subjects had to explicitly attend to the evoked bodily reactions. These experimental outcomes show that transient changes in the sense of the bodily self are responsible for intra-individual variations of felt duration in the seconds range.

What about intra-individual variability in estimating durations lasting several minutes? In one study, we subjected student participants to two different situations of silence lasting 6.5 minutes. The subjective experience was then compared between the two conditions. In our study, a Depth Relaxation Music Therapy (DRMT; Decker-Voigt, 2007) induction formed the experimental condition for university students (Pfeifer, Sarikaya, & Wittmann, 2016). DRMT can be subsumed as relaxation induction comprising elements of mindfulness body awareness. It concerns sensitization to the body, physical comfort, sensitization to emotions and feelings, and sensitization to mental images and auditory perception. We compared two identical periods of 6.5 minutes of silence, one after a 16-minute DRMT induction and one after a 16-minute university seminar focusing on aspects of silence in counseling and therapy. Subjective ratings of 60 participants undergoing both conditions in a group setting were collected with scales assessing various dimensions of states of consciousness. Overall, the silence after DRMT was experienced as more relaxing than the silence after the seminar. Participants reported a diminished sense of space, time, and future perspective. Subjects estimated the silence in the DRMT condition to have lasted longer (10.3 min.) than that in the seminar condition (6.8 min.) ($t = -6.5$, $p < 0.001$; Pfeifer et al., 2016). Figure 9.2 shows that the mode of the distribution (Fig. 9.2a) lies at the average estimate (6.8 min.) with low variance/small range in estimates, following the seminar condition. After the DRMT session, the mode lies above 10 minutes (at the mean value of 10.3 min.) with a considerably higher variance and range.

Regarding the three questions concerning subjective time, silence after DRMT was associated with a decreased intensity of time consciousness and a less pronounced future orientation, but the duration in clock time was relatively overestimated. Paradoxical effects of time passing more quickly on the one hand and duration being judged as seemingly longer on the other hand have been shown in previous investigations (Wearden, 2015; Weiner, Wittmann, Bertschy, & Giersch, 2016). In our study, besides critical methodological concerns about the validity of the scales

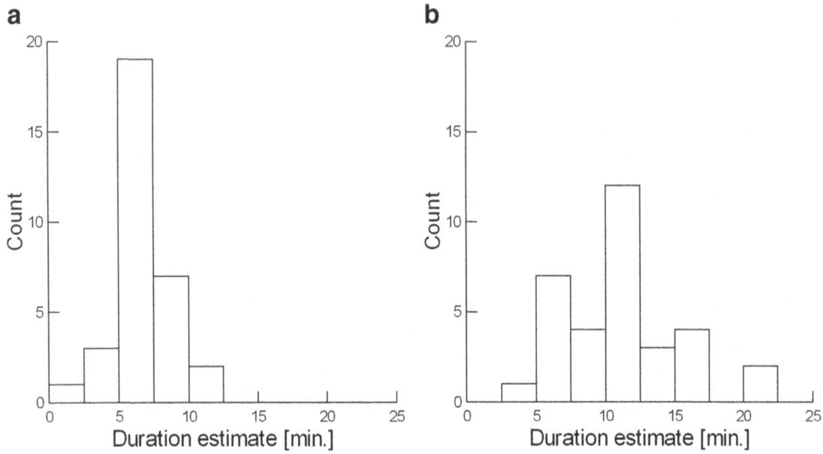

Fig. 9.2 The distribution of felt duration in 60 participants as assessed once after a 6.5-minute period of silence following a seminar (a) and once after a 6.5-minute period of silence following a meditation-type session of Depth Relaxation Music Therapy (b)

used, different mechanisms may underlie how intensely we sense time and how we judge duration explicitly as compared to clock time.

The point we want to make here is that the participants in our study were exposed to two different contextual situations (body-focused intervention vs. university seminar on silence in therapy and counseling) followed by the same period of silence lasting 6.5 minutes. The contextual settings before the interval varied considerably and led to clear intra-individual differences in the judgment of duration. We regard this experimental setup as paradigmatic for different methods to induce intra-individual variations in subjective time. Different induction techniques can be employed surrounding a given interval that can be filled with tasks, such as watching a stage performance (Deinzer, Clancy, & Wittmann, 2017), or unfilled (as in our cases here) and accompanied with measures assessing psychophysiological parameters.

4 Discussion

Recent concepts concerning an embodied view of subjective time (Craig, 2009; Wittmann, 2013; Wittmann & Meissner, 2018) result from earlier attempts to assess the timing accuracy and precision of bodily processes in relation to external events (Münsterberg, 1889; Pavlov, 1928). The majority of investigations assessing behavioral timing and the experience of time focus on the time range of milliseconds to seconds (Nobre & Coull, 2010). Here we present experimental paradigms which assess experienced duration in the minutes range. Experienced duration over several minutes is ecological, as we regularly face situations where we have to explicitly wait for something to happen or when we live through a period of time that is more or less pleasant. We have shown how an impulsive present orientation is related to feeling more irritated during an empty waiting time lasting 7.5 minutes, which in turn leads to an overestimation of duration. This can be interpreted in the context of the waiting situation where more impulsive individuals readily feel more trapped in time and more strongly experience their bodily selves. We have shown how intra-individual variations in situations are associated with strong variations in duration judgments. A body-focused intervention in a group setting evoked a relative overestimation of a period of silence lasting 6.5 minutes as compared to the silence of the same duration preceded by a seminar. These are the first attempts to assess how humans behave and experience times of longer durations in everyday life.

References

Baumann, A. A., & Odum, A. L. (2012). Impulsivity, risk taking, and timing. *Behavioral Processes, 90*, 408–414.

Berkovich-Ohana, A., Dor-Ziderman, Y., Glicksohn, J., & Goldstein, A. (2013). Alterations in the sense of time, space, and body in the mindfulness-trained brain: A neurophenomenologically-guided MEG study. *Frontiers in Psychology, 4*, 912.

Berkovich-Ohana, A., & Glicksohn, J. (2014). The consciousness state space—A unifying model for consciousness and self. *Frontiers in Psychology, 5*, 341.

Block, R. A. (1979). Time and consciousness. In G. Underwood & R. Stevens (Eds.), *Aspects of consciousness* (Vol. 1, pp. 179–217). London: Academic Press.
Bradley, M. M., & Lang, P. J. (1994). Measuring emotion: The self-assessment manikin and the semantic differential. *Journal of Behavior Therapy and Experimental Psychiatry, 25*, 49–59.
Buhusi, C. V., Oprisan, S. A., & Buhusi, M. (2018). Biological and cognitive frameworks for a mental timeline. *Frontiers in Neuroscience, 12*, 377.
Buzsáki, G. (2006). *Rhythms of the brain*. Oxford: Oxford University Press.
Cellini, N., Mioni, G., Levorato, I., Grondin, S., Stablum, F., & Sarlo, M. (2015). Heart rate variability helps tracking time more accurately. *Brain and Cognition, 101*, 57–63.
Craig, A. D. (2009). Emotional moments across time: A possible neural basis for time perception in the anterior insula. *Philosophical Transactions of the Royal Society B, 364*, 1933–1942.
Craig, A. D. (2015). *How do you feel? An interoceptive moment with your neurobiological self*. Princeton: Princeton University Press.
Csikszentmihalyi, M., & Csikszentmihalyi, I. S. (1988). *Optimal experience: Psychological studies of flow in consciousness*. Cambridge: Cambridge University Press.
Danckert, J. A., & Allman, A. A. A. (2005). Time flies when you're having fun: Temporal estimation and the experience of boredom. *Brain and Cognition, 59*, 236–245.
Decker-Voigt, H.-H. (2007). Methodology of music therapy and depth relaxation (DRMT). In I. Frohne-Hagemann (Ed.), *Zeitpunkt Musik. Receptive music therapy: Theory and practice* (pp. 239–251). Wiesbaden: Dr. Ludwig Reichert Verlag.
Deinzer, V., Clancy, L., & Wittmann, M. (2017). The sense of time while watching a dance performance. *SAGE Open, 7*(4), 1–10.
Dirnberger, G., Hesselmann, G., Roiser, J. P., Preminger, S., Jahanshahi, M., & Paz, R. (2012). Give it time: Neural evidence for distorted time perception and enhanced memory encoding in emotional situations. *Neuroimage, 63*, 591–599.
Droit-Volet, S., Fanget, M., & Dambrun, M. (2015). Mindfulness meditation and relaxation training increases time sensitivity. *Consciousness and Cognition, 31*, 86–97.
Droit-Volet, S., & Gil, S. (2009). The time-emotion paradox. *Philosophical Transactions of the Royal Society B, 364*, 1943–1954.

Droit-Volet, S., Monceau, S., Berthon, M., Trahanias, P., & Maniadakis, M. (2018). The explicit judgment of long durations of several minutes in everyday life: Conscious retrospective memory judgment and the role of affects? *PLoS One, 13*, e0195397.
Fink, M., Ulbrich, P., Churan, J., & Wittmann, M. (2006). Stimulus-dependent processing of temporal order. *Behavioural Processes, 71*, 344–352.
Garfinkel, S. N., Minati, L., Gray, M. A., Seth, A. K., Dolan, R. J., & Critchley, H. D. (2014). Fear from the heart: Sensitivity to fear stimuli depends on individual heartbeats. *Journal of Neuroscience, 34*, 6573–6582.
Gelernter, D. (2016). *The tides of mind. Uncovering the spectrum of consciousness.* New York: Liveright.
Golombek, D. A., Bussi, I. L., & Agostino, P. V. (2014). Minutes, days and years: Molecular interactions among different scales of biological timing. *Philosophical Transactions of the Royal Society B, 369*, 20120465.
Grabot, L., & van Wassenhove, V. (2017). Time order as psychological bias. *Psychological Science, 28*, 670–678.
Hölzel, B. K., Lazar, S. W., Gard, T., Schuman-Olivier, Z., Vago, D. R., & Ott, U. (2011). How does mindfulness work? Proposing mechanisms of action from a conceptual and neural perspective. *Perspectives on Psychological Science, 6*, 537–559.
Jokic, T., Zakay, D., & Wittmann, M. (2018). Individual differences in self-rated impulsivity modulate the estimation of time in a real waiting situation. *Timing & Time Perception, 6*, 71–89.
Kononowicz, T. W., & van Rijn, H. (2015). Single trial beta oscillations index time estimation. *Neuropsychologia, 75*, 381–389.
Kramer, R. S. S., Weger, U. W., & Sharma, D. (2013). The effect of mindfulness meditation on time perception. *Consciousness and Cognition, 22*, 846–852.
Matthias, E., Schandry, R., Duschek, S., & Pollatos, O. (2009). On the relationship between interoceptive awareness and the attentional processing of visual stimuli. *International Journal of Psychophysiology, 72*, 154–159.
Meissner, K., & Wittmann, M. (2011). Body signals, cardiac awareness, and the perception of time. *Biological Psychology, 86*, 289–297.
Merchant, H., Harrington, D. L., & Meck, W. H. (2013). Neural basis of the perception and estimation of time. *Annual Review of Neuroscience, 36*, 313–336.
Moreira, D., Pinto, M., Almeida, F., & Barbosa, F. (2016). Time perception deficits in impulsivity disorders: A systematic review. *Aggression and Violent Behavior, 27*, 87–92.

Mueller, A. K., Berger, J., Tucha, O., & Falter, C. M. (2014). Selective impairment of timing functions in non-clinical impulsiveness. *Timing & Time Perception, 2*, 20–34.

Münsterberg, H. (1889). *Beiträge zur experimentellen Psychologie, Heft 2*. Freiburg: J.C.B. Mohr.

Nobre, K., & Coull, J. (2010). *Attention and time*. Oxford: Oxford University Press.

Northoff, G. (2014). *Unlocking the brain volume 2: Consciousness*. Oxford: Oxford University Press.

Otten, S., Schötz, E., Wittmann, M., Kohls, N., Schmidt, S., & Meissner, K. (2015). Psychophysiology of duration estimation in experienced mindfulness meditators and matched controls. *Frontiers in Psychology, 6*, 1215.

Pavlov, I. (1928). *Conditioned reflexes. An investigation of the physiological activity of the cerebral cortex*. London: Oxford University Press, Humphrey Milford.

Pedrosa, D. J., Nelles, C., Maier, F., Eggers, C., Burghaus, L., Fink, G. R., … Timmermann, L. (2016). Variance of essential tremor patients' time reproduction deficits. *Movement Disorders, 31*, 1428–1429.

Pfeifer, E., Sarikaya, A., & Wittmann, M. (2016). Changes in states of consciousness during a period of silence after a session of depth relaxation music therapy (DRMT). *Music and Medicine, 8*(4), 180–186.

Pollatos, O., Laubrock, J., & Wittmann, M. (2014). Interoceptive focus shapes the experience of time. *PLoS One, 9*(1), e86934.

Pollatos, O., Yeldesbay, A., Pikovsky, A., & Rosenblum, M. (2014). How much time has passed? Ask your heart. *Frontiers in Neurorobotics, 8*, 15.

Retz Lucci, C. (2013). Time, self, and intertemporal choice. *Frontiers in Neuroscience, 7*, 40.

Roenneberg, T., Daan, S., & Merrow, M. (2003). The art of entrainment. *Journal of Biological Rhythms, 18*, 183–194.

Schäfer, T., Fachner, J., & Smukalla, M. (2013). Changes in the representation of space and time while listening to music. *Frontiers in Psychology, 4*, 508.

Shanon, B. (2001). Altered temporality. *Journal of Consciousness Studies, 8*, 35–58.

Studerus, E., Gamma, A., Kometer, M., & Vollenweider, F. X. (2012). Prediction of psilocybin response in healthy volunteers. *PLoS One, 7(2)*, e30800.

Todes, D. P. (2014). *Ivan Pavlov. A Russian life in science*. Oxford: Oxford University Press.

Turel, O., & Bechara, A. (2016). A triadic reflective-impulsive-interoceptive awareness model of general and impulsive information system use: Behavioral tests of neuro-cognitive theory. *Frontiers in Psychology, 7*, 601.

Vaitl, D., Birbaumer, N., Gruzelier, J., Jamieson, G. A., Kotchoubey, B., Kübler, A., … Sammer, G. (2005). Psychobiology of altered states of consciousness. *Psychological Bulletin, 131*, 98–127.

van Wassenhove, V. (2016). Temporal cognition and neural oscillations. *Current Opinion in Behavioral Sciences, 8*, 124–130.

Wackermann, J., Meissner, K., Tankersley, D., & Wittmann, M. (2014). Effects of emotional valence and arousal on acoustic duration reproduction assessed via the 'dual klepsydra model. *Frontiers in Neurorobotics, 8*, 11.

Watt, J. D. (1991). Effect of boredom proneness on time perception. *Psychological Reports, 69*, 323–327.

Wearden, J. H. (2015). Passage of time judgments. *Consciousness and Cognition, 38*, 165–171.

Weiner, L., Wittmann, M., Bertschy, G., & Giersch, A. (2016). Dispositional mindfulness and subjective time in healthy individuals. *Frontiers in Psychology, 7*, 786.

Wittmann, M. (2013). The inner sense of time: How the brain creates a representation of duration. *Nature Reviews Neuroscience, 14*, 217–223.

Wittmann, M. (2015). Modulations of the experience of self and time. *Consciousness and Cognition, 38*, 172–181.

Wittmann, M. (2016). *Felt time. The science of how we experience time*. Cambridge, MA: MIT Press.

Wittmann, M. (2018). *Altered states of consciousness: Experiences out of time and self*. Cambridge, MA: MIT Press.

Wittmann, M., Dinich, J., Merrow, M., & Roenneberg, T. (2006). Social jetlag: Misalignment of biological and social time. *Chronobiology International, 23*, 497–509.

Wittmann, M., & Meissner, K. (2018). The embodiment of time: How interoception shapes the perception of time. In M. Tsakiris & H. de Preester (Eds.), *The interoceptive mind—From homeostasis to awareness* (pp. 63–79). Oxford: Oxford University Press.

Wittmann, M., & Paulus, M. (2016). How the experience of time shapes decision-making. In M. Reuter & C. Montag (Eds.), *Neuroeconomics. Studies in neuroscience, psychology and behavioral economics series* (pp. 133–144). Berlin, Heidelberg: Springer.

Wittmann, M., & Paulus, M. P. (2008). Decision making, impulsivity, and time perception. *Trends in Cognitive Science, 12*, 7–12.

Wittmann, M., & Schmidt, S. (2014). Mindfulness meditation and the experience of time. In S. Schmidt & H. Walach (Eds.), *Meditation–neuroscientific approaches and philosophical implications* (pp. 199–209). Cham: Springer.

Wittmann, M., Simmons, A. N., Aron, J. L., & Paulus, M. P. (2010). Accumulation of neural activity in the posterior insula encodes the passage of time. *Neuropsychologia, 48*, 3110–3120.

Wittmann, M., Simmons, A. N., Flagan, T., Lane, S. D., Wackermann, J., & Paulus, M. P. (2011). Neural substrates of time perception and impulsivity. *Brain Research, 1406*, 43–58.

Zakay, D. (2014). Psychological time as information: The case of boredom. *Frontiers Psychology, 5*, 917.

Zimbardo, P. G., & Boyd, J. N. (1999). Putting time in perspective: A valid, reliable individual-difference metric. *Journal of Personality and Social Psychology, 77*, 1271–1288.

10

Against Illusions of Duration

Sean Enda Power

1 Introduction

When I was in college, I worked as a night security. I worked for eight hours each shift. During the shift, this duration *felt* as if, for each hour on the clock, much longer than an hour passed. Yet, when I finished in the morning, it felt like the whole night flew by.

For the same period of clock time, I had experiences of two kinds of *duration*—one prospective, the other retrospective. The difference was phenomenological, a difference in *apparent* duration. It was not of other properties or their differences, such as of spatial extent, color, or shape. Furthermore, these experienced durations seem to be in some way direct. I am not inferring them, nor the difference between them, from something else; nor do I imagine them.

Are differences in these experienced durations because of real differences in duration? Or is at least one an erroneous, distorted, or illusory experience of duration?

S. E. Power (✉)
Department of Philosophy, University College Cork, Cork, Ireland

V. Arstila et al. (eds.), *The Illusions of Time*,
https://doi.org/10.1007/978-3-030-22048-8_10

One might think that they are obviously not real differences in duration and so are illusory; something is wrong here. I agree something is wrong here. However, I do not think it is because of an illusory duration. It is an error in our interpretation of experienced duration and its relationships to other things, including clocks. This kind of error is not an illusion.

An illusion is an *error*, but it is not just any kind of error. It is not even any error related to experience: I can misunderstand what I'm experiencing without being under an illusion, for example, I can believe I am under a color illusion of a blue banana but I correctly see a blue plantain; there can be a discrepancy between appearance and reality but it's because there's more to reality than there appears to be, for example, I think a room is dark but in fact there's very dim starlight from a skylight that, being night-blind, I cannot see.[1]

An illusion is an experience in which something appears to have properties, for example, color, shape, weight, and motion, that it does not have. As Kalderon puts it, 'an illusory appearance is a state of appearing. Specifically, an illusion is an experience of an object o appearing F where o is not in fact F' (Kalderon, 2011; see also Crane & French, 2016).

For example, in the Ebbinghaus illusion, there is a mismatch between some circles' real size (they are same-sized) and their apparent size (they appear differently sized).

This has the following consequence: if one does not know either (a) the appearance of F or (b) the reality of F, then one does not know if there is an illusion of F. If you do not know that the circles in the Ebbinghaus illusion appear to be different sizes, or you do not know their actual relative sizes, then you do not know if this phenomenon is, in fact, an illusion.

As said, illusions are not the only errors around experience. A false belief about some hidden property of what is experienced is not an illusion of that property. For example, when we see any typical object in space, much of its surface faces away from us; if the surface is not broken

[1] For a detailed discussion on the difference between illusions and other kinds of errors relevant to experience (such as illusions), especially as it relates to experience and time, see Power (2018).

in some way, then its internal structure is not visible. Much of the object is not visible to us, it is hidden. Possibly, we are in error about the object because of what is hidden. For example, we are on a movie set with shop-front facades, with nothing behind them. Or what appears to be a red apple is, in fact, a solid metal sculpture of an apple. We are mistaken about what we see, but this is not a case of visual illusion (see Kalderon, *op.cit.*, for discussion).

2 Empirical Evidence

There seem to be many kinds of experiences that provide empirical evidence for illusions of duration. The following are discussed here: round trips; the 'oddball effect'; chronostasis or the 'stopped clock illusion'; the Zeitraffer phenomenon; the sense that time speeds up as you get older. For the purposes of this chapter, these encompass the major examples of alleged illusory duration.

2.1 Round Trips

This common experience does not need much explanation. To experience it, take a journey somewhere you have never been before; travel in both directions at the same speed and along the same route. It is commonly found that, despite it taking the same clock time, the return trip seems to take less time than the trip out. The difference between the experienced duration going out and back does not correspond to a difference in clock time.

2.2 The Oddball Effect

A subject is presented with a sequence of images, each presented for the same length of time. One of the images is non-temporally different to the others. For example, a subject is presented with the same image of a goldfish several times, then presented with an image of a mango, followed by more instances of the image of the goldfish.

Each image is presented for the same length of time. Yet, the odd one out seems to the subject as if it is seen for a longer time. The image of the mango seems to be presented for a greater length of time (for an illustration of this, see the game on http://www.claudiahammond.com/page5.htm, no.4). 'If I flashed up a sequence of pictures on a screen [...] you would think that the picture of a mango was on the screen for longer. It is as though time very briefly slowed down when you looked at the mango'. (Hammond, 2012, p. 70)

This raises a possible argument for illusions of duration:

1. The durations between the common and uncommon stimuli are identical.
2. Yet, it appears as if the durations between the common and uncommon stimuli are different (it appears as if the uncommon stimulus' duration is longer).
3. Either/Both (a) the experience of the duration of the uncommon stimulus is mistaken or/and (b) the duration of the common stimuli is mistaken.
4. If (a), then there is an illusion of the uncommon stimulus' duration.
5. If (b), then there is an illusion of the duration of common stimuli.
6. If (a) and (b), then there is an illusion of both durations.
7. In any of these three cases, there is an illusion of duration.

2.3 Chronostasis

When we glance at a clock's seconds hand—or indeed a spinning fan, or a similar moving thing—the hand seems to hold its position longer when we first see it than afterward: on first seeing it, it appears to slow, pause, or stop. However, it does not slow, pause, or stop. This has led some theorists to call this *chronostasis* (e.g., Hodinott-Hill, Thilo, Cowey, & Walsh, 2002, p. 1779; Le Poidevin, 2007) an illusion: 'If you have a watch with a ticking second hand, you may experience a recurrent, momentary illusion that your watch has stopped' (Bardon, 2013, p. 46). '[T]he first presentation of the repeated stimulus appears longer in duration than successive presentations. With illusions such as these in mind' (Eagleman & Pariyadath, 2007, p. 1841).

This phenomenon resembles the oddball effect, and a similar argument may be made for its illusory status. The only difference is that one substitutes 'first' for 'uncommon' and 'following' for 'common'. The *first stimulus* is the appearance of the seconds hand on first glance; the *following stimuli* are the later appearances of the seconds hand:

1. The duration between the first and the following stimuli are identical.
2. Yet, it appears as if the duration between the first and the following stimuli are different (it appears as if the first stimulus' duration is longer).
3. Either/Both the experience of (a) the duration of the first stimulus is mistaken or/and (b) the duration of the following stimuli is mistaken.
4. If (a), then there is an illusion of the uncommon stimulus' duration.
5. If (b), then there is an illusion of the duration of common stimuli.
6. If (a) and (b), then there is an illusion of both durations.
7. In any of these three cases, there is an illusion of duration.

2.4 Zeitraffer Phenomenon

In the Zeitraffer phenomenon (e.g., Le Poidevin, 2007), the motions of perceived bodies seem to move more rapidly than they actually move. This can be argued to be a durational illusion as follows:

1. If a body in the Zeitraffer phenomenon appears to move rapidly through space, then the body appears to occupy a certain duration D. This is the duration it appears to take to move rapidly through that space.
2. If the body moves slower through the space than it appears to move, then it takes a longer duration than D to move through that space (call that duration E).

3. The body's motion appears to have duration D, but it has duration E (which is not D).
4. Therefore, in the experience of this body's motion, there is an illusion of duration D.

2.5 Time Speeding Up as You Get Older

It is a common claim that subjects experience the same durations—for example, years and days—as feeling shorter than they did when the subjects were younger. This would be an illusion like the oddball effect. If the duration is the same for both subjects' younger and older selves, then either their younger or older selves have a mistaken experience.

Weardon describes a subtler aspect of this experience for older subjects:

> A prominent psychological feature of ageing is changes in the experience of time. People report that "Christmas comes round quicker every year" as they age, yet "Time lies heavy on their hands," and days may seem to crawl by, in a way they never used to when the person was younger. (Weardon, 2005, p. 137)

That is, in the experience of certain short durations, the same duration (e.g., an hour) seems to take more time while, in the experience of certain longer durations, the same periods (e.g., a week) seem to take less time.

Again, one may argue there are illusions of duration; there is a difference between apparent duration and actual duration. Here, it is an appearance of *no difference* between things which *are different*, for example, a day feels the same as a week feels, yet they are different periods of time. One of these experiences is an illusory experience of duration because it appears to be of a duration of which it is not, for example, the actual day appears to be a week.

This is very similar to my job as a security guard: each 15-minute break seemed to take significantly longer than a typical 15-minute period outside the job. Yet, afterward, the whole eight-hour work period seemed to have rushed past, to take much less than a typical eight hours.

3 Against Illusory Durations

Despite how they are often described, I think one can question whether the above examples are illusions of duration. Experiences of duration are not standard clocks. Clock duration is not the relevant real duration, and it is not clear what is. The difference in apparent duration between different experiences of the same event can be explained without appeal to illusion.

3.1 Experience Is Not a Standard Clock

Something may be held to be different—to *diverge*—from a standard. However, it can only be treated as going *wrong* if it is calibrated or set by that standard. For example, any clock in the modern world—in your pocket, on your wrist, on your church tower—is designed to meet standard clock time; it is calibrated and set to it. If the clock's measure of duration diverges from the correct duration under standard clock time, then, since it is designed specifically to meet standard clock time, it is going wrong; it is failing in its function and is in error.

However, the experience of duration is not something that is so calibrated or set, based on the following plausible premises:

1. *Standard clock time is a convention.* It is a socially determined means of dividing time into hours, minutes, and seconds, and, if we like, weeks, months, and years.
2. *Nothing natural is calibrated or set by a convention.* Even if it makes sense to hold that natural things can somehow go wrong, it does not make sense to say they go right because they meet some socially determined standard. An iceberg does not travel in the wrong direction by drifting into shipping lanes. The tide does not come in at the wrong time.
3. *Experiences of duration are natural.* If we have an experience of duration, it is not an experience of technological or socially determined means of measuring time. These are things that measure duration but are not duration themselves.

These premises form the basis of the following argument:

4. (From 1 and 2) Nothing natural is calibrated or set by standard clock time.
5. (From 3 and 4) No experiences of duration are calibrated or set by standard clock time.

The premises of this argument are plausible (obviously) if one finds it plausible that we do have an experience of time and that the experience of time is natural (premise 3), standard clock time is conventional (premise 1), and that natural processes are not calibrated by conventional standards (premise 2), including experiences of duration (premise 3). I can see no obvious reason to deny any of these premises. As such, I can see no reason to hold that clock time is the standard for experiences of duration.

However, this argument does not mean experience has no standard. Perhaps one can hold that some natural process measures duration, by which experience may be calibrated or set. For example, perhaps one can hold that some biological processes are calibrated and set by circadian rhythms, and that circadian rhythms are a natural measure of duration. In that case, such biological processes are in error when they diverge.

It is unclear to me if biological processes are correctly described as being set, calibrated, or evaluable by a standard, even a natural one. Yet, even if we say in such divergence that a biological process is in error, it would not show the process was in error because it diverged from a standard clock. Similarly, the experience of duration, if also a natural process, is inappropriately evaluated for its accuracy in terms of standard clock time as any other biological process.

First, clock time and its measure of duration is a matter of convention. It is a socially determined standard; its adoption is based on contingent factors that appear to be irrelevant to the kind of biological factors that would determine something like an experience. Clock duration is a division of time into hours, minutes, and seconds.

Second, even if clock time were not a convention, the difference may only be evidence of different ways of measuring time. Say we use some non-experiential natural process to judge the time. For example, we use the tide to judge time. Say we call each turn of the tide *six hours*. If the cycle of the tide diverges from clock time, failing to match six hours on the clock, it is a mistake to say that the tide has gone wrong. This is a mistaken description of the relationship between different ways of counting or measuring duration. In the absence of a reason for thinking a measure in terms of one should be evaluated with respect to the other, there is no distortion or error of either being different to the other. There is only the difference.

Furthermore, experienced duration might lack a standard and even a rhythm against which we can keep time. Perhaps it is like a dripping tap. Say that you count an hour by each drip. Depending on how that tap drips, you get significantly different results compared to clocks, with several periods in which there is no easy way to match one to another. Your judgment of how many hours have passed goes wrong. Yet, the dripping tap does not go wrong. The tap is not calibrated to a standard clock. It is not calibrated to anything (indeed, unlike clocks, there is no reason to even think it should be calibrated to other dripping taps).

However, again, if we *assume* experiences of duration fall under the standard clock time, then we do go wrong if the experiences do not fall under that standard (instead, falling under some other standard or indeed no standard at all). However, our mistake is assuming how what we use to measure time is calibrated; the experiences we use may be fine in relation to their relevant standard (if they have such a standard).

In response, one might hold that we assume experiences are calibrated to standard clock time because experiences *feel* like they are calibrated to the standard clock time. Certain experiences of duration feel like hours, others like minutes, others like years, and so forth.

I agree that there are different feelings of duration (and feelings of different durations); I also agree we in some way associate different feelings of durations with different clock durations, for example, one experienced duration feels like an hour, another feels like a minute. One explanation for this association is that the two are identical: the appearance is of some standard duration. That is, there is an appearance of standard clock time in our experience of time. When that appearance is different to actual standard clock time, then we have an illusion.

However, I think we can explain this association without needing to have actual clock time to be what we experience. Again, clock duration is conventional; if the experience of duration is natural, then it is not of clock duration, which is not natural.

3.2 Is Clock Time Real Time?

Here is a reason why clock duration matters to experience and may be treated as being its standard: clock duration is either real duration or a suitable substitute for real duration.

A suitable substitute for a duration is one that, in the context, is indistinguishable from it; we can treat them as the same. Clock duration is indistinguishable from real duration in the context of experience's accuracy. As such, if the experienced duration is different from clock duration, it is illusory because, in being so different, it is different from real duration. So this reasoning goes.

When referring to clock time, theorists do treat it like real duration. '[T]he ratio of judged duration to *real duration* generally increases as a function of both the number of stimuli that occur over an interval' (Tse, 2010). '[V]erbal estimation of the duration of both auditory and visual stimuli produced output that translated into an approximately linear relation between mean estimate and *real time*' (Weardon, 2005).

If clock duration is a suitable substitute for real duration, then we can argue that experiences of duration diverging from it are illusions of duration:

1. Nothing diverges from itself. If x diverges from y, then x is not y.
2. There is real duration.
3. Experienced duration is an appearance of real duration.
4. Clock duration is real duration (or near enough to it).
5. Experienced duration diverges from clock duration.
6. (1 and 5) Experienced duration is not clock duration.
7. (4 and 6) Experienced duration is not real duration.
8. Experienced duration is an illusion of real duration.

So, it is not that experienced duration is simply different to clock duration. It is that clock duration is real duration, and it is different to real duration.

We can ask then: what is the *real* duration that clock time so closely matches?

First, it might be absolute duration, the duration of Newton's independent and absolute flowing time. If so, one could argue for the possibility of illusory duration:

1. Clock duration suitably substitutes absolute duration.
2. There is an appearance of absolute duration that does not match the actual clock duration.
3. Substituting absolute duration by clock duration in 2, we have an appearance of clock duration that does not match clock duration.
4. This appearance of duration an illusion of duration, where duration is understood either as absolute duration or clock duration.

However, defenders of absolute time, and its related duration, do not argue from appearances for their position. It is generally accepted that there is no appearance of absolute time (e.g., Shoemaker, 1993). Furthermore, there is no obvious reason that such absolute time matches clock time or is suitably substituted by it. That clock time is not a suitable substitute denies 1; that there are no appearances of absolute duration denies 2. The above argument is unsound.

What else might the real duration be that is suitably substituted by clock duration?

One possibility is some natural duration, the duration of some natural process or repetitive change. For example, real duration might be the repetitive change between night and day, the day-night cycle. If that is real duration, then clock duration suitably substitutes real duration because it is indistinguishable from the day-night cycle.

However, clock duration *is not* indistinguishable from the day-night cycle (or indeed any similar natural repetitive change). On the level of a day, there might be some similarity; on average, 24 hours fill each day, so a duration of 24 hours matches the duration of a day. However, at the level of hours or minutes, clocks are different to any day-night standard. There is no constant relationship between the period of daylight and the period of clock time. A clock's hour is distinguishable from 1/12 of daylight and a period of daylight is distinguishable from 12 hours. As such, clock duration is not a suitable substitute for the day-night cycle, or for real duration if this is what such a cycle is.

This does not mean that the experience's difference from a clock is not an illusion. It may be that, when experience diverges from day-night cycles, it does so much more than clocks do, and it does so in such a way that its difference from clocks indicates this divergence. In that case, given the day-night cycle is real duration, this is a divergence by experience from real duration. If the assumptions in this section are correct, this is an illusory duration.

Yet, that experience diverges more than clock time from real duration is not the only possibility. Possibly, experience's difference from clock time is because it matches the day-night cycle, that is, the 'real duration' accepted here. Whether or not it does so is an empirical question: if experience diverges from clock time, does it diverge more than a clock from the day-night cycle (or, again, whatever natural process we decide is real duration)?[2] If instead experience follows the day-night cycle more closely than a clock, then it is not the experience that is diverging from real duration. It is the clock that diverges. That is, clock duration is not a good substitute—indeed, given the argument above, experience is a better substitute than that.

The day-night cycle is a toy example. It does not capture the fine divisions of either clock time or experienced time. Hours and seconds, the felt duration of cooking rice or going to work—these are too brief to be measured by changes in day and night. We need something that corresponds to clocks and experienced duration at all scales on which they are divisible. Yet, what applies here for the day-night cycle applies just as much to any other process or repetitive change that we ascribe the role of real duration. I suspect any such natural, real duration is likely different from clock duration in a way that clocks are not obviously a more suitable substitute than experience.

---

[2] In fact, there is a clock standard that matches the day-night cycle. This is the traditional Japanese standard, and the hybrid that developed once Japan introduced European clocks. Japanese clockmakers later copied and modified the mechanism to suit traditional Japanese time telling. A day in the traditional system had 12 hours: six for day and six for night. The hours changed in length according to the season. In summer, daytime hours were long and nighttime hours short. In winter, daytime hours were short and nighttime hours long. (British Museum, 2019)

A further problem is that, even if we find a fine enough process or repetitive change to be real duration, there seems to be too many candidates. It is not obvious what makes one of these changes the right kind, the *relevant* kind, for the experienced duration.

Since we are talking about experience, one class of changes may seem more plausibly relevant than others. These are psychological (or neurological) changes, such as changing emotions or somatic states (e.g., Craig, 2009; Wittmann, 2016 for discussion). These may seem more plausible candidates because one holds that experience is a case of psychological or neurological activity, and so experiences of duration should in some way accurately match measures of psychological activity.

In any case, returning to the question of clock duration as a standard, psychological activity is unlikely to be appropriately substituted by clock duration; psychological activity does not neatly occur in bouts of seconds, minutes, and hours. Again, so far as we may currently know, experiences of duration may be more appropriate as a substitute for measures of psychological activity than clock time.

I think this search for a real and relevant duration, and especially one indistinguishable from clock duration, so we can have illusions of duration is a step in the wrong direction. We may find such a standard, but it is unlikely to lead to illusions of duration from experience's difference to clock duration.

3.3 Converting Between Different Real Measures of Duration

Here is a different approach. When we judge how much clock time passes based on our experience of duration, there is a mistake. However, the error is not an illusion; it is not because of a discrepancy between experienced duration and real duration. The error is because of how we tend to think about our experience of time and its relationship to other kinds of change, including clock time. Experienced duration appears as it is; the error is our tendency to treat it as it is *not*—for example, as something that should meet the standards of clock time, like a manufactured clock. Our mistake is to evaluate experience by its difference from that standard.

Why do we make this mistake? One possible explanation is an old empiricist and psychological idea: association. At some point in our lives, we learn to associate certain experiences of duration with certain periods of clock time. This association may be an abstraction from multiple experiences or may depend on one or a few concrete experiences. For example, you associate the experienced duration of some event in your childhood (going to the seaside, say) with a whole day.

In any case, through this association, we treat experienced duration as an appropriate substitute for clock duration. It is a stand-in when there are no actual clocks, used through the associations we form between the two kinds of duration.

However, this is not about the experience being correctly judged as a standard clock. Such associations could have been made between anything else that repetitively changes. We might also associate a dripping tap with clock duration ('A tap drips every forty minutes') or experienced duration with a washing cycle ('This is the feeling of how long a washing cycle takes').

I suggest that experienced time's 'failure' to correspond to clock time, although we treat it as if it should correspond to clock time, is not an actual failure of experience. It is evidence that it is not a measure of what clock time measures. If it has any standard, it is something else other than the clock.

Yet, we can still do some conversion between experienced duration and clock duration. If experienced duration has a standard, and that standard is not standard clock time, there can still be a relationship between clocks and experience—such that we can, in principle, understand one in terms of the other. However, this is not a case of both falling under the same standard. It is a case of measures in one standard being converted into measures of the other. We get the conversion wrong. This gives rise to many of the errors we may want to call illusions of duration.

How do we get the conversion wrong? This can vary by the standards and our attitudes to them, but one might be this: we treat the relationship between the two kinds of standards as always being stable. If at a particular time or place, two units of some standard A are equal to one unit of some standard B, then, in all times and places, two units of A are equal to one unit of B.

However, standards of measurement can vary in their relationship. Two units of A might equal one unit of B in Ireland but equal two units of B in Sydney; similar varying relationships might happen in other contexts, including time. Consider the example of money.

Say that I know the price of an umbrella in euros. Also, in the past, I assumed a particular exchange rate between euros and dollars. Now I am asked the price of an umbrella *in dollars*. I accurately judge—I know—an umbrella's price in euros. I convert that price to dollars, using my assumed exchange rate, and then give the price in dollars. After all, that is the price I am being asked to give.

However, the exchange rate is no longer the same, and so I give the wrong price in dollars. I am mistaken about the price of umbrellas in dollars. However, I give the wrong price not because I simply do not know the price of umbrellas. I do know their price in euros. I do not know the exchange rate between euros and dollars.

Similarly, the standard for clock time—the second—may be converted into units of experienced duration, and vice versa. Yet, depending on how the standard for experienced duration varies in relation to the standard for clock time, this conversion can be variable. When, based on my experience of duration, I give the wrong clock duration, my belief about the conversion between the two ways of measuring duration goes wrong.

Such a belief is not itself the experience of duration. It is a belief about how the experience of duration relates to clock time. This is a second-order cognitive state, a belief about experience (and clocks, and their relationship), not an experience itself. If it goes wrong, it is not an illusion of duration, but an erroneous belief about duration. Or rather, an erroneous belief about the relationship between experienced duration and clock duration. There is still a mistake, but it is not an illusion.

3.4 Differences Between Experiences of Duration

Some candidate cases of illusory duration seem based *only* on differences between experiences of duration. For the same events, different experiences seem to be of different durations, yet we have reasons to hold that these experiences are of the *same* duration. The examples of this chapter seem to be such cases—the oddball effect, chronostasis, the round trip, or my time spent working in security. For example, the oddball seems to be presented for a longer duration than the control. Yet, as the experiment is typically described, the oddball and control are actually presented for the same duration. One can give similar accounts for the others; in each, the same duration of events appears to be different during different experiences. This difference does not seem to be a difference between clock time and experienced time. So, how can we account for it?

The same cases seem to have obvious analogues with other kinds of experience which we have good reason to call illusory: spatial illusions. Many spatial illusions are illusions of different spatial extents. For example, in the Ebbinghaus illusion, two circles are presented to a subject. These circles are the same size yet appear to be different sizes (e.g., Knol, Huys, Sarrazin, & Jirsa, 2015). Similarly, for a traveler, journeys that last the same duration appear to be different durations; why is this set of experiences not an illusion of *differences in duration*? Or, if the latter is not an illusion of duration, then why is the former an illusion of size?

One response to analogies with spatial illusion is that we should not draw an analogy: space and time are different; experiences of one (space) should not be used to draw any conclusions about experiences of the other (time). Another response is that, since my denial works against illusions of duration, then it also works for experiences of spatial illusions. What we are tempted to call spatial illusions are not spatial illusions at all. The apparent difference in size in the Ebbinghaus 'illusion' is not an illusion of different sizes at all.

Nevertheless, I assume that we can make analogies between space and time, including experiences of them. Furthermore, I assume that spatial illusions *are* illusions. The Ebbinghaus circles are not different sizes; they only appear to be different sizes. This is an illusion of their relative sizes. As such, I cannot deny spatial illusions. Why, then, do I think illusions of duration can be denied?

To know there is an illusion is to know something about reality. The difference between appearance and reality is what matters. If I do not know the reality of something, I cannot know that an appearance differs from it, or that they are mistaken appearances and illusions. With mistaken spatial appearances, I know that they are illusions because I know about real space against which they can be evaluated. I do not have an equivalent way of determining real duration.

Instead, I propose an alternative explanation for cases of apparent duration typically thought of as illusory. As a way of illustrating this alternative, consider an analogue—a false story about the Ebbinghaus Illusion.

1. Assume (falsely) that both an object's real and apparent size is as follows: it is the object's size in a two-dimensional field with a plane that is perpendicular to the subject's line of sight.[3] Its size is like the size of a two-dimensional shape spread over a flat surface facing the subject, such as a wall the subject is looking at. So, (a) if X is the same or a different size to Y, it is the same or a different size in two dimensions perpendicular to the subject's line of sight; (b) if X appears to be the same or different size to Y, it appears to be the same or different size in two dimensions perpendicular to the subject's line of sight.

   Again (for most observations, anyway) this is a false claim about the reality and appearance of objects; objects neither occupy nor appear to occupy two-dimensional space; the sizes of objects seem and are three-dimensional, not two-dimensional, and certainly not two-dimensional relative to the subject.
2. For some reason, when presented with the circles in what is called the Ebbinghaus illusion, with respect to one of the circles, the following happens: our eyes move closer to it (perhaps they extend out on stalks). The circle they move closer to is always the one that appears to be bigger, rather than the one that appears to be smaller.[4]
3. Because one circle is closer, it is larger in two dimensions as defined by its spatial relationship to the subject. Because its size in those two dimensions is *all there is* to something being larger (again, given this false assumption), then that circle *is* larger.
4. As such (from 2) the circle appears larger and (from 3) is larger. There is no illusion of size. The apparent difference in size corresponds to the actual difference in size.

[3] This is false: things appear three-dimensional (contrary to Berkeley) and the size of an object (even a circle drawn on paper) is not determined merely by its spatial relation to the subject.

[4] This, again, is false and physically impossible with our eyes (although, perhaps, not with all animals' eyes, e.g., a snail's stalked eyes are mobile and may be capable of this, for whatever that possibility is worth in this context).

Again, this is a false account of the Ebbinghaus illusion—or indeed, any other optical illusion. Obviously, nothing like this happens in our visual system—and, in any case, the third dimension is important to both actual and apparent size. Instead, we know, the eye is at the same distance from each circle, the circles are the same size. Neither circle is closer; neither circle appears closer; yet, one circle appears to be larger. This appearance of size difference is mistaken. Hence, it is an illusion.

However, an analogous account of the apparent variation in our experience of *duration* is not so obviously false. We have different experiences of durations such that they appear to be different durations, for example, the job feels longer during it than after it; summer days feel longer the first time we experience them.

In these cases, it is possible for there to be illusions of duration: if, as with spatial illusions, we experience events with the same duration while standing in the same relation to us, yet the events appear to have different durations, then there is a discrepancy between all this real sameness and apparent difference; it is analogous to seeing the equal-sized circles with the same eyes at the same distance. The apparent differences in duration are not the actual differences in duration, and we are under an illusion.

However, unlike the spatial illusions, that we experience the events while being in the same relations to them is not obvious. Consider what is meant by *different* experiences of the *same* event.

First, the *same* in *same event* is not *strictly* meant. It does not mean the exact same event. There are different events, occurring in different places or at different times, with different instances of duration. However, the events are the same by having durations of equal length. This is the case with the oddball effect, chronostasis, round trip, and, as we get older, the feeling of months speeding up and days slowing down.

Second, the *same* in *same event* is strictly meant. Each experience is of the exact same event. For example, my experience of the night shift's duration while I am having it and my retrospective experience of its duration after it ends. Or, indeed, in Weardon's examples, days as they are lived (a long duration) and as they are remembered (a short duration).

Whichever of these is meant by 'same', my account covers both. However, I focus only on the second meaning. What applies to it applies to the first meaning as well. The first meaning differs from the second by involving multiple events or instances of time. Multiple events and instances do not offer additional reasons for the position that there is no actual difference (yet there is an apparent difference). With respect to experience and duration, what accounts for proposed illusions under the second meaning accounts for the first as well.

So, given the strict meaning of 'same event', I suggest the following account:

1. During the event, the experience of duration is of several kinds of changes in the event; this is a way of measuring the duration; call that measure M. It does not matter what those changes are: it might be whatever our experience measures as duration (e.g., Craig's changes in emotion) or it may be some change in the world, for example, the flash of car's headlights on the road outside my post.
2. The retrospective experience of the event's duration is of a different number of changes (N). If M is not the same as N, then the experiences of the duration of the same event is different during and after it.
3. In the experience we have while going through the duration, we are in one relation to the event. For example, we are emotionally involved in getting through the night shift while attending to what's happening around us. What exactly this relation is does not matter. What does matter is that, because of this relation, we measure the event's duration in a certain way.
4. In the retrospective experience of duration, we are in a different relation to the events and their duration than when we are going through them; this different relation means we measure the event in a different way. For example, the shift is now over; as we turn to go home, we don't think about the boredom or what happened around us, and so don't measure the shift through them. Instead, we think only about the sandwich we had halfway through the night; that's all we take from the night's events. (Note the similarity to the false story for spatial illusions, where we are in a different relation to each circle; our eyes move toward one circle and don't move toward the other.)
5. The apparent duration of the shift during it is different to its duration after it.

6. In the experiences (from 3 and 4), the measured durations are different and (from 5) the durations appear different. This is not a case in which there is the same real duration but an apparently different duration. It is a case in which there are different measured durations and apparently different durations.

This account allows there to be illusions of duration, even empirically supported illusions. If, in each experience, an event's duration is measured by the same thing in the event (for example, the same emotional change we go through during it), then if there is a difference between the experiences of the event's duration, there is a reason to hold that there is an illusion here: one of the experiences is wrong.

As a final point, I recommend that, to show that there is an illusion of some duration M:

1. Find the appropriate standard for the experience of duration.
2. Check M against an accurate measure of duration under that standard, call that N.
3. If M does not match N, then the experience of duration is mistaken. There is an illusion of duration.

4 Conclusion

This chapter does not insist that illusions of duration are impossible. However, of the cases usually given in the literature—such as the oddball effect, chronostasis, round trips, changes due to age, and so on—I see no reason yet to hold that different experiences are of the same changes and so of the same duration, even if they have events in common. With no such reason, I see no further reason for such cases to be illusions of duration.

There are a few open questions left by this chapter. Here are two.

First, what is the appropriate standard? As argued here, the most immediate standard is unlikely—clock duration. Clock duration is a measure of change, but it is not at all obvious if it is the same change as measured by experience. The real duration may be the standard, and some clock duration may be a suitable surrogate for it, but this is also not obvious. For we must still discover what is meant by real duration. There seem to be many candidates, not all divide time in the same ways, and not all more closely match clock duration than experience.

Second, what if there is no standard (or we cannot find one) for the experienced duration? In that case, we cannot hold that experienced duration fails to meet a standard. It does not fail to correspond to any appropriate real duration (or we cannot show that it does). There are either no illusions of duration or we cannot show that there are any.

We can still go wrong about duration based on our experience. We can still go wrong about clock duration. However, in most cases, this is because we treat what is not a clock as if it were a clock, not because what is not a clock is a clock gone wrong.

Acknowledgments This paper developed out of two talks: the first was for a session at the 2015 *Towards A Science of Consciousness* conference (in Turku), to which I was invited by Valtteri Arstila; the second at a session at the 2016 Pacific meeting of the American Philosophical Association, to which I was invited by Adrian Bardon. I am grateful to Valtteri and Adrian for their invitations, as well as their feedback and interesting discussions. I am also grateful to the thoughtful and thorough comments from Natalja Deng following the 2016 workshop. Thanks also to the audience at each conference for their helpful responses to the talks.

References

Bardon, A. (2013). *A brief history of the philosophy of time*. Oxford: Oxford University Press.

Craig, A. D. (2009). Emotional moments across time: A possible neural basis for time perception in the anterior insula. *Philosophical Transactions of the Royal Society B: Biological Sciences, 364*(1525), 1933–1942. https://doi.org/10.1098/rstb.2009.0008

Crane, T., & French, C., 2016. The problem of perception. In E. N. Zalta (Ed.), *The Stanford encyclopedia of philosophy* (Spring 2016 ed.). http://plato.stanford.edu/archives/spr2016/entries/perception-problem

Eagleman, D., & Pariyadath, V. (2007). Is subjective duration a signature for coding efficiency? *Philosophical Transactions of the Royal Society B, 364*, 1841–1852.

Hodinott-Hill, I., Thilo, K. V., Cowey, A., & Walsh, V. (2002). Auditory chronostasis: Hanging on the telephone. *Current Biology, 12*(20), 1779–1781.

Kalderon, M. E. (2011). Color Illusions. *Nous, 45*(4), 751–775.

Knol, H., Huys, R., Sarrazin, J., & Jirsa, V.K. 2015. Quantifying the Ebbinghaus figure effect: Target size, context size, and target-context distance determine the presence and direction of the illusion', *Frontiers in Psychology*. Retrieved January 10, 2019, from https://doi.org/10.3389/fpsyg.2015.01679

Le Poidevin, R. (2007). *The images of time: An essay on temporal representation*. Oxford University Press.

Museum label for artist. 2019. Illustrated compendium of clocks, London, British Museum, 19 January 2019

Power, S. E. (2018). *Philosophy of time and perceptual experience*. New York: Routledge.

Shoemaker, S. (1993). Time without change. In R. Le Poidevin & M. MacBeath (Eds.), *Philosophy of time*. Oxford: Oxford University Press.

Tse, P. U. (2010). Attention underlies subjective temporal expansion. In A. C. Nobre & J. T. Coulle (Eds.), *Attention and time* (pp. 137–150). Oxford: Oxford University Press.

Weardon, J. H. (2005). The wrong tree: Time perception and time experience in the elderly. In J. Duncan, L. Phillips, & P. Macleod (Eds.), *Measuring the mind: Speed, control, and age* (pp. 137–151). Oxford: Oxford University Press.

Wittmann, M. (2016). *Felt time*. Cambridge, MA: MIT Press.

Part III

Simultaneity and Temporal Order

11

Causality Guides Time Perception

Andrea Desantis and Marc Buehner

1 Causality and Time Perception

Causal inference is essential for understanding the world. It allows us to predict the occurrence of events and to control our environment in order to produce desired outcomes. In addition, failing to represent causal relations can have severe effects on our mental health. For instance, patients affected by delusion of control fail to recognize themselves as the cause of the actions and effects they generate (Blakemore, Wolpert, & Frith, 2002).

A. Desantis (✉)
Département Traitement de l'Information et Systèmes, ONERA, Salon-de-Provence, France

Integrative Neuroscience and Cognition Center, CNRS, Université Paris Descartes, Paris, France

Institut de Neurosciences de la Timone, CNRS, Aix-Marseille Université, Marseille, France

M. Buehner
School of Psychology, Cardiff University, Cardiff, UK
e-mail: BuehnerM@cardiff.ac.uk

V. Arstila et al. (eds.), *The Illusions of Time*,
https://doi.org/10.1007/978-3-030-22048-8_11

David Hume (1739/1985) suggested that causal relations cannot be directly perceived but only inferred from observable information. He identified two factors, linked to time perception, that contribute to causal inferences: *temporal priority* (causes must precede their effects) and *temporal contiguity* (effects occur following a short delay after their cause). Contemporary studies confirmed Hume's intuitions, showing that causal judgments rely, at least partially, upon observation of temporal correlations between two events categorized as cause and effect (e.g., Buehner, 2005; Michotte, 1941; Shanks, Pearson, & Dickinson, 1989; Young, 1995). In short, all other things being equal, people are more likely to perceive two events as causally connected if they follow each other closely in time than if they are separated by a long delay (Michotte, 1941, 1946).

People, however, do not have a sense dedicated to timing. How then does the brain determine the temporal gap separating two events and how does it decide which one came first? This seems particularly challenging when we also consider that our brain processes different sensory signals (e.g., visual and auditory inputs) at different speeds due to transduction delays and to differences in transmission time of different energies (e.g., light and sound). So, how can the brain perceive that the tactile and auditory sensations associated with the action of snapping one's fingers occurred simultaneously and that the movement preceded them? Research has compellingly showed that the brain meets this challenge by combining and comparing available sensory information to internal representations such as prior expectations (Desantis & Haggard, 2016b; Fujisaki, Shimojo, Kashino, & Nishida, 2004; Stetson, Cui, Montague, & Eagleman, 2006; Vatakis & Spence, 2007). For instance, studies showed that our expectations regarding the onset time and the identity of an event strongly influence the perceived onset time of the event (Cravo, Claessens, & Baldo, 2011; Desantis & Haggard, 2016b; Desantis, Haggard, Ikegaya, & Hagura, 2018; Stetson et al., 2006).

Interestingly, recent research explored another form of expectation that could drive time experience, that is, causal expectations. Notably, it has been suggested that the belief that two events are causally linked leads people to experience the presumed cause as occurring before and in close temporal proximity with the presumed effect (Bechlivanidis & Lagnado, 2016; Buehner, 2012; Desantis, Waszak, Moutsopoulou, & Haggard, 2016). In other words, causality may not only be derived by the perceived timing of two events, but it may, in turn, structure our experience of time. The influence of causality on time experience suggests that the time at which sensations are represented in the mind might not correspond to the time at which they are experienced. This notion might provide some insight to the current philosophical debate about the experience of time that opposes mirroring theories according to which our perception of temporal order *inherits* the order at which individual sensations are represented (Mellor, 1985; Phillips, 2014) and theories viewing the perception of temporal order as a second-order judgment (Dainton, 2017; see also Bechlivanidis & Lagnado, 2016). Furthermore, these findings are important since they can shed further light on psychopathological symptoms such as the delusion of control in which both time and causal perception are affected (Fletcher & Frith, 2009; Thoenes & Oberfeld, 2017).

In this chapter, we review the existing evidence in support of the idea that causal expectations structure our perception of time. First, we discuss research showing that causality enhances subjective *temporal contiguity*. Second, we discuss studies suggesting that the *temporal order* of two events is altered by people's assumptions of the causal link between these two events. Finally, we discuss the way in which causality might modulate time perception in terms of Bayesian models of reduction of uncertainty.

2 Causality Enhances Temporal Contiguity

Faro, Leclerc, and Hastie (2005) investigated the influence of causal beliefs on the estimation of the interval of time between historic events. They observed that when people were induced to believe that two historical events were causally connected, they tended to judge them as having occurred closer in time. Moreover, the stronger the causal relation between the events was thought to be, the less time people thought had elapsed between them. More recent studies investigated the influence of causality on people's perceptual experience (Buehner & Humphreys, 2009; Desantis, Roussel, & Waszak, 2011; Dogge, Schaap, Custers, Wegner, & Aarts, 2012; Haering & Kiesel, 2012; Thanopoulos, Psarou, & Vatakis, 2018). These studies were inspired by the Intentional Binding effect, that is, the compelling finding that people report voluntary actions and the sensory event generated by their actions as occurring closer together in time, compared to when the same sensory event occurs after an involuntary muscle twitch induced by Transcranial Magnetic Stimulation or after another sensory signal (Haggard, Clark, & Kalogeras, 2002). Haggard et al. suggested that acting *intentionally* leads to a shortening of the time interval between actions and outcomes, resulting in a coherent conscious experience of being the author of that outcome. Thus, performing intentional actions would lead to an enhancement of subjective temporal contiguity between action and outcomes, reinforcing our experience of agency (i.e., our experience of being in control of our actions and of the outcomes our actions generate).

Based on this finding, researchers then investigated whether temporal binding of action and outcomes is driven by people's beliefs of causality rather than the fact of performing intentional actions. For instance, Buehner and Humphreys (2009) observed that causal beliefs about the connection between an action and a subsequent sensory event were fundamental to temporal binding. In their study, participants completed a condition in which they were asked to synchronize two key-press responses (r1 and r2) to two successive tones (t1 and t2). In a causal condition, r1 triggered t2 after a fixed interval, and in a non-causal control condition, t2 followed t1 after the same fixed interval. In other words, if participants failed to execute r1 in the causal condition, t2 would not sound, whereas it would sound regardless of r1 in the non-causal control condition, and participants were encouraged to verify this. This means that to a naïve onlooker to the experiment, both conditions appeared identical, since they both involved two intentional actions, timed in synchrony with external sensory signals. Crucially, though, participants engaged in the task would know that their key-press caused the tone in one, but not the other condition. If temporal binding is induced by causality, one would expect that participants experienced causal actions as relatively later than non-causal actions, and that caused outcomes would be experienced as relatively earlier than un-caused ones. This is exactly what Buehner and Humphreys found.

Several studies replicated these findings using different paradigms and consistently showed that people's assumption that their actions *caused* the occurrence of a subsequent stimulus leads to a subjective temporal compression of the interval separating these two events (Buehner, 2012; Buehner & Humphreys, 2010; Desantis et al., 2011; Dogge et al., 2012; Haering & Kiesel, 2012; Thanopoulos et al., 2018). In other words, when we believe that two events are causally bound, we perceive them as temporally bound as well. *How* exactly do causal assumptions influence temporal binding? Is causality sufficient or necessary for this phenomenon? These questions have been addressed by some studies, which are described in the following section.

2.1 The Role of Causality on Temporal Binding

In Buehner and Humphreys (2009), participants performed a voluntary action in all experimental conditions. Crucially, not all of these actions were causal, and temporal binding only occurred when voluntary actions were paired with causality. We can, thus, conclude that (a) causality is necessary for temporal binding and (b) intentional action on its own is not sufficient. We do not know, however, whether causality is also sufficient for binding to occur, or whether intentional action is necessary.

Cravo, Claessens, and Baldo (2009) tried to address this issue and suggested that both intentional actions and causality are necessary for temporal binding to occur, but causality alone is not sufficient. Their study employed a variation of Michotte's (1946) launching paradigm and varied the perceptual causality of an animation display containing two disks. In the collision condition, the left disk moved and collided with the right disk, which in turn moved to the right. In the non-collision condition, the left disk, instead of reaching the right disk, moved in the opposite direction until it reached a vertical line, then the right disk started to move. Moreover, in some blocks, participants could trigger the movement of the left disk; in other blocks, its movement was externally controlled. The authors observed temporal binding only when both intentional actions and causality were present.

However, Cravo et al.'s (2009) findings are in contrast with recent studies showing that temporal binding occurs also in situations in which participants do not execute an intentional action but when causality is implied (Buehner, 2012; Dogge et al., 2012). For instance, Dogge et al.

observed temporal binding also in cases in which participants executed an involuntary movement and were told that their involuntary key-press would cause the occurrence of an effect compared to a condition in which they always executed an involuntary key-press but no additional information concerning causality was provided. Furthermore, Buehner showed that temporal binding also occurs for events that are not generated by a voluntary action but visibly produced by a mechanical device (i.e., participants viewed a lever connected to a motor pressing a button). Taken together, and contrary to Cravo et al., these studies suggest that causality is sufficient for temporal binding to occur and that the fact of performing an intentional action is not a necessary condition (Buehner, 2012; see also Thanopoulos et al., 2018).

If this is the case, then it remains unclear how to explain the discrepancies between the studies of Cravo et al. (2009) and Buehner (2012). We, therefore, offer a hypothesis that might reconcile their findings. Research has shown that temporal binding is influenced also by other processes such as temporal predictability (Cravo et al., 2011; Hughes, Desantis, & Waszak, 2013). It is reasonable to assume that performing an intentional action enhances temporal predictability. Notably, by controlling when an action is executed, an individual would also control the onset time of the outcome generated by that action. Accordingly, temporal predictability would be higher when an event is generated by an intentional action compared to when it is generated externally (Desantis, Hughes, & Waszak, 2012; Haering & Kiesel, 2014; Hughes et al., 2013). In other words, it is not the *intentional* nature of an action that enhances temporal binding but the fact that controlling the timing of the cause improves temporal predictability of the effect. Accordingly, in certain circumstances, the fact of performing an intentional action might enhance temporal predictability, and in turn temporal binding, compared to a case in which only causality is involved and where the causal determinant of the event (e.g., the action of a machine or another agent) is not itself temporally predictable.

In summary, there is converging evidence that causal expectations enhance the perceived temporal contiguity between events categorized as cause and effect. Moreover, temporal binding is not necessarily linked to *intentionality* but rather to causality and causality is both sufficient and necessary for temporal binding to occur. The question now remains to understand *how* causality might influence the perceived temporal contiguity between two events.

As mentioned above, temporal contiguity strongly facilitates the discovery of causal relations between events. Humans use temporal contiguity to guide causal perception as early as 6–7 months of life (Leslie & Keeble, 1987; Oakes, 1994). As a consequence, we learn also that causally related events are more likely to be contiguous. Thus, causal expectations might compress the perceived time gap between events that are regarded as causally related. This is in agreement with Bayesian models of uncertainty reduction, according to which the strength of belief in a causal relation between two events modulates an individual's estimate of the temporal interval between them, given inherent uncertainty of time perception (Buehner, 2012; Eagleman & Holcombe, 2002). These models have been successfully applied to different perceptual domains (Kersten, Mamassian, & Yuille, 2004; Körding et al., 2007; Mamassian, Landy, & Maloney, 2002) and to time perception (Jazayeri & Shadlen, 2010; Petzschner, Glasauer, & Stephan, 2015; Rhodes, 2015; Shi, Church, & Meck, 2013) and may offer an elegant way to formalize the process by which the brain combines prior expectations (i.e., the prior) and sensory information (i.e., the likelihood) to elaborate an optimal interpretation of our perceptual experience.

In line with this framework, causal priors might guide time perception in temporal binding in a similar way as the unity assumption guides temporal integration in multisensory perception (Welch, 1999). According to the unity assumption, stimuli of different sensory modalities that are considered to belong to or being generated by the same event are more likely grouped together in time (Chen & Spence, 2017; Welch & Warren, 1980). For instance, Vatakis and Spence (2007) showed that people temporally integrate visual and auditory speech to a greater extent when they matched (i.e., a female face was presented together with a female voice) compared to when they mismatched (i.e., a female face was presented with a male voice). Computational approaches have formalized the unity assumption in terms of priors in Bayesian causal inference models (Körding et al., 2007). In this framework, a unity/causal prior would reflect how likely an observer considers two sensory events as being generated by the same cause. In a similar way, one might consider causal assumptions in temporal binding as how likely an individual considers movement-related signals (e.g., efferent, tactile, kinematic, proprioceptive and auditory) to belong to the same causal action chain. Thus, for instance, the fact of repeatedly performing the action of snapping our fingers leads our brain to learn that internal (e.g., motor signals) and sensory inputs (e.g., auditory, tactile and visual signals) associated with this action belong to the same causal action chain. As a consequence, these signals would be bound in time. Interestingly, in line with this notion, recent studies suggested that processes underlying multisensory integration are also responsible for the temporal binding effect (Kawabe, Roseboom, & Nishida, 2013).

In sum, causal expectations compress the perceived temporal interval between two events categorized as cause and effect. Causal assumptions bias the sensory system to represent events that belong to the same causal action chain via the Bayes' rule such that they are bound together in time. In the next section, we discuss another surprising way in which causal expectations structure time perception.

3 Causality Reorders Time: The Role of Causality on the Perception of Temporal Order

According to the temporal priority principle, causes must precede effects and effects cannot precede causes (Hume, 1739/1985; Wegner & Wheatley, 1999). Use of this principle is evident from at least the age of three years (Rankin & McCormack, 2013). However, to judge the temporal order of two events, people must estimate when each event occurred. Recent research suggests that time estimations are based on available sensory information combined with internal representations such as prior expectations (Desantis et al., 2018; Grondin, 2010; Stetson et al., 2006). For instance, temporal expectations can modulate the perception of temporal order concerning an action and an ensuing outcome. Notably, the brain "recalibrates" time estimations based on temporal predictability.

For example, if people are repeatedly exposed to a fixed time interval separating their action and a sensory outcome, and then this time interval is unexpectedly shortened, they erroneously perceive the outcome as preceding the action that caused it (Heron, Hanson, & Whitaker, 2009; Keetels & Vroomen, 2012; Rohde & Ernst, 2013; Stetson et al., 2006; Sugano, Keetels, & Vroomen, 2010, 2012). Interestingly, this motor-sensory recalibration seems to require causal action, since it is not observed when people remain passive and tactile stimulation is applied to their finger, mirroring the tactile experience of the causal action, that is, a key-press (Stetson et al., 2006).

As we mentioned above, people tend to not consider sensory outcomes as effects of their actions when they occur after long temporal delays (Farrer, Valentin, & Hupé, 2013). Interestingly, Heron et al. (2009) showed that motor-sensory recalibration decreased when observers were exposed to long action-outcome time intervals, suggesting causality might be crucial for motor-sensory recalibration to occur. Notably, the authors suggested that the brain would not recalibrate motor and sensory inputs when outcomes are not considered to be generated by one's actions (i.e., outcomes occurring after long delays), as this recalibration might potentially lead to erroneous self-attribution of sensory events that are instead generated externally (Heron et al., 2009).

Recent studies investigated the role of causality on the perceived temporal order of events more directly by manipulating people's beliefs of causality. For instance, Desantis et al. (2016) showed that causality structures the perceived temporal order between actions and sensory events, imposing a causes-must-precede-effect principle to time perception. In their study, participants first learned that pressing one key (A) induced a brief coherent upward motion in a Random Dot Kinematogram, while pressing another key (B) induced a brief period of coherent downward motion. In a subsequent test phase, participants performed a temporal order judgment task. In this phase, Desantis et al. manipulated the match/mismatch between actions A and B and the direction of coherent motion. Crucially, sometimes, coherent motion preceded the key-press (A or B) and sometimes it occurred afterward. The authors observed that coherent motion presented immediately *before* the key-press was more often erroneously reported to have followed it, when the motion was in the direction associated with the specific key (predicted outcome trials), but not when it was in the unexpected direction (unpredicted outcome trials). This indicates that learning-specific action–outcome associations led to the familiar cause-precedes-outcome relation, creating an illusory reversal of the order of action and sensory events. Indeed, when a specific outcome was expected to occur as a consequence of a specific action, it was consistently perceived as occurring after the action, even though it was in reality presented before. Interestingly, this effect was observed only when participants had a clear evidence that their actions were causally linked to the presentation of the coherent motion during the temporal order judgment task. In fact, the effect disappeared when participants knew that the coherent motion in the temporal order judgement task was always triggered by the computer and never by their key-presses. Based on this finding, the authors suggested that the influence of action–outcome learning on the perception of the temporal order of action and effect is conditional to observers' understanding of their causal effectiveness. Accordingly, they proposed that causal expectation provides the critical information for time perception, and that predictions based on learned action–outcome associations have a further effect, but only if causality is preserved (see also Thanopoulos et al., 2018).

A similar causality-induced temporal re-ordering of merely observational stimuli (i.e., not pertaining to sensory-motor recalibration) has been demonstrated by Bechlivanidis and Lagnado (2013, 2016). Using a variation of Michotte's launching paradigm, Bechlivanidis and Lagnado (2016) presented participants with three objects (A, B and C), with B being in between A and C. In one experimental condition, A moved and hit B. However, after the collision, B remained static for a short delay, and only after that the motion of C started, B also started to move. The authors observed that even though the temporal order of the events was A, C and B, participants more often reported that the order of events followed the causal structure according to which A hit B, and B hit C. Importantly, this erroneous representation of the causal order of events disappeared when the authors removed A from the sequence or when B remained static throughout the trial. The authors suggested that the causal context provided by A led participants to reinterpret the temporal order of the sequence in a way that matched with their causal expectations, that is, that A hit B and B hit C. Thus, this study corroborates the idea that causal assumptions regarding the link between two events determine the order in which these events are perceived.

In sum, these studies suggest that humans' assumptions of causality do not only shorten the time interval between cause and effect, they also reorder events in subjective experience, imposing a cause-precedes-effect principle to temporal order judgments. Here too, Bayesian models could potentially explain the role of causality in the perception of the temporal order of causes and effects.

Indeed, humans from a young age use the temporal priority principle (i.e., causes must precede effects) to infer causal relations in the environment (Rankin & McCormack, 2013). Through experience, we learn that the presumed cause of an event must occur before the event. From Bayes' rule, it follows that the events presumed to be causally connected must occur in succession with the cause preceding the effect. Consequently, the time estimation of when an event occurred would be based not only on the current sensory information (i.e., the likelihood), but also on people's assumption of causality. Notably, people's causal beliefs would shift the time estimates of the presumed cause and effect so that the presumed cause appears before the effect. Thus, for example, in Desantis et al. (2016), people's assumptions regarding their causal effectiveness led them to report more often the sensory event that they expected to be the consequence of their action, as occurring after the action, even though in reality it preceded it.

Even though the studies of Bechlivanidis and Lagnado (2013, 2016) and Desantis et al. (2016) clearly showed that causal assumptions influence time judgments, it remains unclear what the process modulated by these assumptions is. In particular, it is unclear whether causal expectations influence perceptual experience or simply induce a shift in perceptual decisions. In other words, in Bechlivanidis and Lagnado's and Desantis et al.'s studies, causal expectations might have directly influenced perceptual experience (i.e., causal assumptions led to the *illusory* reversal of temporal order), or might have induced a bias in the participants' decision criterion, that is, participants were more inclined (implicitly or explicitly) to report a temporal order that matched their causal assumptions even though they did not *perceive* it. For instance, in Desantis et al., participants were required to judge whether the sensory event (i.e., the coherent dot motion) occurred before or after their action. Causal assumptions might have biased participants' decision criterion, leading them to judge the event as occurring more often "after" their action rather than "before", instead of creating an illusory reversal of time. Further studies should aim at isolating both perceptual and decisional changes induced by causal assumption during temporal order judgment tasks.

4 Conclusions

In this chapter, we showed that time perception and causality are intimately connected: the temporal relations between events guide our causal inferences. In turn, causal expectations structure temporal perception. Temporal perception seems to emerge from the combination of internal (e.g., causal expectations) and sensory signals, providing a coherent representation of the world in which time experience matches causal assumptions. Hence, time perception of the sensory stimulation caused by the action of snapping our fingers is determined not only by the expectation of *what* outcome we are going to generate (Desantis & Haggard, 2016a, 2016b) and *when* it is going to occur (Cravo et al., 2011; Stetson et al., 2006), but also by the belief that our action is the *cause* of these sensory events. In line with this idea, Bayesian models of uncertainty reduction might offer an exploratory hypothesis to formalize the way causal prior and sensory information are combined.

Acknowledgments We would like to thank Karolina Moutsopoulou for her comments and suggestions.

References

Bechlivanidis, C., & Lagnado, D. (2013). Does the "why" tell us the "when"? *Psychological Science, 24*, 1563–1572.

Bechlivanidis, C., & Lagnado, D. A. (2016). Time reordered: Causal perception guides the interpretation of temporal order. *Cognition, 146*, 58–66. https://doi.org/10.1016/j.cognition.2015.09.001

Blakemore, S. J., Wolpert, D. M., & Frith, C. D. (2002). Abnormalities in the awareness of action. *Trends in Cognitive Sciences, 6*(6), 237–242. https://doi.org/16/S1364-6613(02)01907-1

Buehner, M. J. (2005). Contiguity and covariation in human causal inference. *Animal Learning & Behavior, 33*, 230–238. https://doi.org/10.3758/BF03196065

Buehner, M. J. (2012). Understanding the past, predicting the future: Causation, not intentional action, is the root of temporal binding. *Psychological Science, 23*(12), 1490–1497. https://doi.org/10.1177/0956797612444612

Buehner, M. J., & Humphreys, G. R. (2009). Causal binding of actions to their effects. *Psychological Science, 20*(10), 1221–1228. https://doi.org/10.1111/j.1467-9280.2009.02435.x

Buehner, M. J., & Humphreys, G. R. (2010). Causal contraction. *Psychological Science, 21*(1), 44–48. https://doi.org/10.1177/0956797609354735

Chen, Y.-C., & Spence, C. (2017). Assessing the role of the 'Unity Assumption' on multisensory integration: A review. *Frontiers in Psychology, 8*, 445. https://doi.org/10.3389/fpsyg.2017.00445

Cravo, A. M., Claessens, P. M. E., & Baldo, M. V. C. (2009). Voluntary action and causality in temporal binding. *Experimental Brain Research, 199*, 95–99.

Cravo, A. M., Claessens, P. M. E., & Baldo, M. V. C. (2011). The relation between action, predictability and temporal contiguity in temporal binding. *Acta Psychologica, 136*(1), 157–166. https://doi.org/16/j.actpsy.2010.11.005

Dainton, B. (2017, Fall). Temporal consciousness. In E. N. Zalta (Ed.), *The Stanford encyclopedia of philosophy*. Metaphysics Research Lab, Stanford University. Retrieved from https://plato.stanford.edu/archives/fall2017/entries/consciousness-temporal/

Desantis, A., & Haggard, P. (2016a). Action-outcome learning and prediction shape the window of simultaneity of audiovisual outcomes. *Cognition, 153*, 33–42. https://doi.org/10.1016/j.cognition.2016.03.009

Desantis, A., & Haggard, P. (2016b). How actions shape perception: Learning action-outcome relations and predicting sensory outcomes promote audio-visual temporal binding. *Scientific Reports, 6*. https://doi.org/10.1038/srep39086

Desantis, A., Haggard, P., Ikegaya, Y., & Hagura, N. (2018). Specificity of action selection modulates the perceived temporal order of action and sensory events. *Experimental Brain Research, 236*(8), 2157–2164. https://doi.org/10.1007/s00221-018-5292-5

Desantis, A., Hughes, G., & Waszak, F. (2012). Intentional binding is driven by the mere presence of an action and not by motor prediction. *PLoS One, 7*(1), e29557. https://doi.org/10.1371/journal.pone.0029557

Desantis, A., Roussel, C., & Waszak, F. (2011). On the influence of causal beliefs on the feeling of agency. *Consciousness and Cognition, 20*(4), 1211–1220. https://doi.org/10.1016/j.concog.2011.02.012

Desantis, A., Waszak, F., Moutsopoulou, K., & Haggard, P. (2016). How action structures time: About the perceived temporal order of action and predicted outcomes. *Cognition, 146*(Suppl. C), 100–109. https://doi.org/10.1016/j.cognition.2015.08.011

Dogge, M., Schaap, M., Custers, R., Wegner, D. M., & Aarts, H. (2012). When moving without volition: Implied self-causation enhances binding strength between involuntary actions and effects. *Consciousness and Cognition, 21*(1), 501–506. https://doi.org/10.1016/j.concog.2011.10.014

Eagleman, D. M., & Holcombe, A. O. (2002). Causality and the perception of time. *Trends in Cognitive Sciences, 6*(8), 323–325. https://doi.org/10.1016/S1364-6613(02)01945-9

Faro, D., Leclerc, F., & Hastie, R. (2005). Perceived causality as a cue to temporal distance. *Psychological Science, 16*, 673–677.

Farrer, C., Valentin, G., & Hupé, J. M. (2013). The time windows of the sense of agency. *Consciousness and Cognition, 22*(4), 1431–1441. https://doi.org/10.1016/j.concog.2013.09.010

Fletcher, P. C., & Frith, C. D. (2009). Perceiving is believing: A Bayesian approach to explaining the positive symptoms of schizophrenia. *Nature Reviews Neuroscience, 10*(1), 48–58. https://doi.org/10.1038/nrn2536

Fujisaki, W., Shimojo, S., Kashino, M., & Nishida, S. (2004). Recalibration of audiovisual simultaneity. *Nature Neuroscience, 7*(7), 773–778. https://doi.org/10.1038/nn1268

Grondin, S. (2010). Timing and time perception: A review of recent behavioral and neuroscience findings and theoretical directions. *Attention, Perception, & Psychophysics, 72*(3), 561–582. https://doi.org/10.3758/APP.72.3.561

Haering, C., & Kiesel, A. (2012). Mine is earlier than yours: Causal beliefs influence the perceived time of action effects. *Frontiers in Psychology, 3*, 393. https://doi.org/10.3389/fpsyg.2012.00393

Haering, C., & Kiesel, A. (2014). Intentional Binding is independent of the validity of the action effect's identity. *Acta Psychologica, 152*, 109–119. https://doi.org/10.1016/j.actpsy.2014.07.015

Haggard, P., Clark, S., & Kalogeras, J. (2002). Voluntary action and conscious awareness. *Nature Reviews Neuroscience, 5*(4), 382–385. https://doi.org/10.1038/nn827

Heron, J., Hanson, J. V. M., & Whitaker, D. (2009). Effect before cause: Supramodal recalibration of sensorimotor timing. *PLoS One, 4*(11), e7681. https://doi.org/10.1371/journal.pone.0007681

Hughes, G., Desantis, A., & Waszak, F. (2013). Mechanisms of intentional binding and sensory attenuation: The role of temporal prediction, temporal control, identity prediction, and motor prediction. *Psychological Bulletin, 139*(1), 133–151. https://doi.org/10.1037/a0028566

Hume, D. (1985). *A treatise of human nature*. Penguin Books Limited. Retrieved from https://books.google.fr/books?id=JPCgWYSZCUwC

Jazayeri, M., & Shadlen, M. N. (2010). Temporal context calibrates interval timing. *Nature Neuroscience, 13*(8), 1020–1026. https://doi.org/10.1038/nn.2590

Kawabe, T., Roseboom, W., & Nishida, S. (2013). The sense of agency is action-effect causality perception based on cross-modal grouping. *Proceedings of the Royal Society B: Biological Sciences, 280*(1763), 20130991. https://doi.org/10.1098/rspb.2013.0991

Keetels, M., & Vroomen, J. (2012). Exposure to delayed visual feedback of the hand changes motor-sensory synchrony perception. *Experimental Brain Research. Experimentelle Hirnforschung. Experimentation Cerebrale, 219*(4), 431–440. https://doi.org/10.1007/s00221-012-3081-0

Kersten, D., Mamassian, P., & Yuille, A. (2004). Object perception as Bayesian inference. *Annual Review of Psychology, 55*(1), 271–304. https://doi.org/10.1146/annurev.psych.55.090902.142005

Körding, K. P., Beierholm, U., Ma, W. J., Quartz, S., Tenenbaum, J. B., & Shams, L. (2007). Causal inference in multisensory perception. *PLoS One, 2*(9), e943. https://doi.org/10.1371/journal.pone.0000943

Leslie, A. M., & Keeble, S. (1987). Do six-month-old infants perceive causality? *Cognition, 25*(3), 265–288. https://doi.org/10.1016/S0010-0277(87)80006-9

Mamassian, P., Landy, M., & Maloney, L. T. (2002). Bayesian modelling of visual perception. In *Probabilistic models of the brain: Perception and neural function* (pp. 13–36). Cambridge, MA: The MIT Press.

Mellor, D. H. (1985). *Real time*. CUP Archive.

Michotte, A. (1941). La causalité physique est-elle une donnée phénoménale? *Tijdschrift Voor Philosophie, 3*(2), 290–328.

Michotte, A. (1946). *La perception de la causalité*. Editions de l'Institut Supérieur de Philosophie.

Oakes, L. M. (1994). Development of infants' use of continuity cues in their perception of causality. *Developmental Psychology, 30*(6), 869–879. https://doi.org/10.1037/0012-1649.30.6.869

Petzschner, F. H., Glasauer, S., & Stephan, K. E. (2015). A Bayesian perspective on magnitude estimation. *Trends in Cognitive Sciences, 19*(5), 285–293. https://doi.org/10.1016/j.tics.2015.03.002

Phillips, I. (2014). Experience of and in time. *Philosophy Compass, 9*(2), 131–144. https://doi.org/10.1111/phc3.12107

Rankin, M. L., & McCormack, T. (2013). The temporal priority principle: At what age does this develop? *Frontiers in Psychology, 4*. https://doi.org/10.3389/fpsyg.2013.00178

Rhodes, D. (2015). *Bayesian time perception*. PhD thesis, University of Birmingham.

Rohde, M., & Ernst, M. O. (2013). To lead and to lag—Forward and backward recalibration of perceived visuo-motor simultaneity. *Frontiers in Psychology, 3*. https://doi.org/10.3389/fpsyg.2012.00599

Shanks, D. R., Pearson, S. M., & Dickinson, A. (1989). Temporal contiguity and the judgement of causality by human subjects. *The Quarterly Journal of Experimental Psychology Section B, 41*(2), 139–159. https://doi.org/10.1080/14640748908401189

Shi, Z., Church, R. M., & Meck, W. H. (2013). Bayesian optimization of time perception. *Trends in Cognitive Sciences, 17*(11), 556–564. https://doi.org/10.1016/j.tics.2013.09.009

Stetson, C., Cui, X., Montague, P. R., & Eagleman, D. M. (2006). Motor-sensory recalibration leads to an illusory reversal of action and sensation. *Neuron, 51*(5), 651–659. https://doi.org/10.1016/j.neuron.2006.08.006

Sugano, Y., Keetels, M., & Vroomen, J. (2010). Adaptation to motor-visual and motor-auditory temporal lags transfer across modalities. *Experimental Brain Research. Experimentelle Hirnforschung. Experimentation Cerebrale, 201*(3), 393–399. https://doi.org/10.1007/s00221-009-2047-3

Sugano, Y., Keetels, M., & Vroomen, J. (2012). The build-up and transfer of sensorimotor temporal recalibration measured via a synchronization task. *Frontiers in Psychology, 3*. https://doi.org/10.3389/fpsyg.2012.00246

Thanopoulos, V., Psarou, E., & Vatakis, A. (2018). Robust intentional binding for causally-linked sequences of naturalistic events but not for abstract event sequences. *Acta Psychologica, 190*, 159–173. https://doi.org/10.1016/j.actpsy.2018.08.001

Thoenes, S., & Oberfeld, D. (2017). Meta-analysis of time perception and temporal processing in schizophrenia: Differential effects on precision and accuracy. *Clinical Psychology Review, 54*, 44–64. https://doi.org/10.1016/j.cpr.2017.03.007

Vatakis, A., & Spence, C. (2007). Crossmodal binding: Evaluating the "unity assumption" using audiovisual speech stimuli. *Perception & Psychophysics, 69*(5), 744–756.

Wegner, D. M., & Wheatley, T. (1999). Apparent mental causation: Sources of the experience of will. *American Psychologist, 54*(7), 480–492. https://doi.org/10.1037/0003-066X.54.7.480

Welch, R. B. (1999). Meaning, attention, and the "unity assumption" in the intersensory bias of spatial and temporal perceptions. In G. Aschersleben, T. Bachmann, & J. Müsseler (Eds.), *Advances in psychology* (Vol. 129, pp. 371–387). North-Holland. Retrieved from http://www.sciencedirect.com/science/article/pii/S0166411599800363

Welch, R. B., & Warren, D. H. (1980). Immediate perceptual response to intersensory discrepancy. *Psychological Bulletin, 88*(3), 638–667.

Young, M. E. (1995). On the origin of personal causal theories. *Psychonomic Bulletin & Review, 2*, 83–104. https://doi.org/10.3758/BF03214413

12

Getting Stuck in the Ordered Sequence: Disrupted Temporal Processing in Patients with Schizophrenia and What It Tells Us About the Sense of Time Continuity

Anne Giersch

1 Introduction

The exploration of the sense of time continuity is an old question related to the conceptualization of the mind-body relationship. Time continuity is usually attributed to a physical continuity of time, which is axiomatic, yet corresponds to our subjective feeling of time continuity. Intuitively, it is assumed that subjective time continuity results from the ability of our sensory systems to reflect the environment faithfully. However, even though we take time continuity for granted, we struggle to define it. For example, what is the required temporal resolution for consciousness to allow for a feeling of time continuity to emerge? Which are the disturbances of temporal continuity that we find, for example, in psychopathology? In what follows, we first develop what we know about the

A. Giersch (✉)
INSERM U1114, Pôle de Psychiatrie, Centre Hospitalier Régional
Université de Strasbourg, Strasbourg, France
e-mail: giersch@unistra.fr

V. Arstila et al. (eds.), *The Illusions of Time*,
https://doi.org/10.1007/978-3-030-22048-8_12

temporal resolution of conscious and subconscious processing. We then summarize research data in patients with schizophrenia, which leads us to question the mechanisms underlying the feeling of time continuity and the sense of immersion in the world.

2 Time Resolution at Conscious and Non-conscious Levels

Sensory processing is temporally constrained at the neurophysiological level. Transduced information is transmitted through successive neurons, whatever the sensory channel, inducing some delays relative to the physical world (Thorpe, 1990). Action potentials themselves have a duration and are typically followed by a refractory period. Finally, information may be conveyed by means of neuronal network synchronization, which again needs time to develop (Fries, Nikolić, & Singer, 2007; Schumacher, Wunderle, Fries, Jäkel, & Pipa, 2015). Even if delays in these physiological mechanisms are relatively short (a few milliseconds at the neuronal level, 10–20 ms for an auditory signal to reach auditory cortex and 100–150 ms for the visual information to reach temporal cortex), they still constrain processing temporally, and induce small delays between the occurrence of a signal in the outer world and its processing in the brain. Like when we are looking at stars, we are perceiving events now, which in fact occurred sometime before their perception. Yet, like for stars, we do not consciously realize that our perception is delayed relative to the moment of occurrence of external events. To know about delays between external events and their coding in the brain, we need to record the brain's response and the onset of the external event with the same device. In everyday life, there is no external reference to be compared with, and no perception of delay. This simple temporal constraint already means that our sense of immersion, characterized by the impression that we are in direct contact with the environment, does not fit the physical reality, but instead tolerates some degree of approximation in our subjective judgment of synchrony with the external world.

This is not the whole story, though. Being immersed in the world is not only about processing single events but a series of successive events. Intuitively, we have the impression of following events continuously. Yet again, this may have to be disconfirmed. It has been suggested that attention allows information to be sampled in successive snapshots rather than being processed continuously (James, 1890; Romanes, 1878; VanRullen, Reddy, & Koch, 2005; Varela, Lachaux, Rodriguez, & Martinerie, 2001). This hypothesis is strongly counter-intuitive and hotly debated (Fekete, Van de Cruys, Ekroll, & Leeuwen, 2018). Yet, recent results have shown that attention is optimally oriented toward a stimulus depending on the phase of the electroencephalogram (EEG) signals, meaning that attention is rhythmic rather than continuous (Fiebelkorn, Pinsk, & Kastner, 2018; Helfrich et al., 2018; Kienitz et al., 2018; Spyropoulos, Bosman, & Fries, 2018; VanRullen, 2018). The brain seems ill-equipped for a continuous processing of information at the conscious, attentional level. This is also confirmed in experimental psychology. Our ability to distinguish distinct events in time consciously has been studied for more than one century by asking participants to determine whether two sensory events displayed in different locations are simultaneous or asynchronous (see Elliott & Giersch, 2016, for a review). Our ability in this task is far from perfect. Depending on the modality and complexity of the task, 10 to 100 ms are required between stimulus onsets to order them in time or to integrate information over time (Grabot & van Wassenhove, 2017; van Wassenhove, 2009; Wittmann, 2011), whereas individuating events requires only a few ms (van Wassenhove, 2009). This means that the temporal resolution of our conscious perception is limited, which seems in contradiction with the feeling that time flows (Pöppel, 2004; Wittmann, 2011). Several models have been proposed to overcome this difficulty and explain how a sense of time continuity is reached despite this relatively low temporal resolution (see Dainton, 2017, for a complete review on this question). Some authors propose that the sense of continuity is an illusion. Others have proposed that the present is extended, or that successive moments are overlapping. A most popular model is the one issued from phenomenology (Husserl, 2012). It states that the present moment is related to past and future by means of 'retention' and 'protention' mechanisms, respectively. Taking the example of a melody,

Husserl observes that when listening to the present note, we still have the past one in mind. We also anticipate the future note, implying that at the present moment, we have the past, present and future notes in mind. It is this integration of information in the present moment that would yield a sense of continuity. Yet, the concept of present moments that are extended in time requires to explain how it is possible that we never detect any gap between successive windows. As emphasized by Dainton (2017), no model easily accounts for the strength of the feeling of time continuity. The only model that would answer this concern is the 'sliding window'. The main problem of this model, however, is that it is unclear how the window is sliding continuously, given exactly the same arguments as those listed above. There is no reason why the window should slide more continuously than permitted by the temporal resolution of information processing. It might have been proposed that the processing of information in parallel pathways may help, by providing overlapping temporal windows in different pathways. However, this again does not seem to fit well with phenomenology, which tells us that we can focus on a point in space (i.e., relying on information processing in the same pathway continuously) without any temporal gap in our perceptual experience.

Recently it was proposed that additional, non-conscious mechanisms may participate to the sense of time continuity (Giersch & Mishara, 2017a, 2017b, for a review). As a matter of fact, results have shown that the ability to distinguish events at a non-conscious level is characterized by a higher temporal accuracy than the same ability at a conscious level. Despite this, the flow of time is inescapable at a subjective level. What are the reasons for this discrepancy between a discrete processing and our subjective experience? We have proposed that the solution to that question relies on another counter-intuitive distinction between non-conscious and conscious processing (Giersch & Mishara, 2017a, 2017b; Poncelet & Giersch, 2015). The contrast between non-conscious and conscious processing has been nicely illustrated by recent results, based on the use of transcranial magnetic stimulation (TMS) during a visual perception task. Participants were shown two successive visual events, which were displayed in the same place and for too short a duration to be distinguished in time from one another (30 ms each). The two successive stimuli were fused in time, and only one merged stimulus was consciously perceived. Yet, when TMS was applied at various delays after the stimuli, it erased the current stimulus processing, and showed that the processing of the two stimuli extended over more than 300 ms, and that the two stimuli were processed successively (Herzog, Kammer, & Scharnowski, 2016; Scharnowski et al., 2009). In this experiment, the ability to distinguish events in time is thus markedly contrasted at the non-conscious and conscious levels. At an unconscious level, the first and second stimuli are processed one after another and are, thus, distinguished in time, whereas at a conscious level, the two stimuli are merged and perceived as one single stimulus. Complementary evidence supporting the distinction between segregation and integration of information are growing (Wutz, Weisz, Braun, & Melcher, 2014).

Beyond the contrast between conscious and non-conscious mechanisms, it was proposed that the high-temporal resolution at a non-conscious level was underpinned by predictive coding mechanisms. It can be noted here that the term 'prediction' is misleading, inasmuch it entails a meaning of intentionality (see Fuchs, 2018, pp. 61, 97, for more elaboration on this point). We use the term nonetheless, for lack of a better one, but this does not mean that such prediction is part of a sentient brain, or thinking homunculus. On the contrary, the idea is that there are predictions of sequences at the millisecond level, which are basic, automatic mechanisms involved in the functioning brain. We first summarize findings on the ability to process information in time at a non-conscious level, before explaining how this ability may participate to the sense of time continuity. Non-conscious processing of information in time was explored to find out whether participants were better at discriminating visual stimuli in time than indicated by their subjective responses. During the task, participants were shown two squares on a computer screen and were asked to decide whether the two squares were presented simultaneously or asynchronously. They answered by pressing a left response key in case of simultaneity, and a right response key in case of asynchrony. It was then examined whether they had a tendency to press on the side of the first or second square, irrespective of the square's location. For example, in case of a tendency to press to the side of the second square, participants would press on the right key when the order of the squares is left-right, whereas they would press on the left key when the order of squares is right-left. Healthy participants had a global tendency to press to the side of the second key, irrespective of the asynchrony (Lalanne, Van Assche, & Giersch, 2012a). Most interestingly, they had the same tendency for undetectable asynchronies (i.e., yielding the same rate of asynchronous responses than in case of a perfect simultaneity; Lalanne, Van Assche, & Giersch, 2012a; Lalanne, Van Assche, Wang, & Giersch, 2012b). Recent results replicated these findings with a simple priming effect, and suggested such results may be related to an automatic shift of attention toward the second stimulus (Poncelet & Giersch, 2015). In one of the tasks, participants were presented with two empty squares and one of them was filled in with gray color. Participants were instructed to press on a response key as fast as possible to the side of the filled in square. This is a simple detection task. However, it also entailed an incidental manipulation of the temporal properties of the stimuli. The trick was that the two empty squares displayed at the beginning of the trial appeared successively in half of the trials. In this case, the asynchrony was so short (17 ms) that it was not detected consciously (as checked in all participants). The results showed that participants were systematically faster when it was the second empty square that was filled in. The possibility that this was related to an attention shift to the side of the second empty square was supported by another experiment. We reasoned that an attention shift in the location of one of the two empty squares should induce a prior entry effect. The prior entry effect (Spence & Parise, 2010) refers to the fact that an attentional cue displayed before the first target's onset in a series of two can facilitate temporal order judgment. In our experiment, we hypothesized that if the display of two successive empty squares induces an attention shift to the side of the second empty square, then it should facilitate a temporal order judgment on subsequent targets, when the first target is where attention has been shifted. To check this hypothesis, the two empty squares were successively filled in with gray, and participants were asked to determine which square had been filled in first (or second, depending on the task). A prior entry effect was expected to result in improved performance when it was the second empty square that was filled in first with gray. This is exactly what was observed, consistently with the hypothesis of attention shifts. As a whole, this series of experiments showed that visual stimuli are (unconsciously) distinguished in time. Had they not, there would have been no tendency to press to the side of the second square, and no prior entry effect. These effects, thus, demonstrate that information can be represented with a higher temporal resolution at an unconscious than conscious representation can access to. They also suggest a role for prediction effects: attention can shift at a rate of around 10 Hz at the maximum (i.e., a shift takes around 100 ms; Deubel, 2008). Attention cannot be expected to be displaced over a 17 ms time interval. In addition, the first stimulus requires some time to be processed, as already emphasized above (Herzog et al., 2016), and it would be easy to miss the second stimulus (see Poncelet & Giersch, 2015, for a more exhaustive development on this issue). For attention to be displaced to the side of the second visual stimulus when it appears 17 ms later than the first visual stimulus, it must have been expected in advance (see also Nobre & van Ede, 2018). Trial-to-trial effects recently supported the idea that participants expect a sequence of stimuli on each trial (Giersch, Franck, Martin, & Lalanne, 2017). In all, the results yielded the proposal that processing information in time involves (1) predicting sequences of sensory information and (2) checking the occurrence of sensory information by displacing attention in time and space according to the prediction. There is no direct proof that these mechanisms are involved in the feeling of time continuity, but the impairments of these mechanisms in patients with schizophrenia, together with a correlation of these impairments with the sense of immersion in the world, oriented us toward a reflection on the link between the sense of time continuity, the sense of self and unconscious processing in time. This is summarized next.

3 Processing in Time in Schizophrenia

A disruption of the feeling of time continuity has been described by phenomenologists in the early twentieth century, and since then, clinical descriptions have been numerous (Fuchs, 2007, 2013; Minkowski, 1933; Stanghellini et al., 2016; Vogeley & Kupke, 2007). It has been proposed that 'protention' (i.e., the ability to expect future events) is altered in patients, who would remain stuck in the past and would be unable to connect with the reality of the outer world reality (Fuchs & van Duppen, 2017). The fragmentation of the patients' subjective experience may impede them to anticipate and follow events in the world. It may also impact the perception of oneself in the world. Any perception is experienced from one's own perspective, and this perspective affects how we perceive ourselves in the world, by helping to define the limits between us and the outer world. In particular, the continuity of the outer world perception may be a prerequisite to the sense of being one continuous self in time, and its disruption may affect the sense of self in patients with schizophrenia (Giersch & Mishara, 2017a, 2017b; Martin et al., 2014; Northoff & Stanghellini, 2016; Pienkos et al., 2019). As a whole, the literature justifies that questions about the sense of time continuity have been included in phenomenological scales exploring the sense of bodily self (Fuchs & van Duppen, 2017; Parnas et al., 2005; Sass et al., 2017).

The experimental results appear to be consistent with the proposals of the phenomenologists. Patients have difficulties to consciously detect asynchronies and order stimuli (Capa, Duval, Blaison, & Giersch, 2014; Foucher, Lacambre, Pham, Giersch, & Elliott, 2007; Giersch et al., 2009; Lalanne, Van Assche, & Giersch, 2012a; Martin, Giersch, Huron, & van Wassenhove, 2013; Noel, Stevenson, & Wallace, 2018; Schmidt, McFarland, Ahmed, McDonald, & Elliott, 2011; Stevenson et al., 2017). However, the patients' difficulties go beyond conscious, explicit timing. They were also tested on non-conscious timing. We have described above tasks requiring participants to decide whether two stimuli were simultaneous or asynchronous, in which it was found that participants tended to press to the side of the second stimulus (Lalanne, Van Assche, & Giersch, 2012a). When asynchronies were large enough, the tendency to press to the side of the second square in patients with schizophrenia was similar to the tendency found in healthy volunteers. In contrast, when asynchronies were undetectable, patients tended to press to the side of the first stimulus, whereas controls tended to press to the side of the second stimulus (Lalanne, Van Assche, Wang, & Giersch, 2012b). Inasmuch as the tendency in healthy volunteers is interpreted as the marker of the ability to predict sequences of events, the effects in patients are understood as indexing their difficulty at predicting sequences (Giersch et al., 2015). More recently, time prediction was tested at larger timescales in patients with schizophrenia.

Again, patients were impaired but mainly those who showed a lack of immersion in the world (Martin et al., 2017; Martin, Franck, Cermolacce, Coull, & Giersch, 2018), as evaluated by means of phenomenological scales (i.e., the EASE (Examination of Anomalous Self-Experience); Parnas et al., 2005). These results support the idea that patients with schizophrenia suffer from time disorders (see Ciullo, Spalletta, Caltagirone, Jorge, & Piras, 2016; Thoenes & Oberfeld, 2017, for meta-analyses), and that these disorders affect their ability to connect with the world. However, how do these disorders affect the sense of time continuity in patients with schizophrenia, and what does it tell us about time continuity itself?

4 How Do We Reach a Sense of Time Continuity: A Model Involving Interactions Between the Unconscious and Conscious Levels

As developed in Dainton (2017), no model can accommodate for the limitations of the temporal resolution of consciousness while explaining the strength of our feeling of time continuity. This holds for patients like for healthy participants, and makes it unlikely that the extended temporal windows are enough to explain the loss of the sense of time continuity reported by the patients. If any, a commonsense inference would state that extended windows should lead to a stroboscopic perception. Stroboscopic perception, although rare, has been described in patients, especially before the emergence of full-blown psychosis (Gross, Huber, Klosterkötter, & Linz, 2008). This suggests that extended windows may have some impact on the subjective experience. However, such complaint is by no means frequent, and patients more often report a disappearance of time (Martin et al., 2018). It has been proposed that the high resolution of sensory processing at a non-conscious level may help to compensate for the limitations at a conscious level (Herzog et al., 2016; Poncelet & Giersch, 2015). Especially, we have proposed that the ability to automatically predict sequences of events at a non-conscious level may be used to link events at the millisecond level in the same way as proposed by

Husserl at a larger timescale (Giersch & Mishara, 2017a, 2017b; Poncelet & Giersch, 2015). Conversely, the impairment of these mechanisms would impair the sense of time continuity in patients with schizophrenia. However, this explanation seems incomplete. Why would such mechanisms occurring at a non-conscious level emerge as a conscious sense of continuity, and how can they help to avoid a stroboscopic vision? As a matter of fact, when information is complex and multisensory, it can happen that up to 100 ms are needed to consciously distinguish that auditory and visual information is asynchronous (van Wassenhove, 2009). This time window is then enlarged up to 200 ms in patients with schizophrenia (Martin et al., 2013). When we are in a busy street, the amount of sensory information, whether auditory or visual, is much more than in experimental settings. How is it possible that with such extended temporal windows, we still seem to have such a strong feeling of time continuity? An explanation solely in terms of non-conscious mechanisms may not be enough, and there has to be an interaction between non-conscious and conscious mechanisms to yield our strong sense of time continuity.

What the results depicted above suggest is a displacement of attention in addition to prediction of sequences per se. The suggestion is that predictions guide the automatic displacement of attention. This means that our attention is not only guided by our intention but also coupled with unconscious prediction mechanisms. For example, when we predict a trajectory, we can displace our attention to check at any moment and location that the trajectory is as expected, and we can become aware of a prediction error. The important point here is *any* moment in time. It is because the prediction is at the millisecond level that any moment in time can be selected to check the trajectory. We may choose any moment of time in different instances, and it will work each time because the whole trajectory is predicted. Prediction does not only involve tracking a trajectory, though. More often than not, it is predicted that nothing changes. In that case, it can also be checked that information is still there, exactly the same as it was the instant before. If a prediction error occurs, then we can look for an explanation (i.e., something changed or happened in the environment, we lost balance, etc.). There is no case where a prediction leads to a feeling of time gap. Either events occur as predicted, and attention allows us to check and follow information smoothly (i.e., continuously). Alternatively, there is a prediction error, and in that case, attention helps to detect the error and adapt to it. This closed-loop system is akin to predictive coding, which has been successfully applied to the understanding of psychosis (Sterzer et al., 2018). Predictive coding posits that sensory signals are predicted and checked, allowing for disregarding information that corresponds to the prediction and highlighting unpredicted information that needs processing (Friston, 2005). Predictive coding provides a theoretical framework that fits with the data suggesting that sequences of sensory signals are predicted in advance. What our data brings in addition is the time dimension and the suggestion that sensory information is predicted at the millisecond level (i.e., at a temporal resolution that is higher than the resolution observed at a conscious level). These high-temporal resolution predictions would allow attention to be guided in the right place at the right moment, thus coupling non-conscious and conscious mechanisms. Lower temporal-resolution predictions would automatically relate the present perception with the one before, as suggested by Husserl and his melody metaphor, and automatic predictions at the millisecond level would embed the conscious perception within the sequence of automatically predicted events. Even a prediction error would be embedded since it is coded relative to the prediction. Thanks to these mechanisms, events would never be isolated but related by means of the predicted sequences of events. This may involve both predictive and postdictive mechanisms, as long as the mechanisms are coupled with the chain of events associated with predicted sequences.

Importantly, the closed-loop system involving predictive coding would mean that we are always predicting, checking, and readapting, consistent with the embodiment theory (Tschacher, Giersch, & Friston, 2017). A closed-loop system working on a millisecond basis represents a regular exchange of information between what we predict and what we perceive. It may appear as a costly system at first sight, but it might be necessary to allow for an optimal synchronization and adaptation to a dynamic world, especially as our perception is slightly delayed relative to the physical events. The sense of immersion would follow from both this closed-loop high-temporal resolution system and its close synergy with conscious attention. Examples of how predictions contribute to our sense of immersion are described in what follows.

5 Different Time Resolutions at an Automatic and at a Conscious Level: A Contribution to the Sense of Immersion in the World

As a whole, the ability to predict and to orient attention at the right time and place may help us to feel in contact and to adapt to the outer world. It would, thus, participate to our sense of immersion in the world. A number of examples illustrate this sense of immersion, and underline the need for some tolerance to time errors. The precision of the eye movement system is a powerful demonstration of the efficiency of automatic time prediction in avoiding any sense of instability in our environment (Baird-Gunning & Lueck, 2018). This also applies to very simple manual actions. Whenever we tap on a surface, we have the impression that the contact with the surface is immediate. It can be shown that the moment of contact can be predicted very precisely indeed. Tapping on a surface is composed of two motor elements, that is, approaching the surface with the finger and lifting the finger. Lifting the finger should be initiated after entering in contact with the surface. However, processing the tactile information and sending a motor command back requires about 90 ms (Johansson, Riso, Hager, & Backstrom, 1992). Yet, it has been shown that the contact with the surface can be less than 50 ms (Billon, Semjen, Cole, & Gauthier, 1996; Delevoye-Turrell, Giersch, Wing, & Danion, 2007), meaning that the finger is lifted before the sensory signal has been fully processed. The first point here is that the lifting of the finger demonstrates a prediction that is very precise in time. The second point is that the sensory processing of the contact with the surface is necessarily late relative to the physical contact. In that case, we can compare the prediction of the time of occurrence of the contact with the surface (the one that allows the finger to be lifted early) with the time of contact that is really perceived. These two times do not exactly match, but we are never aware of this mismatch. It is here that approximations in our conscious timing are useful. Imperfect temporal resolution helps us to ignore the gap between the prediction and the sensory feedback. Recent results in multisensory timing also suggest that the system adapts to ignore asynchronies (Harvey, Van der Burg, & Alais, 2014). Some degree of tolerance to asynchronies is required in case of multisensory information, because when the different auditory or visual information originates from the same source, it does not travel at the same speed and, thus, cannot reach our sensory cortices at the same time. Multisensory information is nonetheless bound together to produce one unique percept of, for example, a talking and smiling friend. It may help to ignore spurious asynchronies, which are meaningless, and, thus, allow for integrated and meaningful percepts to emerge.

In contrast to healthy participants, patients with schizophrenia do not always ignore spurious and meaningless asynchronies and have difficulty in integrating information over time in a meaningful way (e.g., Noel et al., 2018). What our data has repeatedly shown is a large discrepancy between automatic and conscious processing in time (Giersch et al., 2015; Martin et al., 2014). Altered automatic predictions may make it difficult to embed conscious information in the automatic flow associated with these predictions, and, thus, to derive integrated and meaningful perception. On the other hand, preserved segregation at the millisecond level would lead to bits of information that are not properly integrated within a meaningful flow of information. This may lead patients to misinterpret the outer world (Giersch & Mishara, 2017a, 2017b; Pienkos et al., 2019). Such difficulties may account for the patients' difficulties to feel as being immersed in the world, consistent with the results showing a relation between time expectation and the altered sense of immersion in patients with schizophrenia (Martin et al., 2017, 2018).

6 Conclusion

In all, the mechanisms proposed to explain the emergence of a sense of time continuity involve both conscious attention and non-conscious 'prediction', and an interaction between the two types of mechanisms. Non-conscious 'prediction' is rooted in brain-motor processes that underlie our interaction with the world and may reveal embodied mechanisms rather than an open-loop understanding of information processing, whereby information processing is supposed to reflect the outer world.

The difference in temporal resolution between non-conscious and conscious mechanisms may help to optimize our interaction with our environment, by avoiding a sensitivity to time lags between the timing of external events and the brain processing of these events. This would be crucial to enable our sensation of immersion in the world, by enabling a sensation of having a direct and immediate access to our environment, both in perception and action. A disturbance of such mechanisms may mediate the lack of a sense of immersion in patients with schizophrenia, thus leading to self-disorders.

Acknowledgments I would like to thank Thomas Fuchs and Virginie van Wassenhove for discussions prior to the writing of this chapter, which have been an important source of inspiration for this chapter.

References

Baird-Gunning, J. J. D., & Lueck, C. J. (2018). Central control of eye movements. *Current Opinion in Neurology, 31*(1), 90–95. https://doi.org/10.1097/WCO.0000000000000514

Billon, M., Semjen, A., Cole, J., & Gauthier, G. (1996). The role of sensory information in the production of periodic finger-tapping sequences. *Experimental Brain Research, 110*, 117–130.

Capa, R. L., Duval, C. Z., Blaison, D., & Giersch, A. (2014). Patients with schizophrenia selectively impaired in temporal order judgments. *Schizophrenia Research, 156*(1), 51–55. https://doi.org/10.1016/j.schres.2014.04.001

Ciullo, V., Spalletta, G., Caltagirone, C., Jorge, R. E., & Piras, F. (2016). Explicit time deficit in Schizophrenia: Systematic review and meta-analysis indicate it is primary and not domain specific. *Schizophrenia Bulletin, 42*(2), 505–518. https://doi.org/10.1093/schbul/sbv104

Dainton, B. (2017). Temporal consciousness. In E. N. Zalta (Ed.), *The Stanford Encyclopedia of Philosophy* (Fall 2017 ed.). Retrieved from https://plato.stanford.edu/archives/fall2017/entries/consciousness-temporal/.

Delevoye-Turrell, Y., Giersch, A., Wing, A. M., & Danion, J.-M. (2007). Motor fluency deficits in the sequencing of actions in schizophrenia. *Journal of Abnormal Psychology, 116*, 56–64. https://doi.org/10.1037/0021-843X.116.1.56

Deubel, H. (2008). The time course of presaccadic attention shifts. *Psychological Research, 72*(6), 630–640. https://doi.org/10.1007/s00426-008-0165-3

Elliott, M. A., & Giersch, A. (2016). What happens in a moment? *Frontiers in Psychology, 6*, 1905. https://doi.org/10.3389/fpsyg.2015.01905

Fekete, T., Van de Cruys, S., Ekroll, V., & van Leeuwen, C. (2018). In the interest of saving time: A critique of discrete perception. *Neuroscience of Consciousness, 2018*(1). https://doi.org/10.1093/nc/niy003.

Fiebelkorn, I. C., Pinsk, M. A., & Kastner, S. (2018). A dynamic interplay within the frontoparietal network underlies rhythmic spatial attention. *Neuron, 99*(4), 842–853. https://doi.org/10.1016/j.neuron.2018.07.038

Foucher, J. R., Lacambre, M., Pham, B.-T., Giersch, A., & Elliott, M. A. (2007). Low time resolution in schizophrenia: Lengthened windows of simultaneity for visual, auditory and bimodal stimuli. *Schizophrenia Research, 97*(1–3), 118–127. https://doi.org/10.1016/j.schres.2007.08.013

Fries, P., Nikolić, D., & Singer, W. (2007). The gamma cycle. *Trends in Neurosciences, 30*(7), 309–316. https://doi.org/10.1016/j.tins.2007.05.005

Friston, K. (2005). A theory of cortical responses. *Philosophical Transactions of the Royal Society of London. Series B, Biological Sciences, 360*(1456), 815–836. https://doi.org/10.1098/rstb.2005.1622

Fuchs, T. (2007). The temporal structure of intentionality and its disturbance in schizophrenia. *Psychopathology, 40*, 229–235.

Fuchs, T. (2013). Temporality and psychopathology. *Phenomenology and the Cognitive Sciences, 12*, 75–104.

Fuchs, T., & van Duppen, Z. (2017). Time and events: On the phenomenology of temporal experience in schizophrenia. *Psychopathology, 50*, 68–74.

Fuchs, T. (2018). *Ecology of the brain. The phenomenology and biology of the embodied mind.* Oxford: Oxford University Press.

Giersch, A., Lalanne, L., Corves, C., Seubert, J., Shi, Z., Foucher, J., & Elliott, M. A. (2009). Extended visual simultaneity thresholds in patients with schizophrenia. *Schizophrenia Bulletin, 35*(4), 816–825. https://doi.org/10.1093/schbul/sbn016

Giersch, A., Poncelet, P. E., Capa, R. L., Martin, B., Duval, C. Z., Curzietti, M., ... Lalanne, L. (2015). Disruption of information processing in schizophrenia: The time perspective. *Schizophrenia Research. Cognition, 2*(2), 78–83. https://doi.org/10.1016/j.scog.2015.04.002

Giersch, A., Franck, N., Martin, B., & Lalanne, L. (2017). Sense of time continuity: Patients with schizophrenia show the way. 40th European Conference on Visual Perception (ECVP) (August 27–31 in Berlin, Germany).

Giersch, A., & Mishara, A. (2017a). Disrupted continuity of subjective time in the milliseconds range in the self-disturbances of schizophrenia: Convergence of experimental, phenomenological, and predictive coding accounts. *Journal of Consciousness Studies, 24*(3–4), 62–87.

Giersch, A., & Mishara, A. L. (2017b). Is schizophrenia a disorder of consciousness? Experimental and phenomenological support for anomalous unconscious processing. *Frontiers in Psychology, 8*, 1659. https://doi.org/10.3389/fpsyg.2017.01659

Grabot, L., & van Wassenhove, V. (2017). Time order as psychological bias. *Psychological Science, 28*(5), 670–678. https://doi.org/10.1177/0956797616689369

Gross, G., Huber, G., Klosterkötter, J., & Linz, M. (2008). *BSABS-Bonn scale for the assessment of basic symptoms: 1st English edition: Manual, commentary, references, index, documentation sheet (Berichte aus der Medizin)*. Aachen: Shaker Verlag.

Harvey, C., Van der Burg, E., & Alais, D. (2014). Rapid temporal recalibration occurs crossmodally without stimulus specificity but is absent unimodally. *Brain Research, 1585*, 120–130. https://doi.org/10.1016/j.brainres.2014.08.028

Helfrich, R. F., Fiebelkorn, I. C., Szczepanski, S. M., Lin, J. J., Parvizi, J., Knight, R. T., & Kastner, S. (2018). Neural mechanisms of sustained attention are rhythmic. *Neuron, 99*(4), 854–865. https://doi.org/10.1016/j.neuron.2018.07.032

Herzog, M. H., Kammer, T., & Scharnowski, F. (2016). Time slices: What is the duration of a percept? *PLoS Biology, 14*(4), e1002433. https://doi.org/10.1371/journal.pbio.1002433

Husserl, E. (2012). *On the phenomenology of the consciousness of internal time (1893–1917)*. Springer Science & Business Media.

James, W. (1890). *The principles of psychology*. New York: Dover.

Johansson, S., Riso, R., Hager, C., & Backstrom, L. (1992). Somatosensory control of precision grip during unpredictable pulling loads. II. Changes in load force rate. *Experimental Brain Research, 89*, 192–203.

Kienitz, R., Schmiedt, J. T., Shapcott, K. A., Kouroupaki, K., Saunders, R. C., & Schmid, M. C. (2018). Theta rhythmic neuronal activity and reaction times arising from cortical receptive field interactions during distributed attention. *Current Biology, 28*(15), 2377–2387.e5. https://doi.org/10.1016/j.cub.2018.05.086

Lalanne, L., Van Assche, M., & Giersch, A. (2012a). When predictive mechanisms go wrong: Disordered visual synchrony thresholds in schizophrenia. *Schizophrenia Bulletin, 38*(3), 506–513. https://doi.org/10.1093/schbul/sbq107

Lalanne, L., Van Assche, M., Wang, W., & Giersch, A. (2012b). Looking forward: An impaired ability in patients with schizophrenia? *Neuropsychologia, 50*(12), 2736–2744. https://doi.org/10.1016/j.neuropsychologia.2012.07.023

Martin, B., Giersch, A., Huron, C., & van Wassenhove, V. (2013). Temporal event structure and timing in schizophrenia: Preserved binding in a longer "now". *Neuropsychologia, 51*(2), 358–371. https://doi.org/10.1016/j.neuropsychologia.2012.07.002

Martin, B., Wittmann, M., Franck, N., Cermolacce, M., Berna, F., & Giersch, A. (2014). Temporal structure of consciousness and minimal self in schizophrenia. *Frontiers in Psychology, 5*, 1175. https://doi.org/10.3389/fpsyg.2014.01175

Martin, B., Franck, N., Cermolacce, M., Falco, A., Benair, A., Etienne, E., ... Giersch, A. (2017). Fragile temporal prediction in patients with schizophrenia is related to minimal self disorders. *Scientific Reports, 7*(1), 8278. https://doi.org/10.1038/s41598-017-07987-y

Martin, B., Franck, N., Cermolacce, M., Coull, J. T., & Giersch, A. (2018). Minimal self and timing disorders in Schizophrenia: A case report. *Frontiers in Human Neuroscience, 12*, 132. https://doi.org/10.3389/fnhum.2018.00132

Minkowski, E. (1933/2005). *Le temps vécu: Etudes phénoménologiques et psychopathologiques* (2nd ed.). Presses Universitaires de France—PUF (1st publication 1933).

Nobre, A. C., & van Ede, F. (2018). Anticipated moments: Temporal structure in attention. *Nature Reviews. Neuroscience, 19*(1), 34–48. https://doi.org/10.1038/nrn.2017.141

Noel, J.-P., Stevenson, R. A., & Wallace, M. T. (2018). Atypical audiovisual temporal function in autism and schizophrenia: Similar phenotype, different cause. *The European Journal of Neuroscience, 47*(10), 1230–1241. https://doi.org/10.1111/ejn.13911

Northoff, G., & Stanghellini, G. (2016). How to link brain and experience? Spatiotemporal psychopathology of the lived body. *Frontiers in Human Neuroscience, 10*, 76. https://doi.org/10.3389/fnhum.2016.00172

Parnas, J., Møller, P., Kircher, T., Thalbitzer, J., Jansson, L., Handest, P., & Zahavi, D. (2005). EASE: Examination of anomalous self-experience. *Psychopathology, 38*(5), 236–258. https://doi.org/10.1159/000088441

Pienkos, E., Giersch, A., Hansen, M., Humpston, C., McCarthy-Jones, S., Mishara, A., ... Rosen, C. (2019). Hallucinations beyond voices: A conceptual review of the phenomenology of altered perception in psychosis. *Schizophrenia Bulletin, 45*, S67–S77. https://doi.org/10.1093/schbul/sby057

Poncelet, P. E., & Giersch, A. (2015). Tracking visual events in time in the absence of time perception: Implicit processing at the ms Level. *PloS One, 10*(6), e0127106. https://doi.org/10.1371/journal.pone.0127106

Pöppel, E. (2004). Lost in time: A historical frame, elementary processing units and the 3-second window. *Acta Neurobiologiae Experimentalis, 64*(3), 295–301.
Romanes, G. J. (1878). Consciousness of time. *Mind, 3*(11), 297–303.
Sass, L., Pienkos, E., Skodlar, B., Stanghellini, G., Fuchs, T., Parnas, J., & Jones, N. (2017). EAWE: Examination of anomalous world experience. *Psychopathology, 50*(1), 10–54. https://doi.org/10.1159/000454928
Scharnowski, F., Rüter, J., Jolij, J., Hermens, F., Kammer, T., & Herzog, M. H. (2009). Long-lasting modulation of feature integration by transcranial magnetic stimulation. *Journal of Vision, 9*(6), 1–10. https://doi.org/10.1167/9.6.1
Schmidt, H., McFarland, J., Ahmed, M., McDonald, C., & Elliott, M. A. (2011). Low-level temporal coding impairments in psychosis: Preliminary findings and recommendations for further studies. *Journal of Abnormal Psychology, 120*(2), 476–482. https://doi.org/10.1037/a0023387
Schumacher, J., Wunderle, T., Fries, P., Jäkel, F., & Pipa, G. (2015). A statistical framework to infer delay and direction of information flow from measurements of complex systems. *Neural Computation, 27*(8), 1555–1608. https://doi.org/10.1162/NECO_a_00756
Spence, C., & Parise, C. (2010). Prior-entry: A review. *Consciousness and Cognition, 19*(1), 364–379. https://doi.org/10.1016/j.concog.2009.12.001
Spyropoulos, G., Bosman, C. A., & Fries, P. (2018). A theta rhythm in macaque visual cortex and its attentional modulation. *Proceedings of the National Academy of Sciences of the United States of America, 115*(24), E5614–E5623. https://doi.org/10.1073/pnas.1719433115
Stanghellini, G., Ballerini, M., Presenza, S., Mancini, M., Raballo, A., Blasi, S., & Cutting, J. (2016). Psychopathology of lived time: Abnormal time experience in persons with Schizophrenia. *Schizophrenia Bulletin, 42*(1), 45–55. https://doi.org/10.1093/schbul/sbv052
Sterzer, P., Adams, R. A., Fletcher, P., Frith, C., Lawrie, S. M., Muckli, L., … Corlett, P. R. (2018). The predictive coding account of psychosis. *Biological Psychiatry, 84*(9), 634–643. https://doi.org/10.1016/j.biopsych.2018.05.015
Stevenson, R. A., Park, S., Cochran, C., McIntosh, L. G., Noel, J.-P., Barense, M. D., … T, M. (2017). The associations between multisensory temporal processing and symptoms of schizophrenia. *Schizophrenia Research, 179*, 97–103. https://doi.org/10.1016/j.schres.2016.09.035
Thoenes, S., & Oberfeld, D. (2017). Meta-analysis of time perception and temporal processing in schizophrenia: Differential effects on precision and accuracy. *Clinical Psychology Review, 54*, 44–64. https://doi.org/10.1016/j.cpr.2017.03.007

Thorpe, S. J. (1990). Spike arrival times: A highly efficient coding scheme for neural networks. In R. Eckmiller, G. Hartmann, & G. Hauske (Eds.), *Parallel processing in neural systems and computers* (pp. 91–94). Amsterdam: Elsevier.

Tschacher, W., Giersch, A., & Friston, K. (2017). Embodiment and schizophrenia: A review of implications and applications. *Schizophrenia Bulletin, 43*, 745–753. https://doi.org/10.1093/schbul/sbw220

VanRullen, R., Reddy, L., & Koch, C. (2005). Attention-driven discrete sampling of motion perception. *Proceedings of the National Academy of Sciences of the United States of America, 102*(14), 5291–5296. https://doi.org/10.1073/pnas.0409172102

VanRullen, R. (2018). Attention cycles. *Neuron, 99*(4), 632–634. https://doi.org/10.1016/j.neuron.2018.08.006

van Wassenhove, V. (2009). Minding time in an amodal representational space. *Philosophical Transactions of the Royal Society of London. Series B, Biological Sciences, 364*(1525), 1815–1830. https://doi.org/10.1098/rstb.2009.0023

Varela, F., Lachaux, J. P., Rodriguez, E., & Martinerie, J. (2001). The brainweb: Phase synchronization and large-scale integration. *Nature Reviews Neuroscience, 2*(4), 229.

Vogeley, K., & Kupke, C. (2007). Disturbances of time consciousness from a phenomenological and a neuroscientific perspective. *Schizophrenia Bulletin, 33*(1), 157–165. https://doi.org/10.1093/schbul/sbl056

Wittmann, M. (2011). Moments in time. *Frontiers in Integrative Neuroscience, 5*, 66. https://doi.org/10.3389/fnint.2011.00066

Wutz, A., Weisz, N., Braun, C., & Melcher, D. (2014). Temporal windows in visual processing: "Prestimulus brain state" and "poststimulus phase reset" segregate visual transients on different temporal scales. *The Journal of Neuroscience: The Official Journal of the Society for Neuroscience, 34*(4), 1554–1565. https://doi.org/10.1523/JNEUROSCI.3187-13.2014

13

When the Perception of a Synchronous World Is—Mostly—Just an Illusion

Nadia Paraskevoudi and Argiro Vatakis

1 Introduction

Organisms need to integrate information from different sensory channels to create a coherent representation of a synchronous multisensory world. This is counterintuitive given that sensory signals arising from the same event may arrive at different multimodal areas at slightly different times (Efron, 1963; Halliday & Mingay, 1964; Keetels & Vroomen, 2012; Pöppel, 1988; Vatakis, 2013), due to differences in the speed of sound and light (i.e., physics) and the differential speeds of neural

N. Paraskevoudi
Brainlab, Cognitive Neuroscience Research Group, Department of Clinical Psychology and Psychobiology, University of Barcelona, Barcelona, Spain

Institute of Neurosciences, University of Barcelona, Barcelona, Spain

A. Vatakis (✉)
Cognitive Systems Research Institute, Athens, Greece

Department of Psychology, Panteion University of Social and Political Sciences, Athens, Greece

V. Arstila et al. (eds.), *The Illusions of Time*,
https://doi.org/10.1007/978-3-030-22048-8_13

transmission between the senses (i.e., biophysics; e.g., auditory information is processed much faster than visual signals due to the faster sound transduction by the hair cells of the inner ear compared to the phototransduction times in the retina; Pöppel, 2004). This discrepancy between physical and perceptual synchrony highlights the complexity of the apparently simple question of when two sensory inputs appear to be temporally aligned and has an interesting implication: The perception of a synchronous world is mostly just an illusion—a distortion resulting from the brain's ability to adapt to asynchronies so as to create a smoother percept of the world. In this chapter, we aim to review the research on the relationship between physical, neural, and subjective synchrony by focusing on how the human perceptual system compensates for the temporal differences of perceptually synchronous events.

2 Measures of Perceptual Synchrony

Attempts to better understand how the brain reaches perceptual synchrony have utilized various psychophysical measures that have been mostly obtained in simultaneity (SJ) and temporal order judgment (TOJ) tasks, where participants are presented with two stimuli of different sensory modalities and have to report whether the inputs are simultaneous or which input was presented first, respectively (Kostaki & Vatakis, 2018). In both tasks, researchers typically introduce temporal asynchronies between the two sensory signals (i.e., stimulus onset asynchronies, SOAs) so as to calculate the point of subjective simultaneity (PSS; i.e., the SOA, where the different sensory signals are perceived as being simultaneous) and the just noticeable difference (JND; i.e., the smallest SOA, where participants can reliably detect the temporal asynchrony). Yet, the most critical measure of perceptual synchrony is the width of the temporal binding window (TBW; i.e., the temporal range within which sensory information from the different modalities are integrated), which typically equals the PSS ± JND, and has been found to be movable and highly plastic depending on several experimental manipulations (for previous reviews, see Keetels & Vroomen, 2012; Vatakis, 2013).

Importantly, the measures of perceptual synchrony have been shown to be dependent on the experimental paradigm. That is, despite the fact that the stimuli may be physically identical in the SJ and TOJ tasks, increasing evidence suggests that judgments of simultaneity and temporal order may depend on and, thus, measure different perceptual processes (e.g., TOJs are considered as measuring the smallest audiovisual asynchrony that can be detected by the perceptual system, while SJs have been implicated as providing a measure of the window of perceived simultaneity; Barnett-Cowan & Harris, 2009, 2011; Basharat, Adams, Staines, & Barnett-Cowan, 2018; Bedard & Barnett-Cowan, 2016; Chan, Pianta, & McKendrick, 2014a; De Niear, Koo, & Wallace, 2016; Love, Petrini, Cheng, & Pollick, 2013; Mitrani, Shekerdjiiski, & Yakimoff, 1986; Van Eijk, Kohlrausch, Juola, & van de Par, 2008; Vatakis, Navarra, Soto-Faraco, & Spence, 2008). The potential perceptual processing dissociation is supported by evidence showing different TBW and PSS mean values for SJs and TOJs of both younger and older adults (Basharat et al., 2018; Bedard & Barnett-Cowan, 2016), differential susceptibility of these mechanisms as a function of age (i.e., the age-related effects on synchrony perception are particularly prominent in TOJs, while SJs seem to be largely preserved across the lifespan; Basharat et al., 2018; Bedard & Barnett-Cowan, 2016; Chan et al., 2014a; Setti, Burke, & Kenny, 2011), and differences between these tasks in their cognitive load demands (De Niear et al., 2016; Love et al., 2013; Mitrani et al., 1986; Yarrow, Shapiro, DiCosta, & Arnold, 2014). The latter findings highlight the importance of disentangling the mechanisms underlying judgments of simultaneity and temporal order and create the need to directly assess the modulatory effects of experimental paradigm on the measures of the perception of synchrony.

3 Factors Affecting the Perception of Synchrony

Perceptual synchrony is a highly flexible process that has been found to be modulated by several factors, including the experimental paradigm (Basharat et al., 2018; Love et al., 2013), stimulus features (e.g., intensity, frequency, or duration; Chan et al., 2014a; Eg & Behne, 2015; Tiippana & Salmela, 2018), stimulus complexity and type (Baumann et al., 2018; De Niear et al., 2016; Love et al., 2013), sensory combinations and their presentation order (Benedetto, Burr, & Morrone, 2018; Bhat, Miller, Pitt, & Shahin, 2015; Eg & Behne, 2015; Eg, Griwodz, Halvorsen, & Behne, 2015; Magnotti, Ma, & Beauchamp, 2013; Navarra, Fernández-Prieto, & Garcia-Morera, 2013; Simon & Wallace, 2017; Stevenson, Wilson, Powers, & Wallace, 2013), and age (Basharat et al., 2018; Brooks, Chan, Anderson, & McKendrick, 2018; Chan et al., 2014a), while it undergoes significant changes in pathological conditions (e.g., autism or schizophrenia [SZ]; Balz et al., 2016; Foss-Feig et al., 2010; Foxe et al., 2015; Kwakye, Foss-Feig, Cascio, Stone, & Wallace, 2010; Martin, Giersch, Huron, & van Wassenhove, 2013; Noel, De Niear, Lazzara, & Wallace, 2017; Noel, Lytle, Cascio, & Wallace, 2017; Noel, Stevenson, & Wallace, 2018; Stevenson, Park, et al., 2017; Stevenson et al., 2014; Su et al., 2015; Vatakis & Bakou, 2015). In this section, we will summarize the factors that modulate synchrony perception by mainly focusing on the research conducted over the last five years (for previous reviews, see Keetels & Vroomen, 2012; Vatakis, 2013).

3.1 Stimulus Features

The low-level features of sensory signals have been found to exert a great influence on our ability to perceive two sensory inputs as synchronous. Due to the slower neural transmission and possibly the relative reduction in signal-to-noise-ratio for near- as compared to supra-threshold stimuli (Chan et al., 2014a; Jaskowski, 1999; Neumann & Niepel, 2004; Roefs, 1963; Sanford, 1971; Smith, 1933), the level of stimulus intensity affects both asynchrony detection and the width of the TBW, leading to higher mean asynchrony detection thresholds and higher variability in temporal binding (i.e., increased width of the TBW) for near- than supra-threshold stimuli (e.g., dim as compared to bright visual stimuli), respectively (Chan et al., 2014a; Krueger Fister, Stevenson, Nidiffer, Barnett, & Wallace, 2016). Stimulus frequency has been also found to affect asynchrony detection, with evidence showing worse detection ability (i.e., higher asynchrony detection thresholds) for higher (4 kHz) as compared to lower (0.5 kHz) frequencies, with this effect being further modulated by the individual's age (Chan et al., 2014a). Specifically, aging individuals exhibit a wider window width at lower frequencies (e.g., 0.5 kHz) compared to younger adults, whereas similar width was found for both age groups at higher frequencies (e.g., 4 kHz; Chan et al., 2014a). Finally, stimulus duration has been recently shown to yield task- and stimulus-dependent effects on multisensory synchrony perception (Eg & Behne, 2015; Tiippana & Salmela, 2018). A wider TBW has been found for shorter as compared to longer audiovisual speech stimuli in an SJ task (i.e., in the range of 1–13 s; Eg & Behne, 2015), whereas in a more recent study utilizing a TOJ task, the duration of simple audiovisual stimuli did not affect neither the JND nor the PSS values (Tiippana & Salmela, 2018). Although the discrepancy in the results of the latter two studies (Eg & Behne, 2015; Tiippana & Salmela, 2018) may be due to the informational content differences of the stimuli utilized (e.g., simple vs. complex audiovisual pairings), it has been hypothesized that contrary to simultaneity judgments, audiovisual temporal order is dependent on comparisons between stimulus onsets only, rather than stimulus duration (Tiippana & Salmela, 2018). Future work is, thus, required to assess the interactive and collaborative effects of stimulus duration, experimental paradigm, and informational content on perceptual synchrony for audiovisual stimuli.

Although most studies on synchrony perception have mainly utilized simple stimuli of low informational content, recent work reports that the type and dynamics of the sensory signals also yield significant effects on perceptual sensitivity for asynchrony detection (Baumann et al., 2018; Eg & Behne, 2015). The TBW has been shown to be highly dependent upon the nature (i.e., complexity) of the sensory inputs, with narrower TBWs noted for simple stimuli (i.e., flashes and beeps) and increasingly larger TBWs for pairings of more complex stimuli (i.e., biological motion and speech; Love et al., 2013; Stevenson & Wallace, 2013; Vatakis & Spence, 2006), particularly during TOJs (Love et al., 2013), while the widening of the window is further modulated by the spectrotemporal fidelity of the audiovisual speech signal (Shahin, Shen, & Kerlin, 2017). The effects of stimulus complexity on perceived synchrony are further supported by evidence showing that training-induced enhancements in temporal acuity do not generalize across stimuli of different complexity, thus indicating that perceptual training can elicit changes in temporal acuity, albeit dependent on stimulus features (De Niear, Gupta, Baum, & Wallace, 2018).

3.2 Temporal Order of Multisensory Pairings

Our ability to detect asynchronies between sensory inputs is highly dependent on the physical asynchronies in the external world. The fact that visual stimuli often arrive earlier in our sensory system (due to physics) compared to their correlated auditory signals shapes our prior knowledge about the contingencies in our environment and may account for the brain's tendency to tolerate larger asynchronies when the visual input precedes the auditory one than vice versa (Benedetto et al., 2018; Bhat et al., 2015; Eg et al., 2015; Eg & Behne, 2015; Magnotti et al., 2013; Navarra et al., 2013; Simon & Wallace, 2017; Stevenson et al., 2013). Indeed, the temporal window for evaluating audiovisual simultaneity depends on the temporal order of unisensory constituents (Benedetto et al., 2018; Bhat et al., 2015; Conrey & Pisoni, 2006; Eg & Behne, 2015; Eg et al., 2015; Magnotti et al., 2013; Navarra et al., 2013; Simon & Wallace, 2017; Stevenson et al., 2013; Stevenson & Wallace, 2013; van Wassenhove, Grant, & Poeppel, 2007). Stimuli with a visual lead (VA) are reported as synchronous more frequently than their auditory-leading counterparts (AV), a finding that has been replicated with stimuli of various degrees of complexity, including continuous speech streams (e.g., a female news anchor presenting a story; Eg & Behne, 2015), isolated words (Conrey & Pisoni, 2006), and syllables (Simon & Wallace, 2017; van Wassenhove et al., 2007; Venezia, Thurman, Matchin, George, & Hickok, 2016). This, consequently, results in an asymmetrical TBW, which is consistent with the statistics of the natural environment and the causal structure of speech signals (Bhat et al., 2015; Navarra et al., 2013; Schwartz & Savariaux, 2014; Van der Burg, Alais, & Cass, 2013; van Wassenhove, Grant, & Poeppel, 2005; Venezia et al., 2016). This asymmetry has been suggested to serve different functions depending on the leading anticipatory input (Cecere, Gross, Willis, & Thut, 2017; Cecere, Romei, Bertini, & Làdavas, 2014). For example, the alerting value of auditory information on visual processing (Cecere et al., 2014; Cecere et al., 2017) is reflected in a narrower TBW in auditory-leading scenarios (approx. 50–80 ms; Navarra et al., 2013), presumably due to the much faster latencies of primary auditory than visual cortex activation (Murray & Spierer, 2009; Musacchia & Schroeder, 2009). Conversely, the wider TBW for visual-leading conditions (approx. 150–250 ms; Navarra et al., 2013) has been linked to the predictive value of visual information (e.g., articulatory movements of the vocal apparatus) that may boost the processing of the dynamic and more rapidly changing auditory speech signal (Bhat et al., 2015; Cecere et al., 2017; Navarra et al., 2005; Schwartz & Savariaux, 2014; Simon & Wallace, 2017; Van der Burg et al., 2013; van Wassenhove et al., 2005; Venezia et al., 2016). Critically, the latter behavioral findings are further supported by electrophysiological evidence showing that audiovisual temporal binding (as indexed by SJs) engages different neural pathways depending on the leading sense (Cecere et al., 2017). That is, evaluating synchrony of auditory-leading (AV) versus visual-leading (VA) audiovisual stimulus pairs is characterized by two distinct patterns of brain activity, thereby showing that audiovisual integration is not a unitary process (Cecere et al., 2017), while also pointing to partly independent processes for synchrony and multisensory temporal binding (for a review on the dissociation between synchrony perception and binding for audiovisual stimuli, see Tsilionis & Vatakis, 2016).

3.3 Perceiving Synchrony Through Senses Other Than Vision and Audition

While most research has been mainly focusing on audiovisual synchrony perception, the perceptual synchrony for combinations including somatosensory or proprioceptive inputs remains largely unexplored. Although the TBW correlates across modality pairings (audio-tactile, visual-tactile, audiovisual) within individuals (Noel, Wallace, Orchard-Mills, Alais, & Van der Burg, 2015), several modality-dependent differences have been recently identified. Specifically, research demonstrates increased synchrony resolution for audio-tactile compared to audiovisual and visual-tactile pairings (Fujisaki & Nishida, 2009), and more rapid and malleable recalibration to temporally offset audiovisual than audio-tactile and visual-tactile pairings (Harrar & Harris, 2008; Van der Burg & Goodbourn, 2015; Van der Burg, Orchard-Mills, & Alais, 2015). Additional work shows a positive correlation between each participant's audio-tactile TBW and the distance between the PSS and physical synchrony (compared to audiovisual binding, which is more strongly related to the perceived synchrony for the stimuli presented), suggesting that audio-tactile binding is anchored closer to physical synchrony (i.e., SOA = 0 ms) than audiovisual binding (Noel et al., 2015). Yet, a more recent study supported that in order to achieve perceptual synchrony in audio-tactile combinations, a tactile lead of approximately 40 to 80 ms is required (Roy, Dalla Bella, & Lagarde, 2017). Further research in the somatosensory domain also reports differential effects of limb orientation on asynchrony detection for visual-tactile versus visual-proprioceptive stimulus pairings (Hoover & Harris, 2012, 2016; Keys, Rich, & Zopf, 2018; Zopf, Friedman, & Williams, 2015). In visual-proprioceptive scenarios, delays between self-generated and seen hand movement can be accurately detected but only when the viewed hands are in an anatomically plausible orientation (Hoover & Harris, 2012; Zopf et al., 2015). In contrast, viewed hand orientation does not affect visual-tactile asynchrony detection (within the range of 40–280 ms) between a visual (an LED flash on the model hand) and a tactile stimulus (a tap to fingertip of the participant's hidden hand; Keys et al., 2018). The differential effects of hand orientation on detecting visual-tactile versus visual-proprioceptive asynchronies may be due to the highly spatiotopic representations in the tactile domain (Noel et al., 2015), which may result in diminishing any effects of limb orientation on synchrony judgments, or due to the information from active movement that may enhance sensory predictions in the visual-proprioceptive compared to the visual-tactile scenario (Hoover & Harris, 2012, 2016; Keys et al., 2018; Zopf et al., 2015). The latter speculation is also supported by a significant correlation between asynchrony detection performance and the sense of embodiment of the viewed body parts (Hoover & Harris, 2016).

To date, research on perceived simultaneity for crossmodal stimuli has been mainly limited to the physical (visual, auditory, and tactile modalities) rather than the chemical (olfactory and gustatory inputs) senses. However, recent attempts to assess whether the different properties of physical and chemical senses (e.g., differences in temporal resolution) modulate intersensory synchrony perception show that temporal resolution of synchrony perception (as measured by SJs) is lower in combinations involving one or two chemical stimuli (e.g., olfactory-visual, visual-gustatory, and olfactory-gustatory; Gotow & Kobayakawa, 2017) than in combinations of two physical stimuli (e.g., auditory-tactile, visual-tactile, and audiovisual; Fujisaki & Nishida, 2009). In line with the hypothesis that temporal resolution of crossmodal synchrony perception is driven by the sensory modality with the lower temporal resolution (Fujisaki & Nishida, 2009), this finding highlights the modality-dependent effects on our ability to perceive synchrony between different sensory inputs and creates the need to further examine the neural and behavioral correlates for crossmodal synchrony perception.

3.4 Perceiving Synchrony Across the Lifespan

As with other perceptual and cognitive phenomena, synchrony perception (as measured by the TBW) undergoes significant changes throughout the lifespan (Bedard & Barnett-Cowan, 2016; Brooks et al., 2018; de Boer-Schellekens & Vroomen, 2014; de Dieuleveult, Siemonsma, van Erp, & Brouwer, 2017; Hillock, Powers, & Wallace, 2011; Hillock-Dunn & Wallace, 2012). While being wider during early childhood, the TBW becomes fine-tuned at the time when additional changes in the development of optimal multisensory integration take place (Chen, Lewis, Shore, Spence, & Maurer, 2018; Gori, 2015). Yet, the age at which the TBW reaches its adult-like levels also depends on the sensory pairing, with audiovisual simultaneity reaching its adult-like levels between 7 and 9 years of age, while the window for visual-tactile simultaneity narrows 2 years later in the development, between 9 and 11 years of age (Chen et al., 2018). Consistent with the modality-dependent effects on perceptual synchrony described previously, the latter findings suggest that multisensory perception of simultaneity for different sensory combinations may have distinct rather than shared mechanisms (Baumgarten, Königs, Schnitzler, & Lange, 2017; Baumgarten, Schnitzler, & Lange, 2015; Benedetto et al., 2018; Cecere, Rees, & Romei, 2015; Chen et al., 2018; Landau & Fries, 2012; Noel et al., 2015; Ronconi, Oosterhof, Bonmassar, & Melcher, 2017; VanRullen, 2016).

At the other end of the developmental spectrum, research shows that as we age, it becomes increasingly difficult to separate auditory and visual signals in time (Bedard & Barnett-Cowan, 2016; Brooks et al., 2018; Diederich, Colonius, & Schomburg, 2008; Laurienti, Burdette, Maldjian, & Wallace, 2006; Mahoney, Holtzer, & Verghese, 2014; Peiffer, Mozolic, Hugenschmidt, & Laurienti, 2007; Setti et al., 2011; Wu, Chen, Yeh, Wu, & Tang, 2015). Healthy aging individuals consistently exhibit a wider TBW, which results in unifying irrelevant sensory events, impairs their ability to discriminate simultaneity, temporal order, or the causal relationships between sensory inputs (Basharat et al., 2018; Bedard & Barnett-Cowan, 2016; Brooks et al., 2018; Chan et al., 2014a; Chan, Pianta, & McKendrick, 2014b; Diederich et al., 2008; Laurienti et al., 2006; Mahoney et al., 2014; Peiffer et al., 2007; Ren, Yang, Nakahashi, Takahashi, & Wu, 2017; Setti et al., 2011; Setti, Burke, Kenny, & Newell, 2013; Wu et al., 2015), and makes them more susceptible to temporal illusions (e.g., temporal ventriloquist effect and sound-induced flash illusion; de Boer-Schellekens & Vroomen, 2014; DeLoss, Pierce, & Andersen, 2013; McGovern, Roudaia, Stapleton, McGinnity, & Newell, 2014; Setti et al., 2011; Stapleton, Setti, Doheny, Kenny, & Newell, 2014; Wu et al., 2015). Importantly, the effects of age-related changes in perceiving synchrony and temporal order are not limited to temporal judgments (for a review, see Paraskevoudi, Balcı, & Vatakis, 2018). Rather, the wider TBW has also been related to difficulties in ignoring irrelevant signals (e.g., distractors; Couth, Gowen, & Poliakoff, 2016; DeLoss et al., 2013; Mishra & Gazzaley, 2014; Teramoto, Honda, Furuta, & Sekiyama, 2017), deficient decision making when inferring event causality (Condello et al., 2017), and reduced performance in spatial working memory (Wu et al., 2015), balance (Setti et al., 2011), and walking tasks (Brooks et al., 2018; Mahoney et al., 2014; Paraskevoudi et al., 2018; Setti et al., 2011). The widening of the temporal window as a function of age has also been supported by recent electrophysiological evidence showing broader (i.e., higher auditory N1 amplitude responses) and more sustained cortical responses in the elderly across different SOAs as compared to the neural responses observed in their younger counterparts (Basharat et al., 2018). Although the neural substrates underlying these changes remain largely unknown, research suggests that the wider TBW in the elderly may be related to peripheral visual or hearing decline (Diederich et al., 2008; but see Chan et al., 2014a), the age-related decreases in the concentration of the inhibitory neurotransmitter gamma-aminobutyric acid (GABA; Gao et al., 2013; Takayama et al., 1992), which may result in difficulties in inhibiting the binding of erroneous inputs (Gao et al., 2013), or to an age-related general cognitive decline due to structural changes and loss of brain mass (Mozolic, Hugenschmidt, Peiffer, & Laurienti, 2012).

3.5 Perceiving Synchrony in Pathology

The importance of multisensory temporal acuity in accurately scaffolding perception is highlighted by the altered temporal functioning in pathological conditions and its links with clinical symptomatology. Individuals with diagnoses of autism spectrum disorder (ASD; Balz et al., 2016; Foss-Feig et al., 2010; Foxe et al., 2015; Kwakye et al., 2010; Noel, De Niear, et al., 2017; Noel, Lytle, et al., 2017; Noel et al., 2018; Stevenson et al., 2014) and SZ (Martin et al., 2013; Stevenson, Park, et al., 2017; Su et al., 2015; Vatakis & Bakou, 2015; Zhou et al., 2018) exhibit atypically wide TBWs, with studies also pointing to a close link between temporal binding and clinical symptoms in both groups (i.e., autistic traits and proneness to hallucinations, respectively; Noel, De Niear, et al., 2017; Noel, Lytle, et al., 2017; Stevenson, Park, et al., 2017; Stevenson, Segers, Ferber, Barense, & Wallace, 2016). Yet, although ASD and SZ cohorts exhibit similar atypicalities in multisensory synchrony judgments across a range of temporal asynchronies (for a review, see Zhou et al., 2018), recent data points to distinct causes behind these deviations from optimal synchrony perception (Balz et al., 2016). That is, although both the ASD and SZ groups exhibit larger TBWs compared to healthy controls, the two clinical populations differ in their degree of sensory noise (i.e., larger in SZ than ASD; Balz et al., 2016). Future work is, though, required to replicate these findings and apply them in training protocols that may be able to target the distinct causes behind the atypicalities of multisensory temporal perception in ASD and SZ.

4 Recalibrating the Window of Perceptual Synchrony

Adapting to the temporal asynchronies in our environment is of great ecological importance in order to deal with lags in the arrival time of the different sensory signals. Over the last years, research has been devoted on examining how the brain is able to adaptively recalibrate the PSS when exposed to temporally offset sensory inputs in order to adapt to crossmodal asynchronies and appropriately bind signals with a common source (Jicol, Proulx, Pollick, & Petrini, 2018; Lange, Kapala, Krause, Baumgarten, & Schnitzler, 2018; Simon, Nidiffer, & Wallace, 2018; Van der Burg et al., 2013; Van der Burg & Goodbourn, 2015). This adaptation phenomenon is widely known as recalibration and results in a shift of the PSS toward the modality that was the leading one during the adaptation phase (i.e., shift toward auditory-leading PSS if overexposed to auditory-leading asynchronies).

Increasing evidence shows that temporal recalibration can be induced even by a single trial (i.e., rapid recalibration; Lange et al., 2018; Simon et al., 2018; Van der Burg et al., 2013; Van der Burg et al., 2015; Van der Burg & Goodbourn, 2015). This rapid recalibration was initially considered as being limited to audiovisual combinations (Alais, Ho, Han, & Van der Burg, 2017; Van der Burg & Goodbourn, 2015) and modulated by the recent sensory history (as indexed by differences in the magnitude of the evoked responses depending on the temporal ordering of the stimulus on the previous trial; Simon, Noel, & Wallace, 2017; Simon et al., 2018). Yet, a recent study demonstrated that this adaptation may be also evident for visual-tactile stimuli, albeit dependent on the SOA (Lange et al., 2018) rather than the temporal ordering (Simon et al., 2017) of the preceding trial. Specifically, rapid recalibration for visual-tactile stimuli can occur when the SOA in the preceding, recalibrating trial is lower than 100 ms, but not with SOAs larger than or equal to 100 ms (Lange et al., 2018), presumably due to the shorter time scales reported for integration windows for tactile stimuli (i.e., approx. 50 ms; Baumgarten et al., 2015; Baumgarten et al., 2017). In contrast, the critical SOA for audiovisual recalibration is longer (i.e., equal to or longer than 100 ms; Van der Burg et al., 2013), consistent with the larger temporal integration windows for audiovisual integration (Cecere et al., 2015; Noel et al., 2015; van Wassenhove et al., 2007). Importantly, the effects of SOA on temporal recalibration for different modalities seem to be consistent with evidence implicating different frequency bands as the neural substrate for temporal perception for the visual (alpha frequency with cycle lengths of approx. 100 ms; Benedetto et al., 2018; Cecere et al., 2015; Landau & Fries, 2012; Ronconi et al., 2017; VanRullen, 2016) and tactile (beta-band with cycle lengths of approx. 40–80 ms; Baumgarten et al., 2015; Baumgarten et al., 2017) modalities and suggests that some sensory systems may achieve intersensory synchrony more closely anchored to the physical timing (i.e., SOAs closer to 0) of the pairings (e.g., pairings with a tactile constituent; Noel et al., 2015), while others may be more susceptible to the recent sensory experiences (e.g., audiovisual pairings; Noel et al., 2015; Simon et al., 2018). However, further electrophysiological studies are needed to investigate in more detail the process of dynamically weighting the incoming inputs based on recent sensory history (Noel et al., 2015; Simon et al., 2018) and test the hypothesis that adaptation is contingent on whether the adaptation stimulus triggers the frequency relevant for temporal perception (Simon et al., 2017; Simon et al., 2018).

Rapid recalibration has also been found to be affected by factors related to experimental manipulations, such as the spatial and temporal separation between the sensory signals, the experimental paradigm, and the individual's age. Navarra et al. (2013) assessed whether rapid recalibration can be observed between largely asynchronous signals appearing from clearly different spatial positions in both an SJ and a TOJ task. They obtained a significant temporal shift of the PSS, but only for visual-leading asynchronies in the adaptation phase, in line with evidence reporting the brain's tendency to realign the processing time of visual and auditory signals particularly when the visual stimulus precedes the auditory one (Benedetto et al., 2018; Bhat et al., 2015; Eg et al., 2015; Eg & Behne, 2015; Magnotti et al., 2013; Simon & Wallace, 2017; Stevenson et al., 2013). Critically, this shift was also obtained for stimuli presented at different spatial positions, suggesting that spatial separation yields no significant effects on temporal recalibration. Yet, the recalibration effect was larger in participants that performed SJs (i.e., 37 ms) as compared to those that performed TOJs (i.e., 14 ms), in line with the proposed dissociation between simultaneity and temporal order judgments (Barnett-Cowan & Harris, 2009, 2011; Basharat et al., 2018; Bedard & Barnett-Cowan, 2016; Chan et al., 2014a; De Niear et al., 2016; Love et al., 2013; Mitrani et al., 1986; Van Eijk et al., 2008; Vatakis et al., 2008). Additionally, as with synchrony perception, temporal recalibration undergoes significant changes with age (Chan et al., 2014b). Specifically, although both younger and older adults exhibit an expansion of the synchrony window in the direction of the adapted asynchrony (i.e., sound-lag asynchrony), the degree of expansion has been found to be smaller for the elderly (Chan et al., 2014b). Interestingly, the adaptation effects were not related to the window width in either age group, suggesting that the smaller recalibration effects in the elderly cannot be attributed to age-related differences in the perception of the adaptor stimulus (Chan et al., 2014b).

Although research on recalibration has been mainly focusing on the rapid shift in PSS that can be induced after a few trials, it has recently been shown that rapid recalibration differs from the prolonged and long-term recalibration that can result from training paradigms (De Niear, Noel, & Wallace, 2017; Jicol et al., 2018; Van der Burg et al., 2015; Van der Burg & Goodbourn, 2015). On the one hand, rapid recalibration seems to be an early and fast sensory process (Van der Burg et al., 2013) that occurs almost instantaneously, even after a single exposure to an asynchronous multisensory event, whereas, the prolonged recalibration has been hypothesized to reflect more cognitive processes (Jicol et al., 2018; Van der Burg et al., 2013). Yet, although the two phenomena seem to be partly independent, it has recently been proposed that higher-level cognitive processes (e.g., perceived agency; Rohde & Ernst, 2012, 2016) or long-term training (e.g., music expertise; Jicol et al., 2018) may actually affect rapid recalibration as well.

5 Neural Indicators of Synchrony Perception

The naive view we have about a—seemingly—synchronous world must be reflected in the computations operating at the neural level. Over the last few years, electrophysiological and neuroimaging studies have attempted to elucidate the neural substrates of synchrony perception. Neuroimaging studies support the accumulating psychophysical evidence that dissociates TOJ from SJ (Barnett-Cowan & Harris, 2009, 2011; Basharat et al., 2018; Bedard & Barnett-Cowan, 2016; Chan et al., 2014a; De Niear et al., 2016; Love et al., 2013; Mitrani et al., 1986; Van Eijk et al., 2008; Vatakis et al., 2008) by reporting task-specific activations in distinct areas (Miyazaki et al., 2016). Specifically, temporal order has been suggested to be mediated by activation in the temporal parietal junction, ventral frontal cortices, premotor cortex, cerebellum, and the left temporal and parietal cortices (Adhikari, Goshorn, Lamichhane, & Dhamala, 2013; Binder, 2015; Miyazaki et al., 2016; Takahashi, Kansaku, Wada, Shibuya, & Kitazawa, 2013), thereby partly supporting the view that motor control and TOJs may share the same neural mechanism (Benedetto et al., 2018; Hermosillo, Ritterband-Rosenbaum, & van Donkelaar, 2011). In contrast, SJ-related activation has been found to be limited in the posterior insular cortex (Miyazaki et al., 2016), which has already been implicated as a candidate area underlying the perception of audiovisual simultaneity (Bushara, Grafman, & Hallett, 2001). Collectively, the neuroimaging findings reported to-date not only point to distinct neural mechanisms mediating temporal order and simultaneity judgments, but also support the view that TOJs are more demanding in terms of the resources required and involve more processes than SJs (De Niear et al., 2016; Love et al., 2013; Miyazaki et al., 2016; Yarrow et al., 2014).

Recent electrophysiological studies support the idea that temporal processing is mediated by the synchronized activity of endogenous oscillations (Benedetto et al., 2018; Benedetto & Morrone, 2017; Ronconi et al., 2017; Samaha & Postle, 2015; Simon & Wallace, 2017). Research reports that oscillatory cycles form the neuronal basis of the modality-specific integration windows (Baumgarten et al., 2015; Baumgarten et al., 2017; Cecere et al., 2015; VanRullen, 2016), with the visual and audiovisual windows oscillating within the theta- (~4–7 Hz) and alpha-band (~8–12 Hz; Benedetto et al., 2018; Cecere et al., 2015; Landau & Fries, 2012; Ronconi et al., 2017; VanRullen, 2016), while the beta-band (~13–25 Hz) has been suggested to play a critical role in the somatosensory domain (Baumgarten et al., 2015; Baumgarten et al., 2017). Additional evidence shows a close relationship between gamma-band activity and perceived audiovisual synchrony (Senkowski, Talsma, Grigutsch, Herrmann, & Woldorff, 2007; but see Doesburg, Emberson, Rahi, Cameron, & Ward, 2008, for evidence showing increased synchronization in the gamma range for temporal incongruent relative to congruent continuous audiovisual speech stimuli) and highlights the links between crossmodal interactions and oscillatory synchronization in the delta, theta, and gamma frequencies (Lakatos, Chen, O'Connell, Mills, & Schroeder, 2007), presumably by phase resetting mechanisms that may mediate the effects of signals in one modality on sensory processing in other primary sensory areas (Cecere et al., 2017). Further evidence also supports the coupling of temporal perception and motor rhythms (Benedetto & Morrone, 2017; Benedetto, Spinelli, & Morrone, 2016; Hogendoorn, 2016; Tomassini, Ambrogioni, Medendorp, & Maris, 2017; Tomassini, Spinelli, Jacono, Sandini, & Morrone, 2015; Wutz, Muschter, van Koningsbruggen, Weisz, & Melcher, 2016). In an audiovisual TOJ task, it has recently been found that the PSS (but not the JND) fluctuates in the theta range in synchrony with the execution of a voluntary movement, while most importantly these oscillations precede the action execution by approximately 500 ms (Benedetto et al., 2018). The importance of these low-frequency oscillations in temporal perception and prediction is further highlighted by their role in predicting event timing (Stefanics et al., 2010), integrating temporal information in speech signals (Schroeder, Lakatos, Kajikawa, Partan, & Puce, 2008), as well as by their correlation with perceptual asynchrony (Simon & Wallace, 2017). Collectively, the latter findings suggest that apart from serving as a marker of crossmodal incongruence processing (Simon & Wallace, 2017), theta oscillations seem to operate as predictive priors, thus forming an anticipatory mechanism that drives temporal perception (Benedetto et al., 2018), modulates higher-frequency neuronal synchronization, and facilitates neuronal communication among distant brain areas (Lakatos, Karmos, Mehta, Ulbert, & Schroeder, 2008; Stefanics et al., 2010).

6 Conclusion

Synchrony perception is ubiquitous in creating a unified and coherent representation of the informationally rich and multisensory environment we are living in. The brain's ability to align sensory signals in time has been perplexing researchers working at the intersection between philosophy, psychology, and physics. Inspired by the necessity to understand how the brain deals with lags between the senses, this chapter provided an overview of the factors affecting perceptual synchrony, summarized current research on how short-term exposure to asynchronous signals and training paradigms may modulate our ability to integrate physically asynchronous information into unified percepts, and reviewed the electrophysiological and neuroimaging studies aiming to elucidate the mechanisms underlying our perception of synchrony.

Several conclusions can be drawn based on the research summarized in this chapter. Being highly malleable and adaptive, perceptual synchrony is susceptible to experimental manipulations (e.g., low- and high-level stimulus features, task demands), dependent on the individual's age, prior multisensory experience, and temporal training (e.g., music expertise), sensitive to exposure to intersensory asynchronies (e.g., temporal recalibration), and abnormal in some pathological conditions (e.g., autism, SZ). Research highlights the brain's flexibility to integrate physically asynchronous signals into synchronous and temporally coherent percepts by shifting one sensory signal (typically vision) toward the other, in a way reminiscent of temporal ventriloquism (Keetels & Vroomen, 2012; Vatakis, 2013). Additionally, temporal recalibration studies and perceptual learning protocols have consistently shown that instead of being a static and unchanging process, perceived synchrony can be altered following short- (exposure to asynchronous sensory events or perceptual learning protocols; De Niear et al., 2016; De Niear et al., 2017; Stevenson et al., 2013) and long-term (e.g., music expertise; Jicol et al., 2018; Proverbio, Attardo, Cozzi, & Zani, 2015) training. Collectively, the research conducted so far highlights the inherent plasticity of the TBW, by demonstrating that perceptual training can facilitate asynchrony detection in visuomotor tasks (Rohde & Ernst, 2016), enhance temporal acuity for both simple (flashes and beeps) and complex stimuli (audiovisual speech; De Niear et al., 2016; Setti et al., 2014), and boost older adults' performance in tasks employing audiovisual illusions (Setti et al., 2014). The latter findings, along with evidence reporting correlations between synchrony perception and speech intelligibility (Conrey & Pisoni, 2006) or audiovisual speech integration (Stevenson et al., 2012), raise the possibility of improving temporal acuity via training paradigms. Thus, introducing recalibration or perceptual training protocols as a rehabilitative technique may contribute in fine-tuning multisensory temporal perception and audiovisual speech integration in clinical groups, preventing decision-making errors, and facilitating motor control (Balz et al., 2016; Chan et al., 2014a; Foss-Feig et al., 2010; Merriman, Whyatt, Setti, Craig, & Newell, 2015; Noel, Lytle, et al., 2017; Setti et al., 2014; Stevenson et al., 2012; Stevenson et al., 2014; Stevenson, Park, et al., 2017; Stevenson, Segers, et al., 2017; Stevenson, Toulmin, et al., 2017; Turi, Karaminis, Pellicano, & Burr, 2016).

Several questions are, though, still open for investigation in order to obtain a more comprehensive and thorough understanding of the behavioral and neural correlates of synchrony perception. Evidence remains mixed as concerns the differences between simultaneity and temporal order judgments and the processing stage at which synchrony perception emerges. Preliminary findings suggest that the performance differences in the two tasks may arise either from multisensory integration or higher-level decision-making processes that emerge relatively late in time (Basharat et al., 2018; Simon & Wallace, 2017; Simon et al., 2018; for a review on the distinct mechanisms between integration and synchrony, see Tsilionis & Vatakis, 2016). Additionally, the neural mechanisms that underlie the humans' ability to reach perceptual synchrony are of great importance for future research. Given the proposed links between synchrony perception and oscillatory activity (Benedetto et al., 2018; Benedetto & Morrone, 2017; Ronconi et al., 2017; Samaha & Postle, 2015; Simon & Wallace, 2017), electrophysiological studies may be critical in examining the neural computations that mediate the vivid, albeit illusory, perception of a synchronous world. For example, it has been suggested that measuring the changes in response amplitude, phase locking, and inter-trial coherence may be the first step so as to elucidate the neural processes that give rise to the perception of synchronous events (Dean et al., 2017; Perrodin, Kayser, Logothetis, & Petkov, 2015). Modeling studies would also contribute in examining the mechanisms behind human temporal perception. Recent attempts provide promising results by showing that biologically plausible models based on Bayesian inference can account for the temporal aspects of multisensory integration (e.g., the temporal ventriloquism effect and the percept of unified events as a function of the auditory-visual spatial distance), the inter-subject variability reported in behavioral studies (Stone et al., 2001), and the effects of prior multisensory experience and experimental paradigm on synchrony perception (Cuppini, Shams, Magosso, & Ursino, 2017; Ursino, Cuppini, & Magosso, 2017).

For humans, synchrony is a fundamental, vivid, and essential experience. How the human perceptual system achieves perceptual synchrony in an inherently asynchronous world remains one of the most interesting, yet puzzling, questions within the field of time perception. The increasing research interest in shedding light on the relationship between physical, neural, and perceptual synchrony creates a promising research landscape for unraveling the mysteries behind temporal perception.

References

Adhikari, B. M., Goshorn, E. S., Lamichhane, B., & Dhamala, M. (2013). Temporal-order judgment of audiovisual events involves network activity between parietal and prefrontal cortices. *Brain Connectivity, 3*(5), 536–545. https://doi.org/10.1089/brain.2013.0163

Alais, D., Ho, T., Han, S., & Van der Burg, E. (2017). A matched comparison across three different sensory pairs of cross-modal temporal recalibration from sustained and transient adaptation. *I-Perception, 8*(4), 2041669517718697. https://doi.org/10.1177/2041669517718697

Balz, J., Roa Romero, Y., Keil, J., Krebber, M., Niedeggen, M., Gallinat, J., & Senkowski, D. (2016). Beta/Gamma oscillations and event-related potentials indicate aberrant multisensory processing in schizophrenia. *Frontiers in Psychology, 7*, 1896. https://doi.org/10.3389/fpsyg.2016.01896

Barnett-Cowan, M., & Harris, L. R. (2009). Perceived timing of vestibular stimulation relative to touch, light and sound. *Experimental Brain Research, 198*(2–3), 221–231. https://doi.org/10.1007/s00221-009-1779-4

Barnett-Cowan, M., & Harris, L. R. (2011). Temporal processing of active and passive head movement. *Experimental Brain Research, 214*(1), 27–35. https://doi.org/10.1007/s00221-011-2802-0

Basharat, A., Adams, M. S., Staines, W. R., & Barnett-Cowan, M. (2018). Simultaneity and temporal order judgments are coded differently and change with age: An event-related potential study. *Frontiers in Integrative Neuroscience, 12*, 15. https://doi.org/10.3389/fnint.2018.00015

Baumann, O., Vromen, J. M. G., Cheung, A., McFadyen, J., Ren, Y., & Guo, C. C. (2018). Neural correlates of temporal complexity and synchrony during audiovisual correspondence detection. *eNeuro, 5*(1). https://doi.org/10.1523/ENEURO.0294-17.2018

Baumgarten, T. J., Königs, S., Schnitzler, A., & Lange, J. (2017). Subliminal stimuli modulate somatosensory perception rhythmically and provide evidence for discrete perception. *Scientific Reports, 7*, 43937. https://doi.org/10.1038/srep43937

Baumgarten, T. J., Schnitzler, A., & Lange, J. (2015). Beta oscillations define discrete perceptual cycles in the somatosensory domain. *Proceedings of the National Academy of Sciences of the United States of America, 112*(39), 12187–12192. https://doi.org/10.1073/pnas.1501438112

Bedard, G., & Barnett-Cowan, M. (2016). Impaired timing of audiovisual events in the elderly. *Experimental Brain Research, 234*(1), 331–340. https://doi.org/10.1007/s00221-015-4466-7

Benedetto, A., Burr, D. C., & Morrone, M. C. (2018). Perceptual oscillation of audiovisual time simultaneity. *eNeuro, 5*(3). https://doi.org/10.1523/ENEURO.0047-18.2018

Benedetto, A., & Morrone, M. C. (2017). Saccadic suppression is embedded within extended oscillatory modulation of sensitivity. *Journal of Neuroscience, 37*(13), 3661–3670. https://doi.org/10.1523/JNEUROSCI.2390-16.2016

Benedetto, A., Spinelli, D., & Morrone, M. C. (2016). Rhythmic modulation of visual contrast discrimination triggered by action. *Proceedings of the Royal Society B: Biological Sciences, 283*(1831), 20160692. https://doi.org/10.1098/rspb.2016.0692

Bhat, J., Miller, L. M., Pitt, M. A., & Shahin, A. J. (2015). Putative mechanisms mediating tolerance for audiovisual stimulus onset asynchrony. *Journal of Neurophysiology, 113*(5), 1437–1450. https://doi.org/10.1152/jn.00200.2014

Binder, M. (2015). Neural correlates of audiovisual temporal processing—Comparison of temporal order and simultaneity judgments. *Neuroscience, 300*, 432–447. https://doi.org/10.1016/j.neuroscience.2015.05.011

Brooks, C. J., Chan, Y. M., Anderson, A. J., & McKendrick, A. M. (2018). Audiovisual temporal perception in aging: The role of multisensory integration and age-related sensory loss. *Frontiers in Human Neuroscience, 12*, 192. https://doi.org/10.3389/fnhum.2018.00192

Bushara, K. O., Grafman, J., & Hallett, M. (2001). Neural correlates of auditory-visual stimulus onset asynchrony detection. *Journal of Neuroscience, 21*(1), 300–304. https://doi.org/10.1523/JNEUROSCI.21-01-00300.2001

Cecere, R., Gross, J., Willis, A., & Thut, G. (2017). Being first matters: Topographical representational similarity analysis of ERP signals reveals separate networks for audiovisual temporal binding depending on the leading sense. *Journal of Neuroscience, 37*(21), 5274–5287. https://doi.org/10.1523/JNEUROSCI.2926-16.2017

Cecere, R., Rees, G., & Romei, V. (2015). Individual differences in alpha frequency drive crossmodal illusory perception. *Current Biology, 25*(2), 231–235. https://doi.org/10.1016/j.cub.2014.11.034

Cecere, R., Romei, V., Bertini, C., & Làdavas, E. (2014). Crossmodal enhancement of visual orientation discrimination by looming sounds requires functional activation of primary visual areas: A case study. *Neuropsychologia, 56*, 350–358. https://doi.org/10.1016/j.neuropsychologia.2014.02.008

Chan, Y. M., Pianta, M. J., & McKendrick, A. M. (2014a). Older age results in difficulties separating auditory and visual signals in time. *Journal of Vision, 14*(11), 1–11. https://doi.org/10.1167/14.11.13

Chan, Y. M., Pianta, M. J., & McKendrick, A. M. (2014b). Reduced audiovisual recalibration in the elderly. *Frontiers in Aging Neuroscience, 6*, 226. https://doi.org/10.3389/fnagi.2014.00226

Chen, Y. C., Lewis, T. L., Shore, D. I., Spence, C., & Maurer, D. (2018). Developmental changes in the perception of visuotactile simultaneity. *Journal of Experimental Child Psychology, 173*, 304–317. https://doi.org/10.1016/j.jecp.2018.04.014

Condello, G., Forte, R., Falbo, S., Shea, J. B., Di Baldassarre, A., Capranica, L., & Pesce, C. (2017). Steps to health in cognitive aging: Effects of physical activity on spatial attention and executive control in the elderly. *Frontiers in Human Neuroscience, 11*, 107. https://doi.org/10.3389/fnhum.2017.00107

Conrey, B., & Pisoni, D. B. (2006). Auditory-visual speech perception and synchrony detection for speech and nonspeech signals. *The Journal of the Acoustical Society of America, 119*(6), 4065–4073. https://doi.org/10.1121/1.2195091

Couth, S., Gowen, E., & Poliakoff, E. (2016). Investigating the spatial and temporal modulation of visuotactile interactions in older adults. *Experimental Brain Research, 234*(5), 1233–1248. https://doi.org/10.1007/s00221-015-4431-5

Cuppini, C., Shams, L., Magosso, E., & Ursino, M. (2017). A biologically inspired neurocomputational model for audiovisual integration and causal inference. *Computational Neuroscience, 46*(9), 2481–2498. https://doi.org/10.1111/ejn.13725

de Boer-Schellekens, L., & Vroomen, J. (2014). Multisensory integration compensates loss of sensitivity of visual temporal order in the elderly. *Experimental Brain Research, 232*(1), 253–262. https://doi.org/10.1007/s00221-013-3736-5

de Dieuleveult, A. L., Siemonsma, P. C., van Erp, J. B. F., & Brouwer, A.-M. (2017). Effects of aging in multisensory integration: A systematic review. *Frontiers in Aging Neuroscience, 9*, 80. https://doi.org/10.3389/fnagi.2017.00080

De Niear, M. A., Gupta, P. B., Baum, S. H., & Wallace, M. T. (2018). Perceptual training enhances temporal acuity for multisensory speech. *Neurobiology of Learning and Memory, 147*, 9–17. https://doi.org/10.1016/j.nlm.2017.10.016

De Niear, M. A., Koo, B., & Wallace, M. T. (2016). Multisensory perceptual learning is dependent upon task difficulty. *Experimental Brain Research, 234*(11), 3269–3277. https://doi.org/10.1007/s00221-016-4724-3

De Niear, M. A., Noel, J.-P., & Wallace, M. T. (2017). The impact of feedback on the different time courses of multisensory temporal recalibration. *Neural Plasticity, 2017*, 3478742. https://doi.org/10.1155/2017/3478742

Dean, C. L., Eggleston, B. A., Gibney, K. D., Aligbe, E., Blackwell, M., & Kwakye, L. D. (2017). Auditory and visual distractors disrupt multisensory temporal acuity in the crossmodal temporal order judgment task. *PLoS One, 12*(7), e0179564. https://doi.org/10.1371/journal.pone.0179564

DeLoss, D. J., Pierce, R. S., & Andersen, G. J. (2013). Multisensory integration, aging, and the sound-induced flash illusion. *Psychology and Aging, 28*(3), 802–812. https://doi.org/10.1037/a0033289

Diederich, A., Colonius, H., & Schomburg, A. (2008). Assessing age-related multisensory enhancement with the time-window-of-integration model. *Neuropsychologia, 46*(10), 2556–2562. https://doi.org/10.1016/j.neuropsychologia.2008.03.026

Doesburg, S. M., Emberson, L. L., Rahi, A., Cameron, D., & Ward, L. M. (2008). Asynchrony from synchrony: Long-range gamma-band neural synchrony accompanies perception of audiovisual speech asynchrony.

Experimental Brain Research, 185(1), 11–20. https://doi.org/10.1007/s00221-007-1127-5

Efron, R. (1963). The effect of handedness on the perception of simultaneity and temporal order. *Brain, 86*(2), 261. https://doi.org/10.1093/brain/86.2.261

Eg, R., & Behne, D. M. (2015). Perceived synchrony for realistic and dynamic audiovisual events. *Frontiers in Psychology, 6*, 736. https://doi.org/10.3389/fpsyg.2015.00736

Eg, R., Griwodz, C., Halvorsen, P., & Behne, D. (2015). Audiovisual robustness: Exploring perceptual tolerance to asynchrony and quality distortion. *Multimedia Tools and Applications, 74*(2), 345–365. https://doi.org/10.1007/s11042-014-2136-6

Foss-Feig, J. H., Kwakye, L. D., Cascio, C. J., Burnette, C. P., Kadivar, H., Stone, W. L., & Wallace, M. T. (2010). An extended multisensory temporal binding window in autism spectrum disorders. *Experimental Brain Research, 203*(2), 381–389. https://doi.org/10.1007/s00221-010-2240-4

Foxe, J. J., Molholm, S., Del Bene, V. A., Frey, H.-P., Russo, N. N., Blanco, D., … Ross, L. A. (2015). Multisensory speech integration deficits in high-functioning school-aged children with autism spectrum disorder (ASD) and their resolution during early adolescence. *Cerebral Cortex, 25*(2), 298–312. https://doi.org/10.1093/cercor/bht213

Fujisaki, W., & Nishida, S. (2009). Audio-tactile superiority over visuo-tactile and audio-visual combinations in the temporal resolution of synchrony perception. *Experimental Brain Research, 198*(2–3), 245–259. https://doi.org/10.1007/s00221-009-1870-x

Gao, F., Edden, R. A. E., Li, M., Puts, N. A. J., Wang, G., Liu, C., … Barker, P. B. (2013). Edited magnetic resonance spectroscopy detects an age-related decline in brain GABA levels. *NeuroImage, 78*, 75–82. https://doi.org/10.1016/j.neuroimage.2013.04.012

Gori, M. (2015). Multisensory integration and calibration in children and adults with and without sensory and motor disabilities. *Multisensory Research, 28*(1–2), 71–99. https://doi.org/10.1163/22134808-00002478

Gotow, N., & Kobayakawa, T. (2017). Simultaneity judgment using olfactory-visual, visual-gustatory, and olfactory–gustatory combinations. *PLoS ONE, 12*(4), e0174958. https://doi.org/10.1371/journal.pone.0174958

Halliday, A., & Mingay, R. (1964). On the resolution of small time intervals and the effect of conduction delays on the judgement of simultaneity. *Quarterly Journal of Experimental Psychology, 16*(1), 37–41. https://doi.org/10.1080/17470216408416344

Harrar, V., & Harris, L. (2008). The effect of exposure to asynchronous audio, visual, and tactile stimulus combinations on the perception of simultaneity. *Experimental Brain Research, 186*(4), 517–524. https://doi.org/10.1007/s00221-007-1253-0

Hermosillo, R., Ritterband-Rosenbaum, A., & van Donkelaar, P. (2011). Predicting future sensorimotor states influences current temporal decision making. *Journal of Neuroscience, 31*(27), 10019–10022. https://doi.org/10.1523/JNEUROSCI.0037-11.2011

Hillock, A. R., Powers, A. R., & Wallace, M. T. (2011). Binding of sights and sounds: Age-related changes in multisensory temporal processing. *Neuropsychologia, 49*(3), 461–467. https://doi.org/10.1016/j.neuropsychologia.2010.11.041

Hillock-Dunn, A., & Wallace, M. T. (2012). Developmental changes in the multisensory temporal binding window persist into adolescence. *Developmental Science, 15*(5), 688–696. https://doi.org/10.1111/j.1467-7687.2012.01171.x

Hogendoorn, H. (2016). Voluntary saccadic eye movements ride the attentional rhythm. *Journal of Cognitive Neuroscience, 28*(10), 1625–1635. https://doi.org/10.1162/jocn_a_00986

Hoover, A. E. N., & Harris, L. R. (2012). Detecting delay in visual feedback of an action as a monitor of self recognition. *Experimental Brain Research, 222*(4), 389–397. https://doi.org/10.1007/s00221-012-3224-3

Hoover, A. E. N., & Harris, L. R. (2016). Inducing ownership over an "other" perspective with a visuo-tactile manipulation perception. *Experimental Brain Research, 234*(12), 3633–3639. https://doi.org/10.1007/s00221-016-4760-z

Jaskowski, P. (1999). Reaction time and temporal-order judgment as measures of perceptual latency: The problem of dissociations. In G. Aschersleben, T. Bachmann, & J. Musseler (Eds.), *Cognitive contributions to the perception of spatial and temporal events* (pp. 265–282). North-Holland: Elsevier Science B.V.

Jicol, C., Proulx, M. J., Pollick, F. E., & Petrini, K. (2018). Long-term music training modulates the recalibration of audiovisual simultaneity. *Experimental Brain Research, 236*(7), 1869–1880. https://doi.org/10.1007/s00221-018-5269-4

Keetels, M., & Vroomen, J. (2012). Perception of synchrony between the senses. In M. M. Murray & M. T. Wallace (Eds.), *The neural bases of multisensory processes* (pp. 1–27). Boca Raton, FL: CRC Press.

Keys, R. T., Rich, A. N., & Zopf, R. (2018). Multisensory temporal processing in own-body contexts: Plausibility of hand ownership does not improve visuo-tactile asynchrony detection. *Experimental Brain Research, 236*(5), 1431–1443. https://doi.org/10.1007/s00221-018-5232-4

Kostaki, M., & Vatakis, A. (2018). Temporal order and synchrony judgments: A primer for students. In A. Vatakis, F. Balci, M. Di Luca, & A. Correa (Eds.), *Timing and time perception: Procedures, measures, and applications* (pp. 233–262). Leiden: Brill.

Krueger Fister, J., Stevenson, R. A., Nidiffer, A. R., Barnett, Z. P., & Wallace, M. T. (2016). Stimulus intensity modulates multisensory temporal processing. *Neuropsychologia, 88*, 92–100. https://doi.org/10.1016/j.neuropsychologia.2016.02.016

Kwakye, L. D., Foss-Feig, J. H., Cascio, C. J., Stone, W. L., & Wallace, M. T. (2010). Altered auditory and multisensory temporal processing in autism spectrum disorders. *Frontiers in Integrative Neuroscience, 4*, 129. https://doi.org/10.3389/fnint.2010.00129

Lakatos, P., Chen, C. M., O'Connell, M., Mills, A., & Schroeder, C. E. (2007). Neuronal oscillations and multisensory interaction in primary auditory cortex. *Neuron, 53*(2), 279–292. https://doi.org/10.1016/j.neuron.2006.12.011

Lakatos, P., Karmos, G., Mehta, A. D., Ulbert, I., & Schroeder, C. E. (2008). Entrainment of neuronal oscillations as a mechanism of attentional selection. *Science, 320*(5872), 110–113. https://doi.org/10.1126/science.1154735

Landau, A. N., & Fries, P. (2012). Attention samples stimuli rhythmically. *Current Biology, 22*(11), 1000–1004. https://doi.org/10.1016/j.cub.2012.03.054

Lange, J., Kapala, K., Krause, H., Baumgarten, T. J., & Schnitzler, A. (2018). Rapid temporal recalibration to visuo-tactile stimuli. *Experimental Brain Research, 236*(2), 347–354. https://doi.org/10.1007/s00221-017-5132-z

Laurienti, P. J., Burdette, J. H., Maldjian, J. A., & Wallace, M. T. (2006). Enhanced multisensory integration in older adults. *Neurobiology of Aging, 27*(8), 1155–1163. https://doi.org/10.1016/j.neurobiolaging.2005.05.024

Love, S. A., Petrini, K., Cheng, A., & Pollick, F. E. (2013). A psychophysical investigation of differences between synchrony and temporal order judgments. *PLoS One, 8*(1), e54798. https://doi.org/10.1371/journal.pone.0054798

Magnotti, J. F., Ma, W. J., & Beauchamp, M. S. (2013). Causal inference of asynchronous audiovisual speech. *Frontiers in Psychology, 4*, 798. https://doi.org/10.3389/fpsyg.2013.00798

Mahoney, J. R., Holtzer, R., & Verghese, J. (2014). Visual-somatosensory integration and balance: Evidence for psychophysical integrative differences in aging. *Multisensory Research, 27*(1), 17–42. https://doi.org/10.1163/22134808-00002444

Martin, B., Giersch, A., Huron, C., & van Wassenhove, V. (2013). Temporal event structure and timing in schizophrenia: Preserved binding in a longer now. *Neuropsychologia, 51*(2), 358–371. https://doi.org/10.1016/j.neuropsychologia.2012.07.002

McGovern, D. P., Roudaia, E., Stapleton, J., McGinnity, T. M., & Newell, F. N. (2014). The sound-induced flash illusion reveals dissociable age-related effects in multisensory integration. *Frontiers in Aging Neuroscience, 6*, 250. https://doi.org/10.3389/fnagi.2014.00250

Merriman, N. A., Whyatt, C., Setti, A., Craig, C., & Newell, F. N. (2015). Successful balance training is associated with improved multisensory function in fall-prone older adults. *Computers in Human Behavior, 45*, 192–203. https://doi.org/10.1016/j.chb.2014.12.017

Mishra, J., & Gazzaley, A. (2014). Attentional control of multisensory integration is preserved in aging. In G. R. Mangun (Ed.), *Cognitive electrophysiology of attention. Signals of the mind* (pp. 190–203). San Diego, CA: Academic Press.

Mitrani, L., Shekerdjiiski, S., & Yakimoff, N. (1986). Mechanisms and asymmetries in visual perception of simultaneity and temporal order. *Biological Cybernetics, 54*(3), 159–165.

Miyazaki, M., Kadota, H., Matsuzaki, K. S., Takeuchi, S., Sekiguchi, H., Aoyama, T., & Kochiyama, T. (2016). Dissociating the neural correlates of tactile temporal order and simultaneity judgements. *Scientific Reports, 6*, 23323. https://doi.org/10.1038/srep23323

Mozolic, J. L., Hugenschmidt, C. E., Peiffer, A. M., & Laurienti, P. J. (2012). Multisensory integration and aging. In M. Murray & M. Wallace (Eds.), *The neural bases of multisensory processes* (pp. 381–392). Boca Raton, FL: CRC Press.

Murray, M. M., & Spierer, L. (2009). Auditory spatio-temporal brain dynamics and their consequences for multisensory interactions in humans. *Hearing Research, 258*(1–2), 121–133. https://doi.org/10.1016/j.heares.2009.04.022

Musacchia, G., & Schroeder, C. E. (2009). Neuronal mechanisms, response dynamics and perceptual functions of multisensory interactions in auditory cortex. *Hearing Research, 258*(1–2), 72–79. https://doi.org/10.1016/j.heares.2009.06.018

Navarra, J., Fernández-Prieto, I., & Garcia-Morera, J. (2013). Realigning thunder and lightning: Temporal adaptation to spatiotemporally distant events. *PLoS One, 8*(12), e84278. https://doi.org/10.1371/journal.pone.0084278

Navarra, J., Vatakis, A., Zampini, M., Soto-Faraco, S., Humphreys, W., & Spence, C. (2005). Exposure to asynchronous audiovisual speech extends the temporal window for audiovisual integration. *Cognitive Brain Research, 25*(2), 499–507. https://doi.org/10.1016/j.cogbrainres.2005.07.009

Neumann, O., & Niepel, M. (2004). Timing of "perception" and perception of "time". In C. Kaernbach, E. Schroger, & H. Muller (Eds.), *Psychophysics beyond sensation: Laws and invariants of human cognition* (pp. 245–270). Hillsdale, NJ: Lawrence Erlbaum Associates, Inc.

Noel, J. P., De Niear, M., Lazzara, N., & Wallace, M. T. (2017). Uncoupling between multisensory temporal function and non-verbal turn-taking in Autism Spectrum Disorder. *IEEE Transactions on Cognition and Developmental Systems, 10*, 973–982. https://doi.org/10.1109/TCDS.2017.2778141

Noel, J. P., Lytle, M., Cascio, C., & Wallace, M. T. (2017). Disrupted integration of exteroceptive and interoceptive signaling in autism spectrum disorder. *Autism Research, 11*(1), 194–205. https://doi.org/10.1002/aur.1880

Noel, J.-P., Stevenson, R. A., & Wallace, M. T. (2018). Atypical audiovisual temporal function in autism and schizophrenia: Similar phenotype, different cause. *European Journal of Neuroscience, 47*(10), 1230–1241. https://doi.org/10.1111/ejn.13911

Noel, J.-P., Wallace, M. T., Orchard-Mills, E., Alais, D., & Van der Burg, E. (2015). True and perceived synchrony are preferentially associated with particular sensory pairings. *Scientific Reports, 5*, 17467. https://doi.org/10.1038/srep17467

Paraskevoudi, N., Balcı, F., & Vatakis, A. (2018). "Walking" through the sensory, cognitive, and temporal degradations of healthy aging. *Annals of the New York Academy of Sciences, 1426*(1), 72–92. https://doi.org/10.1111/nyas.13734

Peiffer, A. M., Mozolic, J. L., Hugenschmidt, C. E., & Laurienti, P. J. (2007). Age-related multisensory enhancement in a simple audiovisual detection task. *NeuroReport, 18*(10), 1077–1081. https://doi.org/10.1097/WNR.0b013e3281e72ae7

Perrodin, C., Kayser, C., Logothetis, N. K., & Petkov, C. I. (2015). Natural asynchronies in audiovisual communication signals regulate neuronal multisensory interactions in voice-sensitive cortex. *Proceedings of the National Academy of Sciences of the United States of America, 112*(1), 273–278. https://doi.org/10.1073/pnas.1412817112

Pöppel, E. (1988). *Mindworks. Time and conscious experience*. San Diego: Harcourt Brace Jovanovich.

Pöppel, E. (2004). Lost in time: A historical frame, elementary processing units and the 3-second window. *Acta Neurobiologiae Experimentalis, 64*(3), 295–301.

Proverbio, A. M., Attardo, L., Cozzi, M., & Zani, A. (2015). The effect of musical practice on gesture/sound pairing. *Frontiers in Psychology, 6*, 376. https://doi.org/10.3389/fpsyg.2015.00376

Ren, Y., Yang, W., Nakahashi, K., Takahashi, S., & Wu, J. (2017). Audiovisual integration delayed by stimulus onset asynchrony between auditory and visual stimuli in older adults. *Perception, 6*(2), 205–218. https://doi.org/10.1177/0301006616673850

Roefs, J. A. J. (1963). Perception lag as a function of stimulus luminance. *Vision Research, 3*, 81–91. https://doi.org/10.1016/0042-6989(63)90070-1

Rohde, M., & Ernst, M. O. (2012). To lead and to lag—Forward and backward recalibration of perceived visuo-motor simultaneity. *Frontiers in Psychology, 3*, 599. https://doi.org/10.3389/fpsyg.2012.00599

Rohde, M., & Ernst, M. O. (2016). Time, agency, and sensory feedback delays during action. *Current Opinion in Behavioral Sciences, 8*, 193–199. https://doi.org/10.1016/j.cobeha.2016.02.029

Ronconi, L., Oosterhof, N. N., Bonmassar, C., & Melcher, D. (2017). Multiple oscillatory rhythms determine the temporal organization of perception. *Proceeding of the National Academy of Sciences of the United States of America, 114*, 13435–13440. https://doi.org/10.1073/pnas.1714522114

Roy, C., Dalla Bella, S., & Lagarde, J. (2017). To bridge or not to bridge the multisensory time gap: Bimanual coordination to sound and touch with temporal lags. *Experimental Brain Research, 235*(1), 135–151. https://doi.org/10.1007/s00221-016-4776-4

Samaha, J., & Postle, B. R. (2015). The speed of alpha-band oscillations predicts the temporal resolution of visual perception. *Current Biology, 25*(2), 2985–2990. https://doi.org/10.1016/j.cub.2015.10.007

Sanford, A. J. (1971). Effects of changes in the intensity of white noise on simultaneity judgements and simple reaction time. *Quarterly Journal of Experimental Psychology, 23*(3), 296–303. https://doi.org/10.1080/14640746908401824

Schroeder, C. E., Lakatos, P., Kajikawa, Y., Partan, S., & Puce, A. (2008). Neuronal oscillations and visual amplification of speech. *Trends in Cognitive Sciences, 12*(3), 106–113. https://doi.org/10.1016/j.tics.2008.01.002

Schwartz, J. L., & Savariaux, C. (2014). No, there is no 150 ms lead of visual speech on auditory speech, but a range of audiovisual asynchronies varying from small audio lead to large audio lag. *PLoS Computational Biology, 10*(7), e1003743. https://doi.org/10.1371/journal.pcbi.1003743

Senkowski, D., Talsma, D., Grigutsch, M., Herrmann, C. S., & Woldorff, M. G. (2007). Good times for multisensory integration: Effects of the precision of temporal synchrony as revealed by gamma-band oscillations. *Neuropsychologia, 45*(3), 561–571. https://doi.org/10.1016/j.neuropsychologia.2006.01.013

Setti, A., Burke, K. E., Kenny, R., & Newell, F. N. (2013). Susceptibility to a multisensory speech illusion in older persons is driven by perceptual processes. *Frontiers in Psychology, 4*, 575. https://doi.org/10.3389/fpsyg.2013.00575

Setti, A., Burke, K. E., & Kenny, R. A. (2011). Is inefficient multisensory processing associated with falls in older people? *Experimental Brain Research, 209*(3), 375–384. https://doi.org/10.1007/s00221-011-2560-z

Setti, A., Stapleton, J., Leahy, D., Walsh, C., Kenny, R. A., & Newell, F. N. (2014). Improving the efficiency of multisensory integration in older adults: Audio-visual temporal discrimination training reduces susceptibility to the sound-induced flash illusion. *Neuropsychologia, 61*, 259–268. https://doi.org/10.1016/j.neuropsychologia.2014.06.027

Shahin, A. J., Shen, S., & Kerlin, J. R. (2017). Tolerance for audiovisual asynchrony is enhanced by the spectrotemporal fidelity of the speaker's mouth movements and speech. *Language, Cognition and Neuroscience, 32*(9), 1102–1118. https://doi.org/10.1080/23273798.2017.1283428

Simon, D. M., Nidiffer, A. R., & Wallace, M. T. (2018). Single trial plasticity in evidence accumulation underlies rapid recalibration to asynchronous audiovisual speech. *Scientific Reports, 8*, 12499. https://doi.org/10.1038/s41598-018-30414-9

Simon, D. M., Noel, J.-P., & Wallace, M. T. (2017). Event related potentials index rapid recalibration to audiovisual temporal asynchrony. *Frontiers in Integrative Neuroscience, 11*, 8. https://doi.org/10.3389/fnint.2017.00008

Simon, D. M., & Wallace, M. T. (2017). Integration and temporal processing of asynchronous audiovisual speech. *Journal of Cognitive Neuroscience, 30*(3), 319–337. https://doi.org/10.1162/jocn_a_01205

Smith, W. F. (1933). The relative quickness of visual and auditory perception. *Journal of Experimental Psychology, 16*(2), 239–257. https://doi.org/10.1037/h0071379

Stapleton, J., Setti, A., Doheny, E. P., Kenny, R. A., & Newell, F. N. (2014). A standing posture is associated with increased susceptibility to the sound-induced flash illusion in fall-prone older adults. *Experimental Brain Research, 32*(2), 423–434. https://doi.org/10.1007/s00221-013-3750-7

Stefanics, G., Hangya, B., Hernádi, I., Winkler, I., Lakatos, P., & Ulbert, I. (2010). Phase entrainment of human delta oscillations can mediate the effects of expectation on reaction speed. *Journal of Neuroscience, 30*(41), 13578–13585. https://doi.org/10.1523/JNEUROSCI.0703-10.2010

Stevenson, R. A., Bushmakin, M., Kim, S., Wallace, M. T., Puce, A., & James, T. W. (2012). Inverse effectiveness and multisensory interactions in visual event-related potentials with audiovisual speech. *Brain Topography, 25*(3), 308–326. https://doi.org/10.1007/s10548-012-0220-7

Stevenson, R. A., Park, S., Cochran, C., McIntosh, L. G., Noel, J.-P., Barense, M. D., … Wallace, M. T. (2017). The associations between multisensory temporal processing on symptoms of schizophrenia. *Schizophrenia Research, 179*, 97–103. https://doi.org/10.1016/j.schres.2016.09.035

Stevenson, R. A., Segers, M., Ferber, S., Barense, M. D., & Wallace, M. T. (2016). Keeping time in the brain: Autism spectrum disorder and audiovisual temporal processing. *Autism Research, 9*(7), 720–738. https://doi.org/10.1002/aur.1566

Stevenson, R. A., Segers, M., Ncube, B. L., Black, K. R., Bebko, M., Ferber, S., & Barense, M. D. (2017). The cascading influence of low-level multisensory processing on speech perception in autism. *Autism, 22*(5), 609–624. https://doi.org/10.1177/1362361317704413

Stevenson, R. A., Siemann, J. K., Schneider, B. C., Eberly, H. E., Woynaroski, T. G., Camarata, S. M., & Wallace, M. T. (2014). Multisensory temporal integration in autism spectrum disorders. *Journal of Neuroscience, 34*(3), 691–697. https://doi.org/10.1523/jneurosci.3615-13.2014

Stevenson, R. A., Toulmin, J. K., Youm, A., Besney, R. M., Schulz, S. E., Barense, M. D., & Ferber, S. (2017). Increases in the autistic trait of attention to detail are associated with decreased multisensory temporal adaptation. *Scientific Reports, 7*(1), 14354. https://doi.org/10.1038/s41598-017-14632-1

Stevenson, R. A., & Wallace, M. T. (2013). Multisensory temporal integration: Task and stimulus dependencies. *Experimental Brain Research. Experimentelle Hirnforschung. Experimentation Cerebrale, 227*(2), 249–261. https://doi.org/10.1007/s00221-013-3507-3

Stevenson, R. A., Wilson, M. M., Powers, A. R., & Wallace, M. T. (2013). The effects of visual training on multisensory temporal processing. *Experimental Brain Research, 225*(4), 479–489. https://doi.org/10.1007/s00221-012-3387-y

Stone, J. V., Hunkin, N. M., Porrill, J., Wood, R., Keeler, V., Beanland, M., … Porter, N. R. (2001). When is now? Perception of simultaneity. *Proceedings. Biological Sciences/The Royal Society, 268*(1462), 31–38. https://doi.org/10.1098/rspb.2000.1326

Su, L., Wyble, B., Zhou, L., Wang, K., Wang, Y., Cheung, E. F. C., … Chan, R. C. K. (2015). Temporal perception deficits in schizophrenia: Integration is the problem, not deployment of attentions. *Scientific Reports, 5*, 9745. https://doi.org/10.1038/srep09745

Takahashi, T., Kansaku, K., Wada, M., Shibuya, S., & Kitazawa, S. (2013). Neural correlates of tactile temporal-order judgment in humans: An fMRI study. *Cerebral Cortex, 23*(8), 1952–1964. https://doi.org/10.1093/cercor/bhs179

Takayama, H., Ogawa, N., Yamamoto, M., Asanuma, M., Hirata, H., & Ota, Z. (1992). Age-related changes in cerebrospinal fluid γ-Aminobutyric acid concentration. *European Journal of Clinical Chemistry and Clinical Biochemistry, 30*(5), 271–274. https://doi.org/10.1515/cclm.1992.30.5.271

Teramoto, W., Honda, K., Furuta, K., & Sekiyama, K. (2017). Visuotactile interaction even in far sagittal space in older adults with decreased gait and balance functions. *Experimental Brain Research, 235*(8), 2391–2405. https://doi.org/10.1007/s00221-017-4975-7

Tiippana, K., & Salmela, V. R. (2018). Stimulus duration has little effect on auditory, visual and audiovisual temporal order judgement. *Experimental Brain Research, 236*(5), 1273–1282. https://doi.org/10.1007/s00221-018-5218-2

Tomassini, A., Ambrogioni, L., Medendorp, W. P., & Maris, E. (2017). Theta oscillations locked to intended actions rhythmically modulate perception. *eLife, 6*, e25618. https://doi.org/10.7554/eLife.25618

Tomassini, A., Spinelli, D., Jacono, M., Sandini, G., & Morrone, M. C. (2015). Rhythmic oscillations of visual contrast sensitivity synchronized with action. *Journal of Neuroscience, 35*(18), 7019–7029. https://doi.org/10.1523/JNEUROSCI.4568-14.2015

Tsilionis, E., & Vatakis, A. (2016). Multisensory binding: Is the contribution of synchrony and semantic congruency obligatory? *Current Opinion in Behavioral Sciences, 8*, 7–13. https://doi.org/10.1016/j.cobeha.2016.01.002

Turi, M., Karaminis, T., Pellicano, E., & Burr, D. (2016). No rapid audiovisual recalibration in adults on the autism spectrum. *Scientific Reports, 6*, 21756. https://doi.org/10.1038/srep21756

Ursino, M., Cuppini, C., & Magosso, E. (2017). Multisensory Bayesian inference depends on synapse maturation during training: Theoretical analysis and neural modeling implementation. *Neural Computation, 29*(3), 735–782. https://doi.org/10.1162/NECO_a_00935

Van der Burg, E., Alais, D., & Cass, J. (2013). Rapid recalibration to audiovisual asynchrony. *Journal of Neuroscience, 33*(37), 14633–14637. https://doi.org/10.1523/JNEUROSCI.1182-13.2013

Van der Burg, E., & Goodbourn, P. T. (2015). Rapid, generalized adaptation to asynchronous audiovisual speech. *Proceedings of the Royal Society B: Biological Sciences, 282*(1804), 20143083. https://doi.org/10.1098/rspb.2014.3083

Van der Burg, E., Orchard-Mills, E., & Alais, D. (2015). Rapid temporal recalibration is unique to audiovisual stimuli. *Experimental Brain Research, 233*(1), 53–59. https://doi.org/10.1007/s00221-014-4085-8

Van Eijk, R. L., Kohlrausch, A., Juola, J. F., & van de Par, S. (2008). Audiovisual synchrony and temporal order judgments: Effects of experimental method and stimulus type. *Perception & Psychophysics, 70*(6), 955–968. https://doi.org/10.3758/PP.70.6.955

van Wassenhove, V., Grant, K. W., & Poeppel, D. (2005). Visual speech speeds up the neural processing of auditory speech. *Proceedings of the National Academy of Sciences, 102*(4), 1181–1186. https://doi.org/10.1073/pnas.0408949102

van Wassenhove, V., Grant, K. W., & Poeppel, D. (2007). Temporal window of integration in auditory-visual speech perception. *Neuropsychologia, 45*(3), 598–607. https://doi.org/10.1016/j.neuropsychologia.2006.01.001

VanRullen, R. (2016). Perceptual cycles. *Trends in Cognitive Sciences, 20*(10), 723–735. https://doi.org/10.1016/j.tics.2016.07.006

Vatakis, A. (2013). The role of stimulus properties and cognitive processes in the quality of the multisensory perception of synchrony. In L. Albertazzi (Ed.), *The Wiley-Blackwell handbook of experimental phenomenology: Visual perception of shape, space and appearance* (pp. 243–263). Chichester: Wiley-Blackwell.

Vatakis, A., & Bakou, A. E. (2015). Distorted multisensory experiences of order and simultaneity. In A. Vatakis & M. Allman (Eds.), *Time distortions in mind: Temporal processing in clinical populations* (pp. 1–36). Leiden: Brill.

Vatakis, A., Navarra, J., Soto-Faraco, S., & Spence, C. (2008). Audiovisual temporal adaptation of speech: Temporal order versus simultaneity judgments. *Experimental Brain Research, 185*, 521–529. https://doi.org/10.1007/s00221-007-1168-9

Vatakis, A., & Spence, C. (2006). Audiovisual synchrony perception for music, speech, and object actions. *Brain Research, 1111*(1), 134–142. https://doi.org/10.1016/j.brainres.2006.05.078

Venezia, J. H., Thurman, S. M., Matchin, W., George, S. E., & Hickok, G. (2016). Timing in audiovisual speech perception: A mini review and new psychophysical data. *Attention, Perception & Psychophysics, 78*(2), 583–601. https://doi.org/10.3758/s13414-015-1026-y

Wu, P.-Y., Chen, Y.-C., Yeh, S.-L., Wu, M.-T., & Tang, P.-F. (2015). Multisensory integration ability correlates with spatial working memory and functional mobility in cognitively normal middle-aged and older adults. *Physiotherapy, 101*(1), e1663–e1664. https://doi.org/10.1016/j.physio.2015.03.061

Wutz, A., Muschter, E., van Koningsbruggen, M. G., Weisz, N., & Melcher, D. (2016). Temporal integration windows in neural processing and perception aligned to saccadic eye movements. *Current Biology, 26*(13), 1659–1668. https://doi.org/10.1016/j.cub.2016.04.070

Yarrow, K., Shapiro, S., DiCosta, S., & Arnold, D. H. (2014). A 2AFC simultaneity-judgment task to estimate the point of subjective simultaneity. *Procedia Social and Behavioral Sciences, 126*, 66. https://doi.org/10.3389/fpsyg.2016.00416

Zhou, H. Y., Cai, X. L., Weigl, M., Bang, P., Cheung, E. F. C., & Chan, R. C. K. (2018). Multisensory temporal binding window in autism spectrum disorders and schizophrenia spectrum disorders: A systematic review and meta-analysis. *Neuroscience & Biobehavioral Reviews, 86*, 66–76. https://doi.org/10.1016/j.neubiorev.2017.12.013

Zopf, R., Friedman, J., & Williams, M. A. (2015). The plausibility of visual information for hand ownership modulates multisensory synchrony perception. *Experimental Brain Research, 233*(8), 2311–2321. https://doi.org/10.1007/s00221-015-4300-2

Part IV

Cognition and Representation of Temporal Phenomena

14

Time Opined: A Being in the Moment

Mark A. Elliott

1 In the Beginning There Was a Brief Tale of Time

Undoubtedly, our continued existence, which is the basis for all speculation on time, is, and has always been, marked by regular events. The earth has been spinning on its axis since before sentient life evolved. There is no period in the story of life, or for that matter there is no period in the history of any species able to reflect on 'time' during which the cyclical processes of night and day and of the seasons did not operate. The origin of prototypic timekeeping, whether that be in the mystery traditions that emerged and evolved from regular events in the heavens, or the everyday experiences of most human generations, are based upon the motions of the earth and moon and of the stars and planets.

It is clear from at least prehistory that celestial time has been considered important enough to warrant the building of monuments all over

M. A. Elliott (✉)
School of Psychology, National University of Ireland Galway,
Galway, Republic of Ireland
e-mail: mark.elliott@nuigalway.ie

V. Arstila et al. (eds.), *The Illusions of Time*,
https://doi.org/10.1007/978-3-030-22048-8_14

the prehistoric world. Time reckoning is known to have been undertaken by the use of astronomical monuments and/or calendars since antiquity, and not in directly connected civilizations. Examples of these include Stonehenge and Newgrange in Britain and Ireland, dated conservatively to around 5000 BCE; in Sumer (3500 BCE) and Egypt, where calendars have been used since the eighth to tenth centuries BCE; China since the eighteenth century BCE; while calendars were in use in the pre-Colombian Maya civilization, one of the last major stone age civilizations and a civilization, which to the point of European colonization, existed in geographical isolation of cultural developments Europe, Asia and Africa.

The understanding that things change regularly and may be measured is a general human observation on the nature of existence. From this we may conclude that time as 'object' is also a human observation, and in this sense, time is the condition within which change occurs. However, in this contribution, I do not intend to appeal (much) to philosophical speculation on the nature of time as 'object.' This is partly due to the problem, outlined in the next section, in being able to actually generalize beyond the experience of time as 'subject' to the ontological status of time. It is also because, for the most part and in my opinion, time as 'object' is simply a functional or useful measure of other psychological phenomena, including most particularly, the way psychological content relates to the brain. Again, more on this topic in later sections. It is also because, for the most part, I will discuss the idea of subjective experience of, or in time, and this refers to time 'in the present' or the experience of 'nowness.' This refers to a metaphysical position called 'Presentism' which argues that neither the future nor past exist and that everything exists in the present. A synopsis of the Presentism thesis is beyond the scope of this contribution but it may be found in Crisp (2003), however, when relating to 'time' it is perhaps pertinent to refer Presentism to the deliberations of Augustine of Hippo who proposed (like Zeno before him) the present to be analogous to a sharp edge of a blade placed exactly between the perceived past and the imaginary future. The present, of itself, does not include the concept of time.

I intend to confine my opinion to consideration of the everyday human, subjective experience of time, and the generalizability of this experience to time beyond the subjective. However, this raises the first issue: how generalizable is the human experience of time?

2 Of Fighting Fish, Sea Urchins—and Men

Bergson (1910) considered the experience of time to be related to experience 'in time,' while Husserl considered the immediate present, the 'moment now' to be an amalgamation of future time with the immediate past (Husserl, 1917). Real time, as in the experience of an enduring and continuous flow of events from the future toward the past, Bergson argued, was not directly knowable but may be judged through intuition arising from a series of acts of direct participation in immediate experience. This intuition Bergson referred to as 'lived time,' and it is 'lived time' that is the continuous experience of the moment now, influenced by past events, while influencing and perhaps influenced by events occurring in the future. 'Now' is where we experience events, and it is, at the same time, the nexus of the past and the future.

The original scientific investigations measured the duration of 'now,' known in some scientific circles as the 'psychological moment.' These investigations were undertaken in early schools of theoretical biology, such as that of von Uexküll (1928, 1934, translated in von Uexküll, 1957). From Uexküll's school, experimental work was published in the 1930s, using near identical experimental techniques to delimit the minimum duration of the psychological moment in various creatures, including human beings, fighting fish and snails (Brecher, 1932). While this moment 'quantum' was found to differ between species, in all of the species studied, Brecher found that events separated by time might, given short intervals between their occurrence, combine to form a coexistent experiential content that has meaning by virtue of its presence in the 'now.'

While Brecher showed that both animals and humans could provide an estimate of the experience of present coexistence using similar experimental techniques, he also showed that this experience relates to the 'content' of the temporally separate, but experientially coincident events. At least this is true for the fighting fish who experienced repeated exposures to the image of a conspecific, viewed through slits cut at right angles to one another in a rapidly rotated disk. Above a certain exposure frequency (equivalent to around 30 Hz), the successive images fused to form a continuous image, at which point the fish attempted his attack. Two points can be taken from this example: first, in so far as psychological presence includes content provided by the fusion of temporally separated events, this content may be meaningful. At least, and for creatures other than humans, the past and present are meaningfully unified into an experience of nowness. The second point refers to Uexküll's concept of 'Umwelt,' or 'meaningfully relevant aspects of the environment.' Ultimately, Umwelt determines the sensory, perceptual and subsequent physiological responses of the animal, and in so far as the moment is an index of basic cognitive capacity (i.e. the amount of information we can process in the smallest interval of psychological time), the complexity of Umwelt' responses, and so Umwelt itself determines a given species' moment quantum.

This entails psychological presence to be equated, not with the experience of time flow, but with the 'content structure' of what occurs in time—in this case, the meaningful content of a given moment. Uexküll was clear that while moments may be related to relevant aspects of the environment to which the animal responds, this in turn refers to the number of coordinated reflex arcs in the animal's behavioral repertoire. Uexküll rejected the reflex arc as sufficient to explain either the response itself or the cognitive organization underlying the response. In fact, this may be better conceived as a form of cognitive capacity limit, which may be indexed by the duration, hence, in the content of the moment. Consequently, the moment, in so far as this links past and future, is determined by the content structure that defines the meaningful interaction of the animal with their environment.

The issue raised by the representation of temporality in different species becomes more complex when we take into consideration a further example discussed by Uexküll. He noted that the sea urchin possesses multiple independent reflex arc systems, and while entailing these animals one might experience multiple, but nonetheless concurrent, experiences of Umwelt in time. This clearly differs from the human experience of a single channel flow of time from future to past, and as such, offers one limit to the desired generalizability of the human model. Additionally, and as argued earlier across species, the qualitative content of Umwelten, as well as the timing of the moments within which they present, and the nervous systems within which they are instantiated (their quantitative aspect), are different and this provides a second limit because we do not experience the world as animals do. However, this is not all the animal case tells us. It also tells us that any theory of psychological time, at least a theory that intends to be a complete theory, requires an absolute—that is to say, a non-relative measure of the experience of time passing in each animal species, not only of those who are still living, but also of species that are now extinct. It may be that this line of argument can be extended to include animals and humans, both living and dead, but of course, this is an impossible task even if left at the level of the species.

However, by linking time with the content of Umwelt, Uexküll also showed that one possible solution to this problem lies not in the consideration of psychological time as 'object,' but as 'subject,' and subject to the Umwelt structure experienced by the animal. More on this later.

3 The Human Experience of Time: What's Not Measurable

At best then, we may have to limit our model of psychological time to humans and as many animals as we can examine, without reference of the resulting animal model of experienced time to our own model. Importantly, generalization to support one or other thesis referring to 'time as object' seems impossible. But this is not the topic of this section. Let's instead consider the human experience of time.

For the most part, human beings have not paid much attention to time as object. Consider the common experience of time in the life of a typical individual. First, very few probably considered ontology to any great depth, almost none possessed a timepiece, and almost all measured time implicitly—that is to say, as a part of the event structure in their everyday lives. To do this, they would use both external time markers (the heavenly events mentioned in the introductory paragraphs, as well as more down to earth correlates) and internal events (e.g. regular times for hunger, arousal or sleep, defecation). In addition, there would be task related events, and events related to time, to which their response is a second-order effect. Consider the lives of our ancestors, Hans, Hildegard, John, Elizabeth, Oleg, Katja, Ahmet, Emine, Chang, Chin Chin, Akinyi and Mbuta. Just before cockcrow, he or she would rise ahead of the dawn. Preparing for the day, he or she would check any animals before dawn to feed them or let them out to pasture, prepare lunch and take this to the far field where he or she would work the land until midday (when the sun is at midheaven), until they intuitively felt it was time to eat, or when he or she, and with the others nearby, created consensus that this was the time to eat. After eating, with the time of year in mind, he or she might work until a point of time in the evening at which he or she would stop to eat again, or at a time just ahead of sunset. Everyday life required no particular focus on time as object—while he or she was in the moment—and the requisite information on the personal relevance of a particular point of time on the flow of events is given by a particular event. For the most part we all live in the moment, and although we may have less/more demands on organizing our time than we would have had in the thirteenth or fourteenth century, it is not that often that we sit and reflect on our subjective experience of time passing. For most people, time remains subject.

But the lack of necessity to measure the experience of 'nowness' does not mean it cannot be measured. Since Vierordt's (1868) experiment in time reproduction, the capacity for integrating sequential events into self-contained perceptual units that appear subjectively as occurring in the present is capable of being determined experimentally. This experiment asks participants to reproduce as exactly as possible the duration of a test tone. The results reveal a slope of 1.0 (the reproduced tone is equivalent in duration to the test tone) up to test-tone durations of around three seconds. Beyond this, the slopes flatten because the reproduced estimates are increasingly underestimated relative to increases in the duration of the test tone. Pöppel (1988) interprets this in terms of a temporal limit in information processing beyond which information may be referenced to the temporal limit (leading to underestimations and flattening of the slope of the reproduced estimate). And while Fraisse (1963) refers to this in terms of an upper limit in 'time perception of around three seconds, Pöppel refers to this as the upper limit of 'nowness,' and also makes clear that the experience of 'now' does not allow for a non-extensive boundary between past and future events.

More recently, Wittmann (2015) has summarized psychological 'presence' in the now into three classes: the 'Funktionaler Moment' or 'functional moment' is of duration between 30–300 milliseconds (ms) during which the perceptual and derivative cognitive content structure of events occurring in that interval of time are processed, prior to the deployment of focal attentional mechanisms and prior to any articulate experience of the 'here and now;' an 'Erlebter Moment' or 'lived moment' is of between 300–3000 ms—Pöppel's limit for experienced nowness—during which we are able to experience and coordinate with external event structure, and during which reflexive cognition such as attentional deployment to external events occurs. Finally, Wittmann identifies 'Mentale Präsenz,' a moment ranging between several seconds to a few minutes, which includes the use of explicit cognitive strategies (working memory), and which provides the temporal context (past/future) for a narrative self.

To some extent, measures supporting timing of the lived moment derive from the timing of successive states of introspection on stimuli such as reversible figures (Pöppel, 1988; Wittmann, 2015). These data are subjective, allowing an approximate timing for the moment now and seem to suit the definition of time as subject. The shorter interval includes the interval originally measured by Brecher, who revealed a maximum time for the experience of events as co-temporal as 53 ms for humans (35 ms for fighting fish and around 300 ms for snails). Values between 50 and 60 ms using a paradigm similar to that described by Brecher have also been published from my own work (Elliott, Shi, & Sürer, 2007; Elliott & Giersch, 2016). However, what is interesting with respect to this lowest class of 'nowness' is its identification with the neural process. There is now a very substantial body of evidence that shows that neural 'binding' processes, that is, those processes responsible for the conjoint processing of different perceptual features of a multi-featured object, occur prior to attentional deployment, and may involve the phase synchronization of stimulus-induced neuronal oscillations, which settle at frequencies in the range 30–70 Hz (Singer, 1999), frequencies for which several cycles may be realized during the 30–300 ms time period, argued as necessary to be able to reject the perceptuo-cognitive outcome of spurious, one-off synchronizations.

Wittmann's 'functional moment' includes the period of time during which binding occurs, and at one time several attempts were made to establish whether or not 'perceptual synchrony' could be engineered by stimulus paradigms which could otherwise be interpreted in terms of neural synchrony. The equivalence between a perceived simultaneity and the synchronization of neuronal networks was considered sufficient enough that the former could be used as direct evidence for the latter; however, on closer examination, this leads to a problem that might not confine to the problem of establishing the validity of the binding by synchrony hypothesis. It certainly is a problem for establishing the function of the Funktionaler Moment.

The problem is founded on the fairly common observation that events may be reported as synchronous (or simultaneous, in so far as this is a case of synchrony when events occur at the same time) beyond the case where the two events occur at precisely the same time. As found by Brecher and then myself (Elliott et al., 2007, also Elliott & Giersch, 2016), and depending upon stimulus conditions, two events may be judged to have occurred simultaneously even though they are separated by intervals of a few tens of milliseconds. Additionally, and importantly for this argument, there appears to be little or no difference between the perception of simultaneity following presentation of physically simultaneous events and perceptions that follow presentation of non-simultaneous events. Here, time as 'subject' allows for an interval of 'nowness' during which the physically non-simultaneous are confusable with the simultaneous. Time as subject is different from time as object.

So why is this problematic for the idea of a perceptual synchrony? The synchrony or *simultaneity problem* arises when we ask the question: 'To what extent can we claim to have *experienced events as simultaneous*?' At the crux of the problem are events that are non-simultaneous but which we see as simultaneous, and it might be considered an acceptable claim that irrespective of the actual simultaneity of two events, in perception, events can be considered simultaneous if that is how they seem to the observer (Kelly, 2005). However, the following logic modifies this consideration. Take any two events A_1 and A_2 that appear to the observer to occur simultaneously, in spite of which they are separated by some interval. Then take some third event that occurs at some time later than both A_1 and A_2 but is experienced simultaneous to each. Using this method, a series of events from sets $\{A_1, A_n\}$ may be experienced as simultaneous although a point may be reached at which A_1 and A_n come to be experienced as non-simultaneous.

Perceived simultaneity is, thus, *non-transitive* (i.e. $A_1 = A_2$ and $A_2 = A_3$, but $A_1 \neq A_3$) and non-transitivity should, in principle, preclude a definition that includes *simultaneity* because simultaneity cannot be both *non-transitive* and an *equivalence* relation. It seems an obvious solution that to circumvent the problem of transitivity two events may be defined as perceptually simultaneous if they appear to be simultaneous and if there is no third event with which one event appears to be simultaneous while the other does not. Indeed, this counter argument might be valid if it were not for the following corollary: although two events might *appear to be simultaneous*, it is nonetheless impossible for the observer to conclude that they have *experienced them to be simultaneous* without reference to the third, non-simultaneous event. This event thus marks both the limit of perceived simultaneity and indeed the maximum interval of time between which the two key events would have been perceived as simultaneous. It also means the observer's judgment of what was deemed to be 'simultaneous' is invalid and may only be valid when it takes into account the third, non-simultaneous event, at which point it is a judgment based upon criteria related to successiveness. On these grounds, perceptual synchrony seems little more than a non sequitur, and one which lacks construct validity unless taken to refer to the *interval of time* over which two or more events are seen to occur at the same time. It also means that asking experimental participants to judge the synchrony or asynchrony of events is not possible with any true validity because a priori, they will be unable to make a correct judgment. Thus, and while subject to critical analysis (Arstila, 2012), the problem really does remain that you cannot ask people to report the impossible.

At the lowest level of 'nowness,' at its functional level, it may be difficult to derive measures that allow us to do more than index the intervals of time over which two events appear to be non-simultaneous because we cannot link this perception directly with the process states that are likely to be operating at this lowest level to bring about a state of perceived simultaneity. We have, once again, a problem objectifying time. This may not be a major problem for the models put forward by Pöppel and Wittmann, but again and as with the problem of psychological time as object, the logical impasse may make it impossible to find theoretical closure.

4 Does Anyone Really See the Emperor's New Clothes?

Perhaps the moniker 'time perception' predates Fraisse (1963), but by the time of writing the 'Psychology of Time,' Fraisse developed the idea that we perceive time. So, do we perceive time? The answer is 'no,' we do not perceive time, so why raise this here? One question is whether we really have access to 'time as object' and if we do not, what is it that we are experiencing? There is no need appealing to the scientific literature to make the point that time as subject does not run evenly: if you are bored then things drag on and if you are happy things are over in an inconveniently short time. Time as object would appear somewhat mercurial (it could be that boring things last relatively shorter than enjoyable things), but it is not clear that time is at fault here, or for that matter, any perceptuo-cognitive faculty we possess that is related to the processing of time. There has accumulated a substantial literature on estimates of how long an event or an interval has lasted, which was summarized nearly 20 years ago by Grondin (2001). The reasons to refer to Grondin's, now somewhat dated review, are the following points he makes: (i) In spite of the idea of a single clock mechanism, the results of time estimation experiments have rarely been found to correspond to Weber's law (the law states that the change in a stimulus that is deemed to be just noticeable is at a constant ratio of the original stimulus), and so, time estimation does not scale along a single dimension. Here, one could say that if there is an analog to objective time, it is not in a single processing system. (ii) Related to this perhaps are findings that Weber's law does not hold in situations involving sequences of intervals, explicit counting strategies or extensive training. (iii) Further and most important, the psychophysical functions relating judged to actual durations may be heavily skewed at extremes on the measurement scale. This means, not only is time estimation skewed by affective state, it is also skewed beyond certain estimable limits and because this occurs in the range of a millisecond or tens of milliseconds, it points to an issue of resolution (at least in this case) and possibly the capability of the cognitive system to adequately represent an interval of time that falls outside of the operational capacity of that same system. Time estimation starts to resemble an outcome of an attentional system; retrospective time estimation cannot avoid appeal to the product of our ability to accurately remember the duration of an interval or event. Of course, in the latter case, we need to be reasonably accurate in our 'perception' of time such that the memory can provide an accurate estimate for reproduction, comparison or report. Without going into detail, the take home point from Grondin's summary critique is that there is likely no single perceptuo-cognitive apparatus that provides for time sense, and while the clock metaphor remains substantively elusive, it is not properly accurate to conceive of time sense as a single or dimensionally unitary psychological experience, and certainly not something for which an anatomically singular physiological mechanism should be sought. In addition, it may be more parsimonious and scientifically less prone to mistake to refer to the outcomes of laboratory studies of time perception in terms of memory or attentional processes. Although this may in itself be equivalent to whitewashing the wall by painting it black.

5 If Time Isn't Time, Where Do We Go Now?

Time as object again defies definition in terms of time as subject, or in this instance, time as subject seems prone to enough alternative explanations and measurement issues that a description of what people experience as 'time flow' cannot be easily parceled out from the mixture of experimental instruction set and capacity limits on attentional and perhaps memory processes. None of these are time. In addition, time as subject is essentially the experience of nowness, which has the peculiar property of being by nature atemporal (Zeno's paradox), and if people by and large do not spend their time in a state of awareness of time flow (as I have argued they do not), one might question construct validity in terms of ecological validity, at least as this refers to the experience of time.

So, psychological time (time as subject), for various reasons, is not the same as physical time (time as object). In fact, not only is psychological time not a good measure of physical time, it is often the case that we are interested in it precisely because it varies relative to physical time. Is it then really 'time?' It seems not, and perhaps we should stop referring to various experiential states, or experiential consequences of cognitive performance in terms of time: this may simply serve to mislead. So, we seem to have reached the point at which the argument requires retrieval. Or, put another way, can we measure the effect of physical time in psychological function?

All systems operate in time and the human biobehavioral-interactional-social systems are no exception. It is possible that many of the principles defining dynamics at one level scale to other levels, not surprising in consideration of all such systems being physical, and so describable, systemically, according to the set of physical constraints that define their operation. What does this mean for psychological time? Well, time in psychological systems is undoubtedly one factor that defines the function of all of the dynamic systems that provide for cognitive and other psychological functions: it may be considered to provide a limit in so far as it is the medium within which the several components of any dynamic system interact to bring about a successful outcome. This limit is perhaps better understood though in terms of the extent to which a system or component of that system is able to function and interact regularly, and successfully, with other components of the same system. Referring to one locus of this system, in cognition and the neural systems that support cognition, I have written elsewhere

> the organization of brain circuitry into assemblies defined by their synchrony at particular (and precise) oscillation frequencies is important for the correct correlation of all independent cortical responses to the different aspects of a given complex thought or object. From the point of view of

> anyone operating complex mechanical systems, i.e. those comprising independent components that are required to interact precisely in time, it follows that the precise timing of such a system is essential—not only essential but measurable, and scalable. It must also be reliable over observations to bring about consistent behavior, whatever that behavior is. The catastrophic consequence of an absence of such precision, for instance that required to govern the interference engine in many automobiles, is indicative of how important timing is for the function of dynamical systems at all levels of operation. (Elliott & du Bois, 2017)

Psychological time is, in this view, unequivocally a *measure* of dynamic psychological systems. Given that such systems are now generally accepted to be instrumental in the representation of perceptual and other cognitive events, it is perhaps logical to draw reference back to von Uexküll and the idea that in the moment now, which is the only point at which we have access and may be able to properly measure time, 'nowness' is defined according to the content of Umwelt. In other words, defining time requires defining Umwelt, which in turn, and given its representational nature, entails measuring the dynamic systems that bring Umwelt about. Psychological time, thus considered, is a descriptive analogy for the dynamic systems structure that brings about our immediate and continued experience of the 'moment now.' Taken together with the fact that dynamic systems are defined in terms of regular events, one can conclude that our experience of 'nowness' is the same as our experience of continued existence, both being based upon a structure of regular events, and perhaps ironically, both based upon dynamic systems structures that may share common patterns of regularity.

References

Arstila, V. (2012). Why the transitivity of perceptual simultaneity should be taken seriously. *Frontiers in Integrative Neuroscience, 6*, 3.

Bergson, H. (1910). *Time and free will: An essay on the immediate data of consciousness*. London: Allen & Unwin.

Brecher, G. A. (1932). Die Entstehung und biologische Bedeutung der subject-ktiven Zeiteinheit—des Momentes. *Zeitschrift für Vergleichende Physiologie, 18*, 204–243.
Crisp, T. M. (2003). Presentism. In *The Oxford handbook of metaphysics*. Oxford: Oxford Handbooks.
Elliott, M. A., & du Bois, N. (2017). Dynamical constants and time universals: A first step toward a metrical definition of ordered and abnormal cognition. *Frontiers in Psychopathology, 8*, 332.
Elliott, M. A., & Giersch, A. (2016). What happens in a moment. *Frontiers in Psychology, 6*, 1905.
Elliott, M. A., Shi, Z., & Sürer, F. (2007). The effects of subthreshold synchrony on the perception of simultaneity. *Psychological Research, 71*(6), 687–693.
Fraisse, P. (1963). *The psychology of time*. Oxford: Harper & Row.
Grondin, S. (2001). From physical time to the first and second moments of psychological time. *Psychological Bulletin, 127*(1), 22–44.
Husserl, E. (1917). *On the phenomenology of the consciousness of internal time (1893–1917)* (J. B. Brough, Trans.). Dordrecht: Kluwer Academic Publishers.
Kelly, S. D. (2005). The puzzle of temporal experience. In A. Brook & K. Akins (Eds.), *Cognition and the brain: The philosophy and neuroscience movement* (pp. 208–238). Cambridge: Cambridge University Press.
Pöppel, E. (1988). *Mindworks: Time and conscious experience*. Harcourt Brace Jovanovich.
Singer, W. (1999). Neuronal synchrony: A versatile code for the definition of relations? *Neuron, 24*(1), 49–65.
Vierordt, K. (1868). *Der Zeitsinn nach Versuchen*. H. Laupp.
von Uexküll, J. B. (1928). *Theoretische Biologie* (2nd ed.). Berlin: Springer.
von Uexküll, J. B. (1934). *Streifzüge durch die Umwelten von Tieren und Menschen Ein Bilderbuch unsichtbarer Welten* (Verständliche Wissenschaft, vol. 21). Berlin and Heidelberg: Springer.
von Uexküll, J. B. (1957). A stroll through the worlds of animals and men: A picture book of invisible worlds. In C. H. Schiller (Ed.), *Instinctive behavior: The development of a modern concept* (pp. 5–80). New York: International Universities Press.
Wittmann, M. (2015). *Wenn die Zeit stehen bleibt. Kleine Psychologie der Grenzerfahrungen*. München: C.H. Beck Verlag.

15

Temporal Binding and the Perception/Cognition Boundary

Christoph Hoerl

The term 'temporal binding' refers to a subjective shortening of the interval between two events that occurs when these events are believed to stand in a cause-effect relationship to each other. That is to say, for a certain range of intervals between two events, people will judge those two events to occur closer together in time if they believe the first of them to be the cause of the second, in comparison with two events that they believe to be causally unrelated. Such binding was originally thought to occur only in the context of intentional action (Haggard, Clark, & Kalogeras, 2002)—that is, when the first of the two events is a voluntary movement and the second its consequence—and the effect is therefore also sometimes referred to as 'intentional binding'. However, binding has since also been demonstrated to occur in scenarios that do not involve intentional action, but only the observation of a mechanically produced cause-effect sequence (Buehner, 2012). Hence, I will use the more neutral terminology.

C. Hoerl (✉)
Department of Philosophy, University of Warwick, Coventry, UK
e-mail: C.Hoerl@warwick.ac.uk

V. Arstila et al. (eds.), *The Illusions of Time*,
https://doi.org/10.1007/978-3-030-22048-8_15

Cases such as temporal binding, in which people's judgements about properties of a sequence of events do not correctly reflect the properties the sequence actually has, have played a prominent role in discussions of consciousness.[1] In particular, such cases stand at the heart of an influential argument put forward by Dennett and Kinsbourne (1992) against a view of consciousness they term Cartesian Materialism, and in favour of an alternative view they call the Multiple Drafts Model.[2] The focus in that context has typically been on phenomena other than temporal binding. In what follows, I want to suggest that reflection on the case of temporal binding can help make explicit and undermine a key set of assumptions underlying Dennett and Kinsbourne's argument. In particular, I argue that their argument rests on a representational view of perception—that is, a conception of perception on which perceptual experience is construed as something akin to judgement, thus blurring the boundary between perception and cognition. I then use the case of temporal binding to illustrate an alternative, relational, view of perceptual experience, by providing an explanation of temporal binding on which perceptual experience and judgement play two clearly distinct roles.

1 Dennett and Kinsbourne on Cartesian Materialism

Cartesian Materialism is also described by Dennett and Kinsbourne as a view of the mind as a 'Cartesian Theatre'. This metaphor is meant to evoke the idea of a 'stage' on which mental items must appear in order to become conscious, their appearing on this stage constituting them being conscious. Another metaphor they also use for this view is that of a determinate 'finish line' that information must cross to enter consciousness.

[1] I avoid using the term 'temporal illusions' here because the term 'illusion' typically carries with it the more specific implication that things are not *perceptually experienced* to be the way they really are. To describe temporal binding as a temporal illusion would therefore prejudge one of the key issues at stake in what follows—which is whether temporal binding is a matter of faulty perception or whether it is a matter of erroneous judgement.

[2] See also Dennett (1991). For discussion, see, for example, Akins (1996) and Shoemaker (1993).

The structure of their argument might be summarized in terms of the following *modus tollens*:

(1) If Cartesian Materialism is correct, there are two clearly separable, rival, explanatory hypotheses to account for cases in which people's judgements about a sequence of events differ from how those sequences of events actually unfolded.
(2) No distinction between two such separable explanatory hypotheses can be sustained. The difference between them is a chimera—a 'difference that makes no difference'.
(3) Therefore, Cartesian Materialism is incorrect.

In order to illustrate the reasoning behind (1) and (2), it might help to consider a relatively basic example, namely metacontrast masking.

In a typical metacontrast masking experiment, participants are first asked to focus spatial attention on a fixation point in the centre of a screen. Following this, after a short delay, a circular target is displayed, typically for approx. 40 ms, centred on the same location. Then the screen goes blank again. After a further delay, typically of approx. 20 ms, a mask is displayed on the screen, typically again for approx. 40 ms. This is again centred on the same spot as the circular target, but consists of a ring, whose inner perimeter is just slightly larger than the circular target was.

The typical finding in experiments involving such an arrangement of stimuli is that participants report only seeing the mask—that is, the ring—appearing briefly on an otherwise blank screen. That is, they do not report seeing the circular target that precedes the ring. Yet, in a control experiment in which only that target is displayed, for the same amount of time, but not followed by the ring, participants do report seeing the target. It is in this way that, in metacontrast masking, a temporally later stimulus influences people's awareness of an earlier stimulus.

How might metacontrast masking be explained? It is natural to think that it could be either a misremembering effect or a misperception effect—an idea Dennett and Kinsbourne flesh out by distinguishing between what look like two rival explanatory models, which they label 'Orwellian' and 'Stalinesque', respectively. As they put it, the Orwellian model would have it that '[s]ubjects are indeed conscious of the first stimulus […] but their memory of this conscious experience is almost entirely obliterated by the second stimulus' (Dennett & Kinsbourne, 1992, p. 193). By contrast, the Stalinesque model would have it that 'the second stimulus somehow prevents conscious experience of the first stimulus (it somehow waylays the first stimulus on its way to consciousness). The first stimulus never gets to play on the stage of consciousness' (ibid.).

As the last sentence indicates, Dennett and Kinsbourne interpret the idea that there are two clearly distinguishable rival explanatory hypotheses at issue here to be a symptom of an implicit adoption of a 'Cartesian Theatre' view of the mind, and they think that it needs to be abandoned along with that view. As they argue,

> [b]oth […] models can deftly account for all the data […]. They agree about just where in the brain the mistaken content enters the causal pathways; […] they [also] both account for the subjective data—whatever is obtainable 'from the first-person perspective'—because they agree about how it ought to 'feel' to subjects: Subjects should be unable to tell the difference between misbegotten experiences and immediately misremembered experiences. So, in spite of first appearances, there is really only a verbal difference between the two theories. They tell exactly the same story except for where they place a mythical Great Divide, a point in time (and hence a place in space) whose fine-grained location is nothing that subjects can help them locate, and whose location is also neutral with regard to all other features of their theories. This is a difference that makes no difference. (Dennett & Kinsbourne, 1992, pp. 193f.)

Existing critiques of Dennett and Kinsbourne's argument have typically either focused on whether, contrary to them, there may not in fact be empirical ways to distinguish between the Orwellian and the Stalinesque model, or they have granted that there may not be, but argued that the implied inference from this to the claim that the very idea of such a distinction needs to be rejected is unjustifiably verificationist. My argument will take a somewhat different route. Before setting it out, though, it is worth saying a few words about the model of consciousness advocated by Dennett and Kinsbourne themselves, and isolating in particular one crucial feature of it, which, I will argue, plays a key background role in their argument. Here is their own description of that model, which they call the Multiple Drafts Model:

> All perceptual operations, and indeed all operations of thought and action, are accomplished by multi-track processes of interpretation and elaboration that occur over hundreds of milliseconds, during which time various additions, incorporations, emendations, and overwritings of content can occur, in various orders. Feature-detections or discriminations *have to be made only once*. That is, once a localized, specialized "observation" has been made, the information content thus fixed does not have to be sent somewhere else to be *re*discriminated by some "master" discriminator. In other words, it does not lead to a re-presentation of the already discriminated feature for the benefit of the audience in the Cartesian Theater. [...] These spatially and temporally distributed content-fixations are themselves precisely locatable in both space and time, but their onsets do *not* mark the onset of awareness of their content. (Dennett & Kinsbourne, 1992, p. 185)

The particular feature of this picture I want to highlight is that it explains the operation of the brain in terms of the activity of multiple distinct pieces of mental machinery which, at a high enough level of abstraction, can all be described as performing the same general set of functions, which Dennett and Kinsbourne variously describe as 'content-fixing', 'interpreting' or 'editing' (compare also the metaphor of producing multiple drafts). This basic underlying assumption is not just involved in Dennett and Kinsbourne's own Multiple Drafts Model. It also seems to be in play in the way they construe the Orwellian and Stalinesque models of consciousness. Only, the Orwellian and Stalinesque models, as they see it, also involve an additional (unwarranted) assumption, as indicated in the passage quoted above, of the same input going through a sequence of different such content-fixings with a step in it at which the relevant content becomes conscious, so that we can ask, with respect to a particular content, whether it was ever conscious, or whether it was 'edited out' before reaching consciousness.

The passage quoted above is meant as a description of brain functioning. Note, however, that, as a consequence of construing the basic mental operation as one of content-fixing, this is a view on which perception itself also becomes conceived of in terms of this notion. Consider, for instance, how this approach would construe an 'Orwellian' explanation of metacontrast masking: On such an explanation, at the onset of perceptual consciousness corresponding to the time when the target circle was displayed, the observer is in a state with a representational content in which that circle figures—meaning that he/she is perceptually conscious of the circle. Yet, subsequently, this content is not retained in memory, but is instead converted to one in which the circle does not figure.

Such a view, which conceives of perceptual experience as the upshot of, and as itself a matter of, 'content-fixing', essentially construes perception as something akin to judgement—the arriving at a representation of things as being a certain way. This is true as much of Dennett and Kinsbourne's own Multiple Drafts Model as it is of the Orwellian and Stalinesque models, as characterized by them. The crucial difference between their model and the others lies just in the fact that, on the Multiple Drafts Model, the representational content conscious experience is assumed to possess is not fixed at an arbitrarily fine-grained temporal level. As mentioned before, on that model, 'various additions, incorporations, emendations, and overwritings of content' (Dennett & Kinsbourne, 1992, p. 185) can occur over hundreds of milliseconds after a stimulus is first received by a sensory receptor, and none of them individually, Dennett and Kinsbourne think, should be identified with 'the point' at which the subject becomes conscious of the stimulus (or fails to do so). It is only of that period as a whole that we can say that it constituted the subject's either being conscious of the stimulus or not, where this is construed as the subject being in a state with a certain representational content.

I now want to turn to the case of temporal binding—another phenomenon that involves people making erroneous judgements about a property of a sequence of events. As I will suggest, the case of temporal binding can serve to illustrate an alternative, quite different conception of perceptual experience, which gives perceptual experience and judgement two quite distinct roles. Specifically, this conception allows us to give an explanation of temporal binding that locates the phenomenon firmly on the level of judgement, rather than perception, but an explanation that differs in crucial ways from the Orwellian model, as envisaged by Dennett and Kinsbourne.

2 Temporal Binding

A typical experiment to study temporal binding involves comparing participants' responses across two types of trials, often referred to as *causal trials* and *non-causal trials*, respectively. In each type of trial, participants are asked to make a judgement about the length of an interval between two events (or about the timing of the events). The two types of trial differ in that, in causal trials, there is a plausible causal connection between the relevant events (the first of which is often an action carried out by the participant himself or herself), whereas in non-causal trials, it is not plausible that the first event is the cause of the second. What is taken as a measure of temporal binding is the extent to which the interval between the two events in causal trials is judged to be shorter when compared with the interval between the two events in non-causal trials (in cases where these intervals were in fact of the same length).

There is some evidence that temporal binding is due to a slowing down of the pacemaker in an 'internal clock' used to time events. Models of timing postulating such an internal clock conceive of it as consisting in a pacemaker that emits pulses at a certain rate, which are then accumulated in an accumulator (Gibbon, Church, & Meck, 1984; Wearden, 2001). For a given interval, the stimulus at the onset of that interval causes a switch between the pacemaker and the accumulator to close, so that the latter starts accumulating pulses, until the stimulus at the offset of the interval causes the switch between the pacemaker and the accumulator to open again. Because temporal judgements are based on the number of pulses accumulated in the accumulator, they will be affected by a change in the rate at which the pacemaker emits such pulses. Increases in the rate of pulses will cause the interval to be judged to be longer, and decreases in the rate will cause it to be judged to be shorter.

Evidence that temporal binding may be due to a pacemaker slowing down during causal trials has been reported by Humphreys and Buehner

(2009).[3] In an interval estimation paradigm, they used causal trials in which participants had to press a button at a time of their own choosing, which was then followed, after a set interval, by a tone. They compared this with participants' performance in non-causal trials in which they simply heard a click, which was then again followed, after a set interval, by a tone. (The click was a recording of the noise made by the button press on causal trials.) After each individual trial, participants had to estimate the length of the relevant interval. As expected, Humphreys and Buehner found temporal binding: Participants judged interevent intervals in causal trials to be shorter than corresponding interevent intervals in non-causal trials. However, Humphreys and Buehner also found a difference in the slopes at which interval estimates increased with an increase in the length of the actually presented interval, with the slope for causal trials being shallower than that for non-causal trials. As Humphreys and Buehner point out, such a difference in slopes is a trademark signature of different pacemaker speeds.

How exactly should we interpret an account of temporal binding in terms of a slowed down pacemaker? On the pacemaker-accumulator model, judgements about intervals are based on the pulses accumulated in the accumulator during those intervals. If, during causal trials, the pacemaker slows down, fewer pulses will be accumulated in the accumulator. As a consequence, intervals presented in those trials will thus be judged as being shorter when compared with corresponding intervals in non-causal trials, because the latter involve no slowing down of the pacemaker. This explains the discrepancy of judgements in the causal and non-causal trials measured in temporal binding experiments.

[3] This is likely to be only one element in temporal binding, and is likely to occur only in a context in which participants do not only observe the cause but, at that time, also already expect the effect to happen. (Otherwise it is difficult to see what could trigger the relevant slowing down of the pacemaker during the interevent interval.) For the purposes of this essay, this is the case I will focus on.

3 'Content-fixing' Again

Note that the explanation of temporal binding just sketched locates the phenomenon firmly on the level of judgement. As far as perception is concerned, there is nothing in this explanation that suggests anything other than that participants are perceptually aware of the events as they happen. Indeed, their being thus aware of the events is plausibly what underpins the operation of the switch connecting the pacemaker and the accumulator. That is to say, the switch is closed when participants experience the first event and the switch is opened again when they experience the second event (see also Arstila, 2017). Moreover, if the participants experience the stimulus events as they happen, there is also a natural sense in which it is the case that they perceive the actual interval between them.

On the account of temporal binding I have just sketched, participants are thus, in the causal condition, first perceptually conscious of the relevant events and of the actual interval between them, but when subsequently being asked to make a judgement about that interval, they underestimate it. On the face of it, this might look like a version of what Dennett and Kinsbourne call an Orwellian account. However, there are reasons for thinking that it would be a mischaracterization of the account to give it that label.

I said earlier that one assumption made by Dennett and Kinsbourne that plays a crucial role in their argument is that perception is the upshot of, and indeed itself a matter of, 'content-fixing'. An Orwellian account of temporal binding, as they conceive of it, would have it that, in causal trials, one first has a perceptual experience with a representational content that represents the interevent interval to have the length it actually has, but that this then gets transformed into a content that represents the interval to be shorter than it actually is.

I suggest that the account of temporal binding I have offered should not be understood in this way. In particular, nothing in that account commits us to thinking of perception as the upshot of 'content-fixing', and thus as something that itself has a representational content that represents things as being one way or another. Rather, we can think of the role of perceptual experience here as simply that of putting participants into contact with the relevant stimulus events themselves—something that is a precondition of the interval between them having an impact on the participants' cognition. It is in this sense that the events as they happen themselves figure in the participants' perceptual experience. Similarly, as I said, we can think of participants as being perceptually aware of the interevent interval simply in virtue of their being aware of the relevant two events as they happen. This is different from their forming a representation of that interval as being of a certain length that is then later overwritten by a different representation of it as being of a shorter length, as an Orwellian account, as conceived of by Dennett and Kinsbourne, would have it. Rather, the point at which a representation of that interval is formed is the point at which participants use the number of accumulated pulses to make a judgement about that interval.

The conception of perceptual experience at issue here, which I am contrasting with Dennett and Kinsbourne's conception of perception as the upshot of 'content-fixing', is sometimes referred to as the relational view of experience, as opposed to a representational view of experience.[4] Travis describes it at one point as follows:

> *[P]erception*, as such, simply places our surroundings in view; affords us awareness of them. There is no commitment to their *being* one way or another. It confronts us with what is there, so that, by attending, noting, recognizing, and otherwise exercising what capacities we have, we may, in some respect or other, make out what is there for what it is—or again, fail to. (Travis, 2004, p. 65, emphases in original)

It is by adopting a relational view of perception along these lines, I want to suggest, that we can avoid the conception of the mind as a Cartesian Theatre, whilst at the same time hanging on to the idea of a

[4] The particular account I have given of temporal binding does not straightforwardly generalize to other postdiction effects, but these too might be explained in ways compatible with a relational view of experience. For instance, in Hoerl (2015), I give an account of apparent motion that is compatible with a relational view (which could also be extended to the case of metacontrast masking, as discussed in Sect. 1). This account makes use of the idea that temporal experience is not, as Soteriou (2013, p. 104) puts it, homogeneous down to instants. An earlier version of something like that idea already figures in Shoemaker's (1993) response to Dennett.

clear difference between perception and cognition. Take again the case of temporal binding. The relationalist can, of course, allow that there are a number of things that have to happen in participants' brains in order for them to perceive the interevent interval in a typical temporal binding experiment. But just like Dennett and Kinsbourne, the relationalist will consider it a mistake to identify one of these brain events with the state of being perceptually conscious of the interval, in a way that would allow for raising the question as to what length that interval is represented as having by that state.[5] Rather, what the relevant events in the brain do is enable the interval itself to have an influence on the participants' cognition by putting those participants into the relation of awareness to that interval.

The relational view of experience is sometimes criticized as leaving obscure the relationship between perception and judgement.[6] It should be clear, though, that this is not a criticism that can be levelled at the account of temporal binding I have given. As I said, participant's perceptual experience of the two events in the causal and non-causal trials can be seen to play a clear causal role in the operation of the internal clock: It allows the operation of the switch that determines the onset and offset of the interval during which pulses are accumulated in the accumulator to be governed by these events themselves. In this way, perceptual experience grounds the ability to make judgements about the interval between the events, based on these accumulated pulses. In Travis's words, this is how, by confronting us with what is there—the events—perceptual experience enables us to make out the interevent interval for what it is—or sometimes for a different, shorter, interval, as it happens in the case of temporal binding.

[5] On related issues, see also Soteriou (2013, p. 145ff.).

[6] Questions have also been raised, however, about the way this relationship is construed on the alternative representational view of experience. On it, making a judgement based on a perceptual experience is conceived of a transition between one representational state and another, akin to the drawing of an inference. Yet, this arguably leaves out the more fundamental sense in which perception provides us with a grasp of what it is our judgements are about in the first place.

4 Conclusion

I have argued that Dennett and Kinsbourne's argument trades crucially on the idea of 'context-fixing' as the basic mental operation. This idea also leads them to adopt a view of perceptual experience as the forming of a representation, thus blurring the boundary between perception and judgement. What I have argued, in effect, is that it is this assumption that we should reject. I have tried to show how adopting instead a relational view of perceptual experience lets us provide an explanation of temporal binding that invokes neither what Dennett and Kinsbourne call an Orwellian nor a Stalinesque explanation of temporal binding, but one which can nevertheless deliver a clear verdict as to whether such binding should be seen as being a matter of faulty perception or mistaken judgement.[7]

References

Akins, K. (1996). Lost the plot? Reconstructing Dennett's multiple drafts theory of consciousness. *Mind & Language, 11*(1), 1–43.

Arstila, V. (2017). Experience and the pacemaker-accumulator model. *Journal of Consciousness Studies, 24*(3–4), 14–36.

Buehner, M. J. (2012). Understanding the past, predicting the future: Causation, not intentional action, is the root of temporal binding. *Psychological Science, 23*(12), 1490–1497.

Dennett, D. C. (1991). *Consciousness explained.* Boston, MA: Little, Brown and Company.

Dennett, D. C., & Kinsbourne, M. (1992). Time and the observer: The where and when of consciousness in the brain. *Behavioral and Brain Sciences, 15*(2), 183–201.

Gibbon, J., Church, R. M., & Meck, W. H. (1984). Scalar timing in memory. *Annals of the New York Academy of Sciences, 423*, 52–77.

Haggard, P., Clark, S., & Kalogeras, J. (2002). Voluntary action and conscious awareness. *Nature Neuroscience, 5*(4), 382–385.

[7]Work on this chapter was supported by a grant from the Leverhulme Trust for the project 'Time and Causality in Cognitive Development', Grant No. RPG-2015-267.

Hoerl, C. (2015). Seeing motion and apparent motion. *European Journal of Philosophy, 23*(3), 676–702.

Humphreys, G. R., & Buehner, M. J. (2009). Magnitude estimation reveals temporal binding at super-second intervals. *Journal of Experimental Psychology: Human Perception and Performance, 35*(5), 1542.

Shoemaker, S. (1993). Lovely and suspect ideas. *Philosophy and Phenomenological Research, 53*(4), 905–910.

Soteriou, M. (2013). *The mind's construction: The ontology of mind and mental action*. Oxford: Oxford University Press.

Travis, C. S. (2004). The silence of the senses. *Mind, 113*(449), 57–94.

Wearden, J. H. (2001). Internal clocks and the representation of time. In C. Hoerl & T. McCormack (Eds.), *Time and memory: Issues in philosophy and psychology* (pp. 37–58). Oxford: Oxford University Press.

16

Spatiotemporal Illusions Involving Perceived Motion

Timothy L. Hubbard

A moving object occupies different spatial locations at different times. Thus, an illusion involving differences in the perceived relative timing or duration of a moving target frequently also involves differences in the perceived spatial location of that target. In other words, an illusion of time often involves an illusion of space; indeed, it is difficult to separate spatial components and temporal components of some illusions, and so such illusions are often referred to as *spatiotemporal illusions*. In many of the spatiotemporal illusions considered in this chapter, researchers usually focused on the spatial components of those illusions; however, spatiotemporal illusions clearly have a temporal component, and this temporal component has implications for the nature of mental representation (i.e., for the ways in which information is structurally and functionally encoded by the brain). Some well-known and lesser-known spatiotemporal illusions related to perceived motion of a

T. L. Hubbard (✉)
Arizona State University, Tempe, AZ, USA

Grand Canyon University, Phoenix, AZ, USA
e-mail: timothyleehubbard@gmail.com

V. Arstila et al. (eds.), *The Illusions of Time*,
https://doi.org/10.1007/978-3-030-22048-8_16

target are considered in this chapter, and the focus is on spatiotemporal illusions related to the localization of a moving target or other stimulus at the end of target presentation, during target presentation, or at the beginning of target presentation. Different spatiotemporal illusions are suggested to share many characteristics but to differ in other characteristics, and the existence of spatiotemporal illusions is suggested to show that temporal information is a necessary and intrinsic aspect of the dynamic mental representation that underlies perceptual and cognitive processing.

1 End of Target Presentation

Five spatiotemporal illusions that involve judgment of location and occur at the end of target presentation are considered. These include representational momentum, representational gravity, representational friction, representational centripetal force, and the looming effect.

1.1 Representational Momentum

The judged final location of a moving target is often displaced in the direction of anticipated target motion, and this is referred to as *representational momentum* (e.g., Freyd & Finke, 1984; for review, see Hubbard, 2005, 2014a, 2018a). Representational momentum is increased with increases in target velocity (Hubbard & Bharucha, 1988), horizontal motion relative to vertical motion and descending motion relative to ascending motion (Hubbard, 1990), if motion is toward a landmark (Hubbard & Ruppel, 1999), if the target is a type of object that typically exhibits the observed motion (Reed & Vinson, 1996), if attention is divided between the target and another stimulus or task (Hayes & Freyd, 2002), and if a target is considered threatening (Greenstein, Franklin, Martins, Sewack, & Meier, 2016). Representational momentum is decreased but not eliminated if the vanishing point is cued (Hubbard, Kumar, & Carp, 2009) or observers receive feedback (Ruppel, Fleming, & Hubbard, 2009) or instruction regarding representational momentum and are asked to compensate (Courtney & Hubbard, 2008). Representational momentum peaks after a few hundred milliseconds and then declines (Freyd & Johnson, 1987) and can displace the judged location of a nearby stationary object in the direction of target motion (Hubbard, 2008). If a target vanishes at the moment it is expected to change direction, then displacement is in the direction of expected rather than previous motion (Hubbard & Bharucha, 1988). Representational momentum can be elicited with a variety of stimulus formats (smooth motion, implied motion, single frozen-action stimuli) and assessed with a variety of response measures (cursor-positioning, probe judgment, reaching).

Several theories have been proposed for representational momentum (for review, see Hubbard, 2010). One theory suggested forward displacement resulted from visual persistence and inertia of smooth pursuit eye movements, but findings of representational momentum with single frozen-action stimuli and implied motion stimuli, as well as findings of representational momentum for auditory stimuli and tactile stimuli, rule out a primary causal role for eye movements (Hubbard, 2005, 2014a). A more cognitively oriented theory suggests representational momentum compensates for delays in perception due to the time required for neural processing. When a moving object is first sensed, a cascade of sensory, cognitive, and perhaps motor responses is initiated. Although fast, this processing is not instantaneous (e.g., transmission of information from retina to visual cortex takes at least 100 milliseconds); by the time the cortex has processed information related to the initially seen location, the target has already moved beyond that location, and so the cortical representation of location lags behind the actual location. If an observer is to interact with a moving target in real time, the represented target location should reflect where the target would be when a response would reach that target and not where the target was when it was initially sensed. Observers need some way to compensate for the lag in perception due to neural processing times. By correcting for a lag in time, representational momentum also corrects for a lag in space. Seen in this perspective, such an illusion is adaptive, as it allows more effective interaction with moving targets.

1.2 Representational Gravity

The judged final location of a moving target or a stationary object is displaced in the direction of implied gravitational attraction, and this has been referred to as *representational gravity* (e.g., Hubbard, 1997; for review, see Zago, 2018). Representational gravity is related to target size, as larger horizontally moving targets exhibit larger downward displacement (in addition to forward displacement), larger ascending targets exhibit relatively smaller forward displacement, and larger descending targets exhibit relatively larger forward displacement (Hubbard, 1997). Just as physical gravity influences stationary objects as well as moving targets, representational gravity influences judgments of the locations of stationary objects as well as judgments of the locations of moving targets. For example, in a display in which a stationary object is depicted as supported by another object (e.g., a flowerpot sitting on a table or hanging from a hook), that object is judged to be slightly lower in the display if the object is subsequently shown without the supporting object (Freyd, Pantzer, & Cheng, 1988). The magnitude of representational gravity increases with increases in retention interval between when the target vanishes and judgments of location are made (de sá Teixeira, 2016; de sá Teixeira, Hecht, & Oliveira, 2013), at least for relatively brief retention intervals. Even though a stationary target does not actually move, representational gravity is a spatiotemporal illusion in that displacement in judged location is influenced by duration of the retention interval.

Representational gravity can be combined with representational momentum in the determination of target location (e.g., a horizontally moving target is displaced downward as well as forward; if representational momentum and representational gravity operate in the same direction [descending motion], they combine and forward displacement is relatively large, whereas if representational momentum and representational gravity operate in opposite directions [ascending motion], they partially cancel and forward displacement is relatively small). However, the extent to which representational momentum and representational gravity are similarly influenced by similar variables is not clear (e.g., limiting smooth pursuit eye movements for a smoothly moving target decreases representational momentum but does not influence representational gravity, de sá Teixeira et al., 2013). There are relatively few theories for representational gravity, with most investigators at least partially supporting the notion of an internal model or internalization of gravity (e.g., de sá Teixeira, 2016; de sá Teixeira et al., 2013; Lacquaniti et al., 2015; Zago, McIntyre, Senot, & Lacquaniti, 2008). Representational gravity has been suggested to reflect implied weight of the target (Hubbard, 1997; but see de sá Teixeira & Oliveira, 2014). Representational gravity might be related to vestibular function, as the time course of representational gravity can be disrupted by vestibular stimulation (de sá Teixeira et al., 2017).

1.3 Representational Friction and Representational Centripetal Force

Representational momentum and representational gravity have been the focus of the majority of investigation directed at effects of invariant physical principles on judged location. In addition to momentum and gravity, two other invariant physical principles that influence judged location but have received less investigation are friction and centripetal force. Forward displacement of a moving target is decreased when that target contacts another object, and this has been referred to as *representational friction* (e.g., Hubbard, 1995a, 1998). Interestingly, such decreases in forward displacement occur even when the velocity of targets experiencing implied friction does not decrease upon encountering implied friction and does not differ from that of an otherwise identical target in isolation.[1] Increases in friction also decrease the effect of target velocity on displacement. Judged location of a target moving along a circular trajectory is displaced forward (due to effects of implied momentum) and inward (due to effects of implied centripetal force), and the latter of these components is referred to as *representational centripetal force* (Hubbard, 1996). Representational

[1] Just as representational momentum and representational gravity combine to determine the ultimate displacement of a horizontally or vertically moving target, so too might representational friction and representational momentum combine. Representational friction operates in the direction opposite to representational momentum, and so representational friction partially cancels representational momentum, resulting in a smaller overall forward displacement of the target (Hubbard, 1995a, 1995b).

centripetal force is increased with increases in radius length of the target's trajectory or the target's angular velocity. Relatedly, when a moving target exits a spiral tube, forward displacement is larger for a target that continues to move in a curved (physically incorrect) path than for a target that moves in a straight (physically correct) path (Freyd & Jones, 1994), suggesting that displacement does not reflect objective Newtonian physical principles (for discussion, see Hubbard, 2005, 2006).

1.4 Looming Effect

The time-to-contact for an approaching object is usually underestimated, and this has been referred to as a *looming effect* (e.g., Neuhoff, 2001; for review, see Neuhoff, 2018). Whereas studies of many spatiotemporal illusions usually involved target motion in the picture plane, the looming effect involves target motion in depth (typically along or near the line-of-sight). Looming effects are larger for women than for men for auditory (Neuhoff, Planisek, & Seifritz, 2009) and visual (Schiff & Oldak, 1990) targets, larger when the moving target is interpreted as threatening (e.g., snakes, Vagnoni, Lourenco, & Longo, 2012; threatening faces, Brendel, DeLucia, Hecht, Stacy, & Larsen, 2012), and larger when the target leads to higher arousal (Brendel, Hecht, DeLucia, & Gamer, 2014). Looming stimuli automatically attract attention even if they are not relevant to the current task (Franconeri & Simons, 2003), and providing feedback regarding accuracy of time-to-contact judgments diminishes but does not eliminate the looming effect (Rosenblum, Gordon, & Wuestefeld, 2000). Looming effects can be found with human infants (Morrongiello, Hewitt, & Gotowiec, 1991) and with nonhuman primates (Ghazanfar, Neuhoff, & Logothetis, 2002). The looming effect is smaller for visual stimuli than for auditory stimuli; for multimodal stimuli in which both visual information and auditory information are available, the magnitude of the looming effect is similar to that obtained when only visual information is available (DeLucia, Preddy, & Oberfeld, 2016).

Changes in the projected retinal size, binocular disparity, and other depth cues regarding the target provide information regarding visual looming (Tresilian, 1999), and changes in intensity, interaural time differences, and a Doppler shift provide information regarding auditory looming (Rosenblum, Carello, & Pastore, 1987). Increases in size and in intensity for visual and auditory stimuli, respectively, are referred to as *visual tau* (e.g., Savelsbergh, Whiting, & Bootsma, 1991) and *acoustic tau* (e.g., Shaw, McGowan, & Turvey, 1991), respectively. One theory of the looming effect stresses evolutionary adaptiveness, as there is minimal cost in underestimating time-to-collision (and thus avoiding collision with time to spare) but significant cost in overestimating time-to-collision (and thus not avoiding collision), and Neuhoff (2018) presents a detailed discussion of whether the looming effect is adaptive. Underestimation of time-to-collision suggests perceived distance to the approaching target is also underestimated; indeed, underestimation of time-to-collision is consistent with representational momentum for the approaching object, as representational momentum for an approaching object would shift its position forward and decrease the judged distance between the target and observer (cf. Gray & Thornton, 2001). Auditory looming effects are larger under high cognitive load (McGuire, Gillath, & Vitevitch, 2016), and this is consistent with the larger representational momentum that occurs if attention is divided between the target and some other stimulus or task.

2 During Target Presentation

Three spatiotemporal illusions that involve judgments of location and occur during target presentation are considered. These include the flash-lag effect, the kappa effect, and the tau effect.

2.1 Flash-Lag Effect

When a briefly presented (i.e., flashed) stationary object is aligned with the location of a moving target, the location of the stationary object is perceived to lag behind the target, and this has been referred to as the *flash-lag effect* (e.g., Nijhawan, 1994; for review, see Hubbard, 2014b, 2018b). The flash-lag effect is increased with increases in target velocity

(Nijhawan, 1994), attending to multiple targets or tasks (Sarich, Chappell, & Burgess, 2007), target motion toward fixation (Brenner, van Beers, Rotman, & Smeets, 2006) or the observer (Harris, Duke, & Kopinska, 2006), increased uncertainty regarding target location (Kanai, Sheth, & Shimojo, 2004) or when the stationary object will appear (Vreven & Verghese, 2005), increased contrast of the moving target and stationary object with the background (Kanai et al., 2004) or of the moving target with the stationary object (Arnold, Ong, & Roseboom, 2009), decreased luminance of the stationary object (Öğmen, Patel, Bedell, & Camuz, 2004), and alignment of the stationary object with the leading edge of the target (Watanabe, Nijhawan, Khurana, & Shimojo, 2001). The flash-lag effect is found for stationary objects aligned with the middle or beginning, but not the end, of target motion (see Hubbard, 2014b). There are several related spatiotemporal illusions that can be obtained by varying parameters of a standard flash-lag effect display, and these include the flash-lead (the flashed object appears to lead, rather than lag, the moving target), flash-drag (the location of the flashed object is shifted in the direction of target motion), and flash-jump (a change in a moving target [e.g., color, size] is judged to occur later [farther along the trajectory] than it did) effects (see Hubbard, 2018b).

Several theories have been proposed for the flash-lag effect, and different theories emphasize either the temporal aspects or the spatial aspects of the effect. The earliest and most well-known theory involves motion extrapolation (Nijhawan, 1994, 2008), in which the location of the stationary object is veridically perceived, but the location of the moving target is extrapolated forward.[2] This extrapolation seems similar to representational momentum, and so the flash-lag effect might be a special case of representational momentum in which location of the target is compared to the location of a different (stationary) object rather than to the actual location of that target (Hubbard, 2013, 2014b). A second theory suggests

[2] Subsequent studies have suggested perception of the stationary object is not necessarily veridical. Shi and de'Sperati (2008) measured judged location for both the moving target and the flashed object. Judged locations for both stimuli were shifted forward, with the shift being larger for the moving target, and the combination of shifts being equivalent to the flash-lag effect. Hubbard (2008) demonstrated a similar pattern within a representational momentum study, in which both the target and a stationary object briefly presented near the end of the target trajectory were shifted forward, and the forward displacement of the target was larger than that for the stationary object.

information regarding moving targets is processed faster than information regarding stationary objects, and so by the time information regarding the stationary object has reached perceptual awareness, the target has moved some distance (Purushothaman, Patel, Bedell, & Öğmen, 1998). A third theory suggests attention is initially on the moving target, and when the stationary object is presented, attention is shifted to that object. By the time the stationary object is perceived, the target has moved some distance, and so location of the stationary object is perceived as lagging behind the moving target (Baldo & Klein, 1995; although see Baldo & Klein, 2010). A fourth theory suggests location information for the moving target at the time the stationary object was presented is based on an averaging of positions during a brief duration after the stationary object is presented, and this averaging shifts the represented location of the moving target forward (Eagleman & Sejnowski, 2000).

2.2 Kappa Effect and Tau Effect

Perhaps the most well-known examples of spatiotemporal illusions are the kappa effect and the tau effect. When differences between successive physical distances influence judgments of successive elapsed times, this is referred to as the *kappa effect* (e.g., Cohen, Hansel, & Sylvester, 1953), and when differences between successive elapsed times influence judgments of successive physical distances, this is referred to as the *tau effect* (e.g., Helson & King, 1931). More specifically, in the kappa effect, observers overestimate elapsed time in traveling between two successive stimuli when the physical distance between the stimuli is larger and underestimate elapsed time when the physical distance between two successive stimuli is smaller. In other words, if the distance between A and B is larger than the distance between B and C, observers will perceive that it takes longer to travel between A and B than to travel between B and C, even if the travel times are identical. Similarly, in the tau effect, observers overestimate physical distance between two successive stimuli when the elapsed time in traveling between stimuli is longer and underestimate physical distance when the elapsed time is shorter. In other words, if the elapsed time to travel between A and B is longer than the elapsed time to travel between B and C, observers perceive the distance between A and B to be larger than the distance between B and C, even if the physical distances are identical. A physical target is not necessary for the kappa effect or tau effect; the beginning and ending of a presented or implied motion can be indicated by flashing lights, brief tones, or brief tactile stimuli.

Both the kappa effect and the tau effect can be found in visual, auditory, and tactile modalities, as well as cross-modally. The kappa effect increases as velocity of motion increases (Henry & McAuley, 2009; Newman & Lee, 1972; Ono, 1969) and is larger for upward motion than for downward motion (Cohen, Hansel, & Sylvester, 1955), and similar effects of velocity and direction are found for the tau effect (Henry, McAuley, & Zaleha, 2009; Ono, 1969). Both the kappa effect and the tau effect have been attributed to observers assuming a constant velocity for the moving stimulus (e.g., Jones & Huang, 1982) and to surrounding context (Alards-Tomalin, Leboe-McGowan, & Mondor, 2013; Masuda, Kimura, Dan, & Wada, 2011), and both effects have been suggested to occur at a relatively early stage of processing (Shigeno, 1986). Masuda et al. (2011) speculated that the larger kappa effect with ascending motion than with descending motion was consistent with expectations that rising objects decelerate and falling objects accelerate and with representational gravity. The kappa and tau effects are similar to the filled duration illusion (a temporal duration containing more events is judged as longer than an equal temporal duration containing fewer events, e.g., Thomas & Brown, 1974) and clutter effect (a physical distance containing more objects is judged as longer than an equal physical distance containing fewer objects, e.g., Thorndyke, 1981), as additional duration or distance would be analogous to more events or objects. However, the filled duration illusion and clutter effect are usually viewed as temporal or spatial, respectively, rather than as spatiotemporal.

3 Beginning of Target Presentation

Four spatiotemporal illusions that involve judgments of location and occur at the beginning of target presentation are considered. These include the Fröhlich effect, onset repulsion effect, tandem effect, and illusory line motion.

3.1 Fröhlich Effect and Onset Repulsion Effect

The judged initial location of a moving target is often displaced forward in the direction of motion or backward in the direction opposite to motion, and this has been referred to as the *Fröhlich effect* (e.g., Müsseler & Aschersleben, 1998) or the *onset repulsion effect* (e.g., Thornton, 2002), respectively. The Fröhlich effect, similar to representational momentum, involves a forward displacement; however, studies in which observers indicated the initial location and the final location of the target suggested that the Fröhlich effect is not simply representational momentum for the initial location of the target or a forward displacement of the entire trajectory (Actis-Grosso & Stucchi, 2003; Hubbard & Motes, 2002). The Fröhlich effect increases with increases in target velocity (Müsseler & Aschersleben, 1998) and trajectory length (Geer & Schmidt, 2006), but effects of cueing the initial location are not clear (cf. Müsseler & Aschersleben, 1998; Müsseler & Tiggelbeck, 2013). The onset repulsion effect has received less investigation than the Fröhlich effect. The onset repulsion effect is increased with rightward or upward motion or with smooth motion rather than implied motion (Thornton, 2002), but findings regarding whether increases in velocity increase the onset repulsion effect are mixed (cf. Actis-Grosso & Stucchi, 2003; Hubbard & Motes, 2002; Kerzel, 2002; Thornton, 2002). The onset repulsion effect can displace judged locations of stationary objects near the initial location of the target in the direction of target displacement (Hubbard & Courtney, 2008) and is decreased by cueing the initial location (Hubbard & Ruppel, 2011a).

Several theories have been proposed for the Fröhlich effect (for review, see Kerzel, 2010; Müsseler & Kerzel, 2018). Early theories involved sensation time and metacontrast masking, but it was not clear why such accounts would apply only to the initial location and not the entire trajectory. Subsequent theories suggested visual attention to subsequent positions along the trajectory resulted in the initial position being masked but subsequent positions being visible (e.g., Kirschfeld & Kammer, 1999). Other theories involving attention focused on the shift of attention toward the target. When a target appears, attention is shifted toward the target. However, this shift takes time, and while attention is shifting, the target continues to move. A perceptual representation is not available before the end of the attention shift, and so the initial portion of the trajectory is not perceived. However, attentional theories cannot account for some findings (e.g., effects of trajectory length). Theories of the Fröhlich effect cannot account for onset repulsion because the target never occupied the location where it was perceived to appear (Müsseler & Kerzel, 2018), and so theories of the onset repulsion effect usually appealed to non-perceptual factors (e.g., overcompensation, see Thornton, 2002). The relationship between the Fröhlich effect and the onset repulsion effect is not clear, although some findings suggest an effect of context (e.g., a Fröhlich effect is more likely if the initial location is more predictable or closer to a boundary, whereas an onset repulsion is more likely if the initial location is less predictable or farther from a boundary, Hubbard & Motes, 2005; Müsseler & Kerzel, 2004).

3.2 Tandem Effect

If two moving targets emerge from behind a barrier in close succession and continue moving at the same velocity and in the same direction, the perceived distance between those targets is judged as smaller than the physical distance, and this is referred to as a *tandem effect* (e.g., Müsseler & Neumann, 1992). The tandem effect has received relatively little investigation. The magnitude of the tandem effect increases (i.e., perceived distance between the targets decreases) as physical distance between the two targets increases and as target velocities increase. One theory of the tandem effect proposed by Müsseler and Neumann (1992) and Müsseler (1999) suggests attention is initially focused on the first target. Once the second target appears, attention is shifted to that target. However, the second target continues to move while attention is being shifted. The location of the second target at the time the shift reaches that target is compared to where the first target was when the shift began. Given that the second target continued to move during the shift, its location is relatively farther forward, and so the distance between the first and second targets is perceived as smaller. Such an explanation is similar to the attention-shift mechanism suggested for the Fröhlich effect, and so the tandem effect might reflect a Fröhlich effect for the second target. Alternatively, the trailing edge of a target exhibits larger representational momentum than the leading edge (Hubbard & Ruppel, 2018b), and such anisotropy created by the first target might extend through perceived space to the second target, thus displacing the second target farther forward.

3.3 Illusory Line Motion

If a specific location is cued, then a stationary line subsequently presented near that cue is often perceived to "unfold" or "be drawn" from the end of the line closest to the cue to the end of the line farthest from the cue. This occurs even though the entirety of the line is presented simultaneously, and so this is referred to as *illusory line motion* (also referred to as *line motion illusion* or *shooting line illusion*, e.g., Downing & Treisman, 1997; Christie & Klein, 2005; Hamm & Klein, 2002). Illusory motion begins when the line (i.e., target) is presented; the end of the line closest to the cue is stationary, and the end of the line farthest from the cue is perceived to move away from the cue until the entire length is visible. Most studies of illusory line motion used cues involving changes in luminance, color, or shape. Studies of illusory line motion have involved observers rating velocity or other aspects of illusory motion (e.g., Christie & Klein, 2005), adjusting the stimulus so that no motion is perceived (e.g., Fuller & Carrasco, 2009), or making judgments regarding stimuli at different locations relative to the line (e.g., Hamm & Klein, 2002). Increases in the distance of the far end of the line from the cued location do not influence relative velocity of illusory line motion, but increases in line length result in faster apparent velocity (Hubbard & Ruppel, 2011b). Increases in latency between onset of the cue and onset of the line decreases relative velocity and strength of illusory line motion (Hubbard & Ruppel, 2018a). Perceived strength of illusory motion is influenced by elements that distinguish the line from the background (e.g., contrast and luminance; von Grünau & Faubert, 1994).

Several theories have been proposed for illusory line motion. The most well-known theory suggests illusory line motion is due to an attentional gradient. Attention is greater near the cued location, and so the part of the line closest to the cued location enters conscious awareness more quickly (Hikosaka, Miyauchi, & Shimojo, 1993); however, whether differences in attention occur before or after illusory motion has been debated (e.g., see Hamm & Klein, 2002). There is also debate regarding whether illusory line motion involves or is dependent upon exogenous attention or endogenous attention (cf. Blanco & Soto, 2009; Christie & Klein, 2005; Schmidt, 2000). Some demonstrations claiming to elicit illusory line motion have been suggested to actually elicit polarized gamma motion or transformational apparent motion (Ha, Li, Patten, & Hamm, 2018; Hamm, 2017). In polarized gamma motion (e.g., Kanizsa, 1979), a stationary object that suddenly appears is perceived to appear from the center outward (i.e., expanding from the center), whereas a stationary object that suddenly disappears is perceived to disappear from the periphery inward (i.e., contracting toward the center). In transformational apparent motion (e.g., Tse, 2006), an object that abruptly changes its shape is perceived to exhibit a continuous change between the initial and final shapes. In polarized gamma motion and transformational apparent motion, there is no physical motion, but observers perceive a smooth change extending across space and time and connecting the initial and final states (e.g., appearance and disappearance of the object in gamma motion, different shapes in transitional apparent motion).

4 Summary and Conclusions

There are several similarities of the spatiotemporal illusions considered here. A given variable can similarly affect multiple spatiotemporal illusions (e.g., increased perceived target velocity increases representational momentum, flash-lag effect, kappa effect, tau effect, Fröhlich effect, tandem effect; increased perceived threat from a target increases representational momentum and looming effect); however, there are data on only a few variables across multiple spatiotemporal illusions, and so the extent to which different variables have similar effects on different spatiotemporal illusions is an important area for further research. Displacements in judged location of a moving target or other stimulus can occur regardless of whether judgment occurs at the beginning, middle, or end of target presentation. Shifts or gradients of attention from one area of the visual field to another might be a component or contributor to several spatiotemporal illusions (e.g., flash-lag effect, Fröhlich effect, tandem effect, illusory line motion). Differences in neural processing latencies for different types or lengths of stimuli might be a component or contributor to several spatiotemporal illusions (e.g., flash-lag effect, kappa effect, tau effect), and relatedly, some spatiotemporal illusions might reflect compensation for neural processing latencies (e.g., spatiotemporal illusions reflecting invariant physical principles). Also, many spatiotemporal illusions involve a likely future location of a target (e.g., representational momentum, looming effect), and this might increase effectiveness of interactions with moving targets and be evolutionarily adaptive.

There are several differences between the spatiotemporal illusions considered here. One potential difference is that some spatiotemporal illusions appear to require physical motion (e.g., onset repulsion effect, tandem effect), whereas other spatiotemporal illusions do not require physical motion (e.g., representational gravity, illusory line motion). A second potential difference involves the time courses of different spatiotemporal illusions (e.g., representational momentum peaks more quickly than representational gravity); however, there are data on the time courses of only some spatiotemporal illusions, and the extent to which different spatiotemporal illusions exhibit a similar time course is an important area for further research (especially in light of the suggestions regarding dynamic mental representation below). Intriguingly, some apparent differences might reflect a deeper underlying similarity. For example, increased uncertainty of initial location increases the Fröhlich effect and decreases the onset repulsion effect, but this might reflect a general increase in forward displacement with increases in uncertainty of the initial location of the target (as an increased forward shift would increase the Fröhlich effect and decrease the onset repulsion effect), and this would be consistent with findings that increases in blurriness (and hence uncertainty) increase forward displacement in final target location (Fu, Shen, & Dan, 2001). Similarly, increased stationary object brightness increases the flash-lead effect and decreases the flash-lag effect, but this might reflect a more general facilitation of stimulus processing with increases in stimulus intensity.

It is unlikely that a single mechanism is responsible for all (or even most) spatiotemporal illusions. For example, some spatiotemporal illusions might primarily involve extrapolation of a likely future position (e.g., representational momentum, looming effect), differences in processing times between different types of stimuli (e.g., flash-lag effect, tau effect, kappa effect), some type of averaging processes (e.g., flash-lag effect), or properties of attention (e.g., Fröhlich effect, illusory line motion). However, the different potential mechanisms that could contribute to spatiotemporal illusions need not be mutually exclusive, and effects of some spatiotemporal illusions might be combined to influence the overall judged location (e.g., Hubbard, 1997, 2005, suggests representational momentum and representational gravity combine to influence displacement of a target; Baldo & Klein, 2010, suggest that multiple mechanisms combine to produce the flash-lag effect). Some spatiotemporal illusions (e.g., kappa effect, tau effect) might also share mechanisms with more purely temporal (e.g., filled duration illusion) or spatial (e.g., clutter effect) illusions. Alternatively, some spatiotemporal illusions might reflect special cases of other spatiotemporal illusions (e.g., the flash-lag effect might reflect representational momentum when target location is measured relative to a stationary object rather than relative to the actual target location; the looming effect might reflect representational momentum for an approaching object; the tandem effect might reflect a Fröhlich effect for the second of two moving objects).

Spatiotemporal illusions highlight how spatial processing and temporal processing are intimately interconnected. More broadly, changes in time are usually related to changes in space. Changes in judged location, usually considered as reflecting changes in spatial information, might actually reflect or be driven by changes in temporal information (i.e., experimental results suggested to reflect properties of spatial representation might actually reflect properties of temporal representation such as consistent directionality and continuity, cf. Freyd, 1987). Such a view supports the notion that mental representation can be dynamic or exhibit dynamic qualities. Indeed, dynamic mental representation includes temporal information as a necessary and intrinsic part of the representation (for discussion, see Freyd, 1987; Hubbard, 2015, 2017a, 2017b); therefore, temporal information is not merely an added tag or procedure applied to a static representation, but is an essential part of a dynamic representation. Indeed, if temporal information were removed from the representation of the stimulus, such an illusion would not be likely to occur, but the representation would also not be as useful. Furthermore, spatiotemporal illusions occur even if observers are knowledgeable about the illusions or try not to see the illusions, thus suggesting that such dynamics are not necessarily under the control of higher cognitively penetrable processes but might be part of the functional architecture of mental representation. By revealing the dynamic aspects of our representational system, spatiotemporal illusions highlight a critical aspect of our perceptual and cognitive experience.

References

Actis-Grosso, R., & Stucchi, N. (2003). Shifting the start: Backward mislocation of the initial position of a motion. *Journal of Experimental Psychology: Human Perception & Performance, 29*, 675–691.

Alards-Tomalin, D., Leboe-McGowan, L. C., & Mondor, T. A. (2013). Examining auditory kappa effects through manipulating intensity differences between sequential tones. *Psychological Research, 77*, 480–491.

Arnold, D. H., Ong, Y., & Roseboom, W. (2009). Simple differential latencies modulate, but do not cause the flash-lag effect. *Journal of Vision, 9*(5), 4. https://doi.org/10.1167/9.5.4

Baldo, M. V. C., & Klein, S. A. (1995). Extrapolation or attention shift? *Nature, 378*, 565–566.

Baldo, M. V. C., & Klein, S. A. (2010). Paying attention to the flash-lag effect. In R. Nijhawan & B. Khurana (Eds.), *Space and time in perception and action* (pp. 396–407). Cambridge, UK: Cambridge University Press.

Blanco, M. J., & Soto, D. (2009). Unconscious perception of a flash can trigger line motion illusion. *Experimental Brain Research, 192*, 605–613.

Brendel, E., DeLucia, P. R., Hecht, H., Stacy, R. L., & Larsen, J. T. (2012). Threatening pictures induce shortened time-to-contact estimates. *Attention, Perception, & Psychophysics, 74*(5), 979–987.

Brendel, E., Hecht, H., DeLucia, P. R., & Gamer, M. (2014). Emotional effects on time-to-contact judgments: Arousal, threat, and fear of spiders modulate the effect of pictorial content. *Experimental Brain Research, 232*, 2337–2347.

Brenner, E., van Beers, R. J., Rotman, G., & Smeets, J. B. J. (2006). The role of uncertainty in the systematic spatial mislocalization of moving objects. *Journal of Experimental Psychology: Human Perception and Performance, 32*, 811–825.

Christie, J., & Klein, R. M. (2005). Does attention cause illusory line motion? *Perception & Psychophysics, 67*(6), 1032–1043.

Cohen, J., Hansel, C. E. M., & Sylvester, J. D. (1953). A new phenomenon in time judgment. *Nature, 172*, 901.

Cohen, J., Hansel, C. E. M., & Sylvester, J. D. (1955). Interdependence in judgments of space, time and movement. *Acta Psychologica, 11*, 360–372.

Courtney, J. R., & Hubbard, T. L. (2008). Spatial memory and explicit knowledge: An effect of instruction on representational momentum. *Quarterly Journal of Experimental Psychology, 61*, 1778–1784.

de sá Teixeira, N. A. (2016). How fast do objects fall in visual memory? Uncovering the temporal and spatial features of representational gravity. *PLoS One, 11*(2). https://doi.org/10.1371/journal.pone.0148953

de sá Teixeira, N. A., Hecht, H., Diaz Artiles, A., Seyedmadani, K., Sherwood, D. P., & Young, L. R. (2017). Vestibular stimulation interferes with the dynamics of an internal representation of gravity. *Quarterly Journal of Experimental Psychology, 70*, 2290–2305.

de sá Teixeira, N. A., Hecht, H., & Oliveira, A. M. (2013). The representational dynamics of remembered projectile locations. *Journal of Experimental Psychology: Human Perception and Performance, 39*(6), 1690–1699.

de sá Teixeira, N. A., & Oliveira, A. M. (2014). Spatial and foveal biases, not perceived mass or heaviness, explain the effect of target size on representational momentum and representational gravity. *Journal of Experimental Psychology: Learning, Memory, and Cognition, 40*(6), 1664–1679.

DeLucia, P. R., Preddy, D., & Oberfeld, D. (2016). Audiovisual integration of time-to-contact information for approaching objects. *Multisensory Research, 29*, 365–395.

Downing, P. E., & Treisman, A. M. (1997). The line-motion illusion: Attention or impletion? *Journal of Experimental Psychology: Human Perception and Performance, 23*, 768–779.

Eagleman, D. M., & Sejnowski, T. J. (2000). Motion integration and postdiction in visual awareness. *Science, 287*, 2036–2038.

Franconeri, S. L., & Simons, D. J. (2003). Moving and looming stimuli capture attention. *Perception & Psychophysics, 65*(7), 999–1010.

Freyd, J. J. (1987). Dynamic mental representations. *Psychological Review, 94*, 427–438.

Freyd, J. J., & Finke, R. A. (1984). Representational momentum. *Journal of Experimental Psychology: Learning, Memory, and Cognition, 10*, 126–132.

Freyd, J. J., & Johnson, J. Q. (1987). Probing the time course of representational momentum. *Journal of Experimental Psychology: Learning, Memory, and Cognition, 13*, 259–268.

Freyd, J. J., & Jones, K. T. (1994). Representational momentum for a spiral path. *Journal of Experimental Psychology: Learning, Memory, and Cognition, 20*, 968–976.

Freyd, J. J., Pantzer, T. M., & Cheng, J. L. (1988). Representing statics as forces in equilibrium. *Journal of Experimental Psychology: General, 117*, 395–407.

Fu, Y. X., Shen, Y., & Dan, Y. (2001). Motion-induced perceptual extrapolation of blurred visual targets. *Journal of Neuroscience, 21*, RC172.

Fuller, S., & Carrasco, M. (2009). Perceptual consequences of visual performance fields: The case of the line motion illusion. *Journal of Vision, 9*, 1–17. https://doi.org/10.1167/9.4.13

Geer, M., & Schmidt, W. C. (2006). Perception of initial moving target signals: Support for a cumulative lateral inhibition theory. *Journal of Experimental Psychology: Human Perception and Performance, 32*(5), 1185–1196.

Ghazanfar, A. A., Neuhoff, J. G., & Logothetis, N. K. (2002). Auditory looming perception in rhesus monkeys. *Proceedings of the National Academy of Sciences of the USA, 99*(24), 15755–15757.

Gray, R., & Thornton, I. M. (2001). Exploring the link between time to collision and representational momentum. *Perception, 30*, 1007–1022.

Greenstein, M., Franklin, N., Martins, M., Sewack, C., & Meier, M. A. (2016). When anticipation beats accuracy: Threat alters memory for dynamic scenes. *Memory & Cognition, 44*, 633–649.

Ha, H., Li, D., Patten, B., & Hamm, J. P. (2018). The relationship between flash based illusory line motion and exogenous visual attention. *Canadian Journal of Experimental Psychology, 72*(1), 38–47.

Hamm, J. P. (2017). A comparison of colour, shape, and flash induced illusory line motion. *Attention, Perception, & Psychophysics, 79*, 911–928.

Hamm, J. P., & Klein, R. M. (2002). Does attention follow the motion in the "shooting line" illusion? *Perception & Psychophysics, 64*, 279–291.

Harris, L. R., Duke, P. A., & Kopinska, A. (2006). Flash lag in depth. *Vision Research, 46*, 2735–2742.

Hayes, A. E., & Freyd, J. J. (2002). Representational momentum when attention is divided. *Visual Cognition, 9*, 8–27.

Helson, H., & King, S. M. (1931). The tau effect: An example of psychological relativity. *Journal of Experimental Psychology, 14*, 202–217.

Henry, M. J., & McAuley, J. D. (2009). Evaluation of an imputed pitch velocity model of the auditory kappa effect. *Journal of Experimental Psychology: Human Perception and Performance, 35*(2), 551–564.

Henry, M. J., McAuley, J. D., & Zaleha, M. (2009). Evaluation of an imputed pitch velocity model of the auditory tau effect. *Attention, Perception, & Psychophysics, 71*, 1399–1413.

Hikosaka, O., Miyauchi, S., & Shimojo, S. (1993). Focal visual attention produces illusory temporal order and motion sensation. *Vision Research, 33*, 1219–1240.

Hubbard, T. L. (1990). Cognitive representation of linear motion: Possible direction and gravity effects in judged displacement. *Memory & Cognition, 18*, 299–309.

Hubbard, T. L. (1995a). Cognitive representation of motion: Evidence for friction and gravity analogues. *Journal of Experimental Psychology: Learning, Memory, and Cognition, 21*, 241–254.

Hubbard, T. L. (1995b). Environmental invariants in the representation of motion: Implied dynamics and representational momentum, gravity, friction, and centripetal force. *Psychonomic Bulletin & Review, 2*, 322–338.

Hubbard, T. L. (1996). Representational momentum, centripetal force, and curvilinear impetus. *Journal of Experimental Psychology: Learning, Memory, and Cognition, 22*, 1049–1060.

Hubbard, T. L. (1997). Target size and displacement along the axis of implied gravitational attraction: Effects of implied weight and evidence of representational gravity. *Journal of Experimental Psychology: Learning, Memory, and Cognition, 23*, 1484–1493.

Hubbard, T. L. (1998). Some effects of representational friction, target size, and memory averaging on memory for vertically moving targets. *Canadian Journal of Experimental Psychology, 52*, 44–49.

Hubbard, T. L. (2005). Representational momentum and related displacements in spatial memory: A review of the findings. *Psychonomic Bulletin & Review, 12*, 822–851.

Hubbard, T. L. (2006). Bridging the gap: Possible roles and contributions of representational momentum. *Psicologica, 27*, 1–34.

Hubbard, T. L. (2008). Representational momentum contributes to motion induced mislocalization of stationary objects. *Visual Cognition, 16*, 44–67.

Hubbard, T. L. (2010). Approaches to representational momentum: Theories and models. In R. Nijhawan & B. Khurana (Eds.), *Space and time in perception and action* (pp. 338–365). Cambridge, UK: Cambridge University Press.

Hubbard, T. L. (2013). Do the flash-lag effect and representational momentum involve similar extrapolations? *Frontiers in Psychology, 4*, 290. https://doi.org/10.3389/fpsyg.2013.00290

Hubbard, T. L. (2014a). Forms of momentum across space: Representational, operational, and attentional. *Psychonomic Bulletin & Review, 21*, 1371–1403.

Hubbard, T. L. (2014b). The flash-lag effect and related mislocalizations: Findings, properties, and theories. *Psychological Bulletin, 140*, 308–338.

Hubbard, T. L. (2015). Forms of momentum across time: Behavioral and psychological. *Journal of Mind and Behavior, 36*, 47–82.

Hubbard, T. L. (2017a). Toward a general theory of momentum-like effects. *Behavioural Processes, 141*, 50–66.

Hubbard, T. L. (2017b). Momentum in music: Musical succession as physical motion. *Psychomusicology: Music, Mind, and Brain, 27*, 14–30.

Hubbard, T. L. (2018a). Influences on representational momentum. In T. L. Hubbard (Ed.), *Spatial biases in perception and cognition* (pp. 121–138). Cambridge, UK: Cambridge University Press.

Hubbard, T. L. (2018b). The flash-lag effect. In T. L. Hubbard (Ed.), *Spatial biases in perception and cognition* (pp. 139–155). Cambridge, UK: Cambridge University Press.

Hubbard, T. L., & Bharucha, J. J. (1988). Judged displacement in apparent vertical and horizontal motion. *Perception & Psychophysics, 44*, 211–221.

Hubbard, T. L., & Courtney, J. R. (2008). The onset-repulsion effect and motion-induced mislocalization of a stationary object. *Perception, 37*, 1386–1398.

Hubbard, T. L., Kumar, A. M., & Carp, C. L. (2009). Effects of spatial cueing on representational momentum. *Journal of Experimental Psychology: Learning, Memory, and Cognition, 35*, 666–677.

Hubbard, T. L., & Motes, M. A. (2002). Does representational momentum reflect a distortion of the length or the endpoint of a trajectory? *Cognition, 82*, B89–B99.

Hubbard, T. L., & Motes, M. A. (2005). An effect of context on whether memory for initial position exhibits a Fröhlich effect or an onset repulsion effect. *Quarterly Journal of Experimental Psychology, 58A*, 961–979.

Hubbard, T. L., & Ruppel, S. E. (1999). Representational momentum and the landmark attraction effect. *Canadian Journal of Experimental Psychology, 53*, 242–256.

Hubbard, T. L., & Ruppel, S. E. (2011a). Effects of spatial cuing on the onset repulsion effect. *Attention, Perception, & Psychophysics, 73*, 2236–2248.

Hubbard, T. L., & Ruppel, S. E. (2011b). Effects of temporal and spatial separation on velocity and strength of illusory line motion. *Attention, Perception, & Psychophysics, 73*, 1133–1146.

Hubbard, T. L., & Ruppel, S. E. (2018a). Does allocation of attention influence relative velocity and strength of illusory line motion? *Frontiers in Psychology, 9*, 147. https://doi.org/10.3389/fpsyg2018.00147

Hubbard, T. L., & Ruppel, S. E. (2018b). Representational momentum and anisotropies in nearby visual space. *Attention, Perception, & Psychophysics, 80*, 94–105.

Jones, B., & Huang, Y. L. (1982). Space-time dependencies in psychophysical judgment of extent and duration: Algebraic models of the tau and kappa effects. *Psychological Bulletin, 91*(1), 128–142.

Kanai, R., Sheth, B. R., & Shimojo, S. (2004). Stopping the motion and sleuthing the flash-lag effect: Spatial uncertainty is the key to perceptual mislocalization. *Vision Research, 44*, 2605–2619.

Kanizsa, G. (1979). *Organization in vision: Essays on Gestalt perception*. New York, NY: Praeger.

Kerzel, D. (2002). Different localization of motion onset with pointing and relative judgments. *Experimental Brain Research, 145*, 340–350.

Kerzel, D. (2010). The Fröhlich effect: Past and present. In R. Nijhawan & B. Khurana (Eds.), *Space and time in perception and action* (pp. 321–337). Cambridge, UK: Cambridge University Press.

Kirschfeld, K., & Kammer, T. (1999). The Fröhlich effect: A consequence of the interaction of visual focal attention and metacontrast. *Vision Research, 39*(22), 3702–3709.

Lacquaniti, F., Bosco, G., Gravano, S., Indovina, I., La Scaleia, B., Maffei, V., & Zago, M. (2015). Gravity in the brain as a reference for space and time perception. *Multisensory Research, 28*(5-6), 397–426.

Masuda, T., Kimura, A., Dan, I., & Wada, Y. (2011). Effects of environmental context on temporal perception bias in apparent motion. *Vision Research, 51*, 1728–1740.

McGuire, A. B., Gillath, O., & Vitevitch, M. S. (2016). Effects of mental resource availability on looming task performance. *Attention, Perception, & Psychophysics, 78*, 107–113.

Morrongiello, B. A., Hewitt, K. L., & Gotowiec, A. (1991). Infants' discrimination of relative distance in the auditory modality: Approaching versus receding sound sources. *Infant Behavior & Development, 14*(2), 187–208.

Müsseler, J. (1999). Perceiving and measuring of spatiotemporal events. In J. S. Jordan (Ed.), *Modeling consciousness across the disciplines* (pp. 95–112). University Press of America.

Müsseler, J., & Aschersleben, G. (1998). Localizing the first position of a moving stimulus: The Fröhlich effect and an attention-shifting explanation. *Perception & Psychophysics, 60*(4), 683–695.

Müsseler, J., & Kerzel, D. (2004). The trial context determines adjusted localization of stimuli: Reconciling the Fröhlich and onset repulsion effects. *Vision Research, 44*(19), 2201–2206.

Müsseler, J., & Kerzel, D. (2018). Mislocalizations at the onset position of moving stimuli. In T. L. Hubbard (Ed.), *Spatial biases in perception and cognition* (pp. 109–120). Cambridge, UK: Cambridge University Press.

Müsseler, J., & Neumann, O. (1992). Apparent distance reduction with moving stimuli (Tandem Effect): Evidence for an attention-shifting model. *Psychological Research, 54*, 246–266.

Müsseler, J., & Tiggelbeck, J. (2013). The perceived onset position of a moving target: Effects of trial contexts are evoked by different attentional allocations. *Attention, Perception, & Psychophysics, 75*, 349–357.

Neuhoff, J. G. (2001). An adaptive bias in the perception of looming auditory motion. *Ecological Psychology, 13*(2), 87–110.

Neuhoff, J. G. (2018). Adaptive biases in visual and auditory looming perception. In T. L. Hubbard (Ed.), *Spatial biases in perception and cognition* (pp. 180–190). Cambridge, UK: Cambridge University Press.

Neuhoff, J. G., Planisek, R., & Seifritz, E. (2009). Adaptive sex differences in auditory motion perception: Looming sounds are special. *Journal of Experimental Psychology: Human Perception and Performance, 35*(1), 225–234.

Newman, C. V., & Lee, S. G. (1972). The effect of real and imputed distance on judgments of time: Some experiments on the Kappa phenomenon. *Psychonomic Science, 29*(4-A), 207–211.

Nijhawan, R. (1994). Motion extrapolation in catching. *Nature, 370*, 256–257.

Nijhawan, R. (2008). Visual prediction: Psychophysics and neurophysiology of compensation for time delays. *Behavioral and Brain Sciences, 31*, 179–198.

Öğmen, H., Patel, S. S., Bedell, H. E., & Camuz, K. (2004). Differential latencies and the dynamics of the position computation process for moving targets, assessed with the flash-lag effect. *Vision Research, 44*, 2109–2128.

Ono, A. (1969). Interdependence in successive judgments of the duration, distance and speed of a manual movement. *Tohoku Psychologica Folia, 28*(1-2), 29–53.

Purushothaman, G., Patel, S. S., Bedell, H. E., & Öğmen, H. (1998). Moving ahead through differential visual latency. *Nature, 396*, 424.

Reed, C. L., & Vinson, N. G. (1996). Conceptual effects on representational momentum. *Journal of Experimental Psychology: Human Perception and Performance, 22*, 839–850.

Rosenblum, L. D., Carello, C., & Pastore, R. E. (1987). Relative effectiveness of three stimulus variables for locating a moving sound source. *Perception, 16*(2), 175–186.

Rosenblum, L. D., Gordon, M. S., & Wuestefeld, A. P. (2000). Effects of performance feedback and feedback withdrawal on auditory looming perception. *Ecological Psychology, 12*(4), 273–291.

Ruppel, S. E., Fleming, C. N., & Hubbard, T. L. (2009). Representational momentum is not (totally) impervious to error feedback. *Canadian Journal of Experimental Psychology, 63*, 49–58.

Sarich, D., Chappell, M., & Burgess, C. (2007). Dividing attention in the flash-lag illusion. *Vision Research, 47*, 544–547.

Savelsbergh, G. J., Whiting, H. T., & Bootsma, R. J. (1991). Grasping tau. *Journal of Experimental Psychology: Human Perception and Performance, 17*(2), 315–322.

Schiff, W., & Oldak, R. (1990). Accuracy of judging time to arrival: Effects of modality, trajectory, and gender. *Journal of Experimental Psychology: Human Perception and Performance, 16*(2), 303–316.

Schmidt, W. C. (2000). Endogenous attention and illusory line motion reexamined. *Journal of Experimental Psychology: Human Perception and Performance, 26*, 980–996.

Shaw, B. K., McGowan, R. S., & Turvey, M. T. (1991). An acoustic variable specifying time-to-contact. *Ecological Psychology, 3*, 253–261.

Shi, Z., & de'Sperati, C. (2008). Motion-induced positional biases in the flash-lag configuration. *Cognitive Neuropsychology, 25*, 1027–1038.

Shigeno, S. (1986). The auditory tau and kappa effects for speech and non-speech stimuli. *Perception & Psychophysics, 40*(1), 9–19.

Thomas, E. A., & Brown, I. (1974). Time perception and the filled-duration illusion. *Perception & Psychophysics, 16*(3), 449–458.

Thorndyke, P. W. (1981). Distance estimation from cognitive maps. *Cognitive Psychology, 13*(4), 526–550.

Thornton, I. M. (2002). The onset repulsion effect. *Spatial Vision, 15*(2), 219–243.

Tresilian, J. R. (1999). Visually timed action: Time-out for 'tau'? *Trends in Cognitive Sciences, 3*(8), 301–310.

Tse, P. U. (2006). Neural correlates of transformational apparent motion. *NeuroImage, 31*, 766–773.

Vagnoni, E., Lourenco, S. F., & Longo, M. R. (2012). Threat modulates perception of looming visual stimuli. *Current Biology, 22*, R826–R827.

von Grünau, M., & Faubert, J. (1994). Intraattribute and interattribute motion induction. *Perception, 23*, 913–928.

Vreven, D., & Verghese, P. (2005). Predictability and the dynamics of position processing in the flash-lag effect. *Perception, 34*, 31–44.

Watanabe, K., Nijhawan, R., Khurana, B., & Shimojo, S. (2001). Perceptual organization of moving stimuli modulates the flash-lag effect. *Journal of Experimental Psychology: Human Perception and Performance, 27*, 879–894.
Zago, M. (2018). Perceptual and motor biases in reference to gravity. In T. L. Hubbard (Ed.), *Spatial biases in perception and cognition* (pp. 156–166). Cambridge, UK: Cambridge University Press.
Zago, M., McIntyre, J., Senot, P., & Lacquaniti, F. (2008). Internal models and prediction of visual gravitational motion. *Vision Research, 48*, 1532–1538.

17

Perceptual Illusions Caused by Discrete Sampling

Rodika Sokoliuk and Rufin VanRullen

1 Human Visual Perception—Discrete or Continuous?

We believe what we see, and what we cannot see with our own eyes is sometimes hard to believe. Therefore, the notion of discrete visual perception might seem odd. How could our own perception, which provides us with a continuous stream of visual information, actually be based on discrete snapshots of the world around us? How would we not be aware of that? In the following, we will elaborate on the discrete perception theory and present the existing evidence for it. Importantly, we will review neuroscientific and psychophysical work investigating different

---

R. Sokoliuk (✉)
Centre for Human Brain Health, University of Birmingham, Birmingham, UK
e-mail: r.sokoliuk@bham.ac.uk

R. VanRullen
Centre de Recherche Cerveau et Cognition, Université Paul Sabatier, Toulouse, France
e-mail: rufin.vanrullen@cnrs.fr

V. Arstila et al. (eds.), *The Illusions of Time*,
https://doi.org/10.1007/978-3-030-22048-8_17

optical illusions that could reveal the perceptual consequences of discrete perception.

Although it remains a controversial debate, the idea of discrete visual perception arose a long time ago in theories (James, 1890) and was also underpinned by behavioral evidence. As early as 1875, Exner reported evidence for discrete visual perception (Exner, 1875). In a behavioral experiment, participants were exposed to two succeeding light stimuli that were presented at the same location with varying inter-stimulus intervals. Participants were asked to report the number of perceived light stimuli, which Exner found was correlated with the length of the interval between them. Shorter intervals (<44 ms) thereby led to the perception of only one stimulus whereas longer intervals (>44 ms) resulted in two perceived stimuli. These results were replicated by many studies, several of which are presented below. This experimental paradigm was extended by separating the light sources spatially, and judgments of temporal order of appearance of the stimuli were investigated. Kristofferson hereby found that participants' performance to judge the order of appearance increased with the length of the inter-stimulus interval. Thereby, 80–120 ms were found to be the optimal condition for subjects to clearly separate two succeeding stimuli in time (Kristofferson, 1967). Stroud worked on similar psychophysical experiments in visual perception and integrated his findings into a theory—his "perceptual moment theory" in 1956 (Stroud, 1956); he compared human visual perception to a cinematograph showing individual snapshots that build a continuous visual stream when played in fast succession. He came to this theory when replicating Exner's findings (Exner, 1875) by observing a critical inter-stimulus interval of 62.5 ms on an average that either led to the simultaneous or sequential perception of two successive light stimuli presented at the same location. He concluded that stimuli that were perceived simultaneously would fall into the same "moment" and thus be perceived as simultaneous; stimuli that were presented with the same inter-stimulus interval but perceived sequentially would reflect a "late arrival" of the first stimulus in the perceptual moment so that the two stimuli would fall into two succeeding moments (see Fig. 17.1). His assumption was that a mechanism in the hierarchical processing of visual

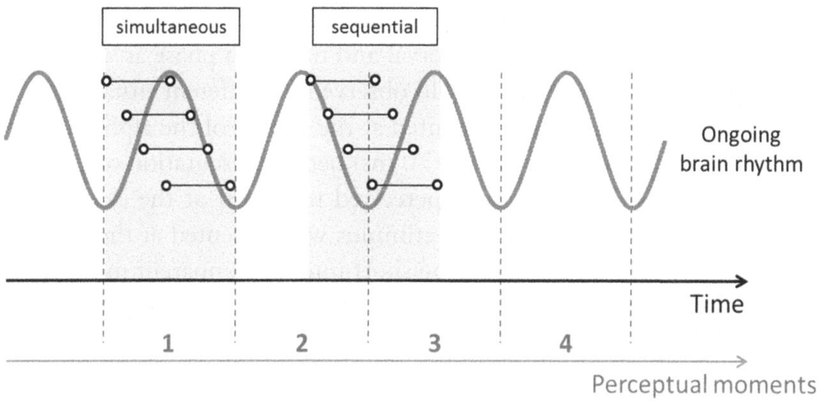

Fig. 17.1 The perceptual moment theory. Visual perception is divided into separate perceptual moments, shown here as the oscillatory cycles of an ongoing brain oscillation (gray). Two visual stimuli (black circles) that are presented successively with a certain delay seem to appear simultaneously, if they fall into the same perceptual moment ("simultaneous"). With the same delay, if the first stimulus is presented at a different phase and the two stimuli fall into separate perceptual moments, they are perceived one after the other ("sequential") (This figure was inspired by Baumgarten, Schnitzler & Lange, 2015)

information is separating the continuous visual stream into "moments" of approximately 100 ms and thus he used the term "perceptual moments".

All these studies had a specific timing in common that led several researchers to the assumption that the human alpha rhythm (8–12 Hz) could be the underlying neuronal mechanism, providing perceptual "moments" or "cycles" of ~100 ms length. The perceptual moment theory is the foundation of numerous studies in visual perception that tried to solve the issue of whether alpha oscillations are the underlying brain rhythm that separates the visual stream into individual snapshots. A very convincing, but to date not replicated study, was presented by Varela and colleagues in 1981 (Varela, Toro, John, & Schwartz, 1981). Including electroencephalography (EEG) into the equation, the authors provided an actual measure of the human alpha rhythm. Their paradigm used LED lights at two close spatial locations which were presented sequentially with different inter-stimulus delays. Importantly, Varela et al. measured the phase of ongoing alpha oscillations at occipital electrode locations and used it to control online the moment of stimulus presentation. Modulating the inter-stimulus interval and the alpha phase at which the stimuli were presented, Varela could observe four different situations: (i) When the first stimulus was presented at the trough of the alpha oscillation followed by a short interval (<70 ms) before presentation of the second stimulus, both stimuli were perceived to appear at the exact same time. (ii) However, when the first stimulus was presented at the peak of the oscillation followed by the same short interval, "apparent motion" of the first stimulus to the location of the second stimulus could be observed. (iii) Presenting the first stimulus at the trough of the alpha cycle, followed by a slightly longer interval (>80 ms) before presenting the second stimulus led as well to the perception of apparent motion. (iv) However, when the first stimulus was presented at the peak of the alpha oscillation followed by the same long interval, observers could clearly identify sequential presentation of the two stimuli. Importantly, the critical inter-stimulus delay leading to different percepts differed between subjects, which suggests that it might depend on the individual alpha frequency (IAF) of each participant. As mentioned above, this highly interesting result has not been replicated so far in vision (there is one recent report in somatosensory perception; Baumgarten et al., 2015), which did not help in the acceptance of the discrete perception theory; nevertheless, it inspired a lot of vision scientists in their research directions.

Given the different possible visual percepts for two succeeding stimuli with a fixed interval—be it simultaneous, sequential (Exner, 1875; Stroud, 1956; Kristofferson, 1967) or apparent motion from one stimulus location to the other (Varela et al., 1981)—it seems that the exact moment at which a stimulus is presented is of crucial importance. In terms of oscillations, this moment would reflect a specific oscillatory phase, which changes over time (from 0 to 2π; cf. Fig. 17.1). Recently, a distinction between "discrete" and "rhythmic" perception has been suggested (VanRullen, 2016): discrete perception describes a situation where perception is temporally divided into individual snapshots, whereas rhythmic perception merely shows continuous periodic modulation of perception, providing phases that lead to higher or lower efficiency in the processing of sensory information. The underlying brain mechanism for both could be ongoing oscillations, and the crucial component that would determine different perceptual outcomes would be their phase. Discrete perception would result in temporal phenomena as described above, where the oscillatory phase would modulate the timing of perception of two succeeding events—if an event arrives "too late", it would be delayed up to the next moment or snapshot (cf. Fig. 17.1). In rhythmic perception, oscillatory phase would rather influence whether low-intensity visual stimuli would reach perceptual threshold or not. Thus, in rhythmic perception, the probability or intensity of perception of a visual stimulus would be modulated, rather than its timing in discrete perception (VanRullen, 2016).

Recent progress in the field has revealed that the phase of 7–15 Hz oscillations influences visual perception crucially. Here, a specific phase value determines the way an incoming visual stimulus will be processed. In terms of rhythmic visual perception, oscillatory phase modulates excitability and therefore the probability or intensity with which a visual stimulus will be perceived. For a brief light stimulus of very low intensity, it can happen that the stimulus will not be perceived at all, whereas a stimulus with the same properties (same luminance and presentation duration) would be perceived if it arrived at a different moment in time, and thus at a different phase of an ongoing 7–15 Hz oscillation (Busch, Dubois, & VanRullen, 2009; Busch & VanRullen, 2010; Dugué, Marque, & VanRullen, 2011; Mathewson, Gratton, Fabiani, Beck, & Ro, 2009; Scheeringa, Mazaheri, Bojak, Norris, & Kleinschmidt, 2011). Therefore, there seem to be "good" and "bad" phases for visual perception; phases that reflect high and phases that reflect low excitability states. In terms of discrete visual perception in contrast, oscillatory phase modulates the timing with which a visual stimulus is perceived. An interesting phenomenon in this regard is patients suffering from akinetopsia. This neuropsychological disorder prevents patients from perceiving a smooth stream of visual motion; instead, these patients perceive individual snapshots of the movement (for a review, see Zeki, 1991). A similar phenomenon has been reported to exist upon the use of the hallucinogenic drug LSD (Lysergic acid diethylamide) which led Dubois and VanRullen to the idea of conducting an online study with a large number of LSD users. In the study, participants were exposed to short video clips that were "discretized" at different sampling rates, mimicking different lengths of "snapshots". Observers would then compare the video clips with their own experience of visual motion perception in the past upon LSD intake. Interestingly, they found that video clips that presented the snapshots at a frequency of 15–20 Hz showed the strongest concordance with the visual experience of the drug users (Dubois & VanRullen, 2011) and concluded that this phenomenon was probably linked to the discrete character of visual perception. Fortunately, healthy individuals in their daily life are not aware of the fluctuations in excitability (rhythmic visual perception), nor can they identify the individual "snapshots" (discrete visual perception); otherwise, the world around us would seem to flicker at around 10 Hz. However, some visual illusions can trick our minds and lead to the conscious perception of our own brain rhythms. In the following, different optical illusions will be presented that all share a specific property. All illusions can be explained by or seem to be related to ongoing brain oscillations, which can be measured using electroencephalography (EEG); we will present different visual illusions that reflect rhythmic visual perception (the flickering wheel illusion (Sokoliuk & VanRullen, 2013) and the triple-flash illusion (Gulbinaite, İlhan, & VanRullen, 2017)), discrete visual perception (the continuous wagon wheel illusion (c-WWI) (e.g. VanRullen et al., 2006) and the flash-lag effect (FLE) (Chota & VanRullen, 2019)) and visual illusions that can be explained by both rhythmic and discrete visual perception (the fluttering heart illusion (Minami & Amano, 2017) and the saltation effect (Nakayama, Motoyoshi, & Sato, 2018)).

2 The Flickering Wheel Illusion

The flickering wheel illusion (Sokoliuk & VanRullen, 2013) can be experienced when a specific black-and-white wheel stimulus (see Fig. 17.2 A) is viewed in the visual periphery. A regular flicker can be perceived emerging from the stimulus center; this illusory flicker increases in intensity while eye movements are performed. Importantly, the observer should not look at the stimulus itself. To explore this illusion, read the legend of Fig. 17.2 while covertly attending (without looking at the stimulus) to the center of the wheel stimulus in panel A of Fig. 17.2. Another way to

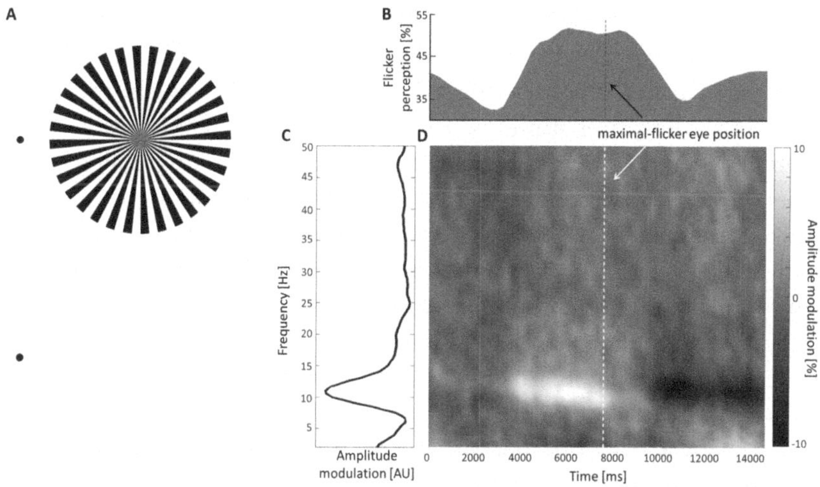

Fig. 17.2 The flickering wheel illusion. (a) When fixating the dot to the left of the wheel stimulus or when moving your eyes (like while you are reading these lines) while you are paying attention to the center of the wheel stimulus, you should be able to experience a regular flicker emerging from the center of the stimulus. (b) When following a dot slowly circling around the wheel stimulus, observers had a "preferred phase", a location of the dot with respect to the wheel stimulus where illusory flicker perception was the strongest. Observers responded to flicker perception by pressing a button (gray area). Data is centered on the mean preferred phase of the illusion (dashed vertical black line). (c and d) EEG correlate of the illusory flicker, centered on the mean preferred phase for the illusion (dashed vertical white line). A strong increase in alpha power (brighter colors; ~10 Hz, cf. (c)) occurs simultaneously to an increase in illusory flicker perception, followed by a decrease in alpha power (darker colors) (Figure adapted from Sokoliuk & VanRullen, 2013). (Color figure online)

experience this illusion is to fixate the black dot (Fig. 17.2a) to the left of the stimulus for ~10 seconds. As you change fixation to the dot below, you should be able to see the flicker illusion in the afterimage of the stimulus on the white background.

After conducting a behavioral experiment which gave insight into the optimal stimulus parameters in order to perceive the illusion (32 spokes, maximum luminance contrast) as well as information about the estimated perceived flicker frequency (averaged over ten participants: 9.1 Hz), we were interested in the potential oscillatory neuronal correlates of the illusion given the regular rhythmicity observed in the illusion. We performed an experiment where we measured brain activity using EEG while participants were following with their gaze a dot which was orbiting around the wheel stimulus. Participants indicated via button press the time when they were perceiving the illusion. Interestingly, all participants showed a preferred phase, that is, a specific location of the dot relative to the wheel stimulus, where the flicker was significantly stronger than at all other locations of the orbit (Fig. 17.2b). When comparing behavioral and EEG results, an increase in alpha power could be observed at the preferred phase for the illusion, which was followed by a decrease in alpha power (Figs. 17.2c and 17.2d). We believe that this increase in alpha power that temporally coincides with the illusory perception is a direct correlate of the perception of our own alpha oscillations. Interestingly, in a control condition, where a similar wheel stimulus with a lower spatial frequency (eight black-and-white spokes) that did not lead to illusory flicker perception was presented instead, a similar power modulation in the alpha band could be observed. However, this power increase was significantly lower than in the illusory condition. The alpha increase in the control condition thus seems to be related to a specific gaze position. Indeed, such alpha increases when directing the gaze to locations in the upper visual field have been reported much earlier (Dewan, 1967; Fenwick & Walker, 1968; Mulholland & Evans, 1965). This prevalent alpha power increase at this specific gaze location might thus reflect a certain alpha power level that is needed to induce the illusory percept; importantly, this illusion only happens in an interaction with a stimulus of specific spatial frequency (it does not happen for the control stimulus). The illusory flicker that was estimated to occur in the alpha band would then provoke a positive feedback loop, increasing the predominant alpha power, as observed in the EEG. But how can our alpha oscillations be "translated" into an illusory visual flicker? Speculatively, the rhythmic flicker perception could represent periodically recurring "bad oscillatory phases" for perception. This hypothesis builds on the "pulsed inhibition theory" brought forward by Mathewson and colleagues (Mathewson et al., 2011) and Jensen and Mazaheri (Jensen & Mazaheri, 2010). According to this theory, alpha oscillations are modulating cortical excitability. When alpha power is high enough, this rhythmic modulation at

~10 Hz would influence visual perception which would periodically drop below the detection threshold, every ~100 ms, at a specific phase of the alpha rhythm. These moments, when visual perception drops below the perceptual threshold, could reflect the moments when we perceive the flicker—one "gap" in the incoming visual stream, recurring every 100 ms. This could lead to the perceived illusory flicker, here, at ~10 Hz. We believe that this illusion is one of the rare examples that makes us consciously aware of the rhythmic character of our visual perception, oscillating in the alpha frequency band and providing fluctuations of excitability.

3 The Triple-Flash Illusion

Another illusion that reflects rhythmic visual perception is the triple-flash illusion, originally discovered by Bowen (Bowen, 1989). In this illusion, a third flash is occasionally perceived upon the presentation of only two brief succeeding visual flash stimuli (Fig. 17.3a). Bowen theoretically explained the origin of the illusion by assuming the superposition of the brain's oscillatory responses (so-called impulse response functions (IRFs) to two succeeding flash stimuli. A superposition of the two IRFs would only occur when the stimuli are presented with a specific delay (~100 ms) between them, which matches the period of the IRF (Bowen, 1989). Such oscillatory IRFs as a response to a single visual flash have been shown to exist; importantly, they are reverberating at ~10 Hz (VanRullen & MacDonald, 2012). Gulbinaite et al. re-investigated the triple-flash illusion in an EEG study, and could confirm Bowen's theoretical explanation: they reported that the optimal stimulus onset asynchrony (SOA) between two flashes in order to perceive an illusory third flash matched the period of each participant's individual IRF (Gulbinaite et al., 2017). Furthermore, they could extend Bowen's findings by revealing that illusory flash perception depended on the individual alpha frequency (IAF) of each participant, measured over parietal and occipital areas. Tailoring the inter-stimulus intervals between the two flashes to the period of the IAF, however, still only led to illusory flash perception in about 50% of the trials. Gulbinaite and colleagues found that the cause of this limited

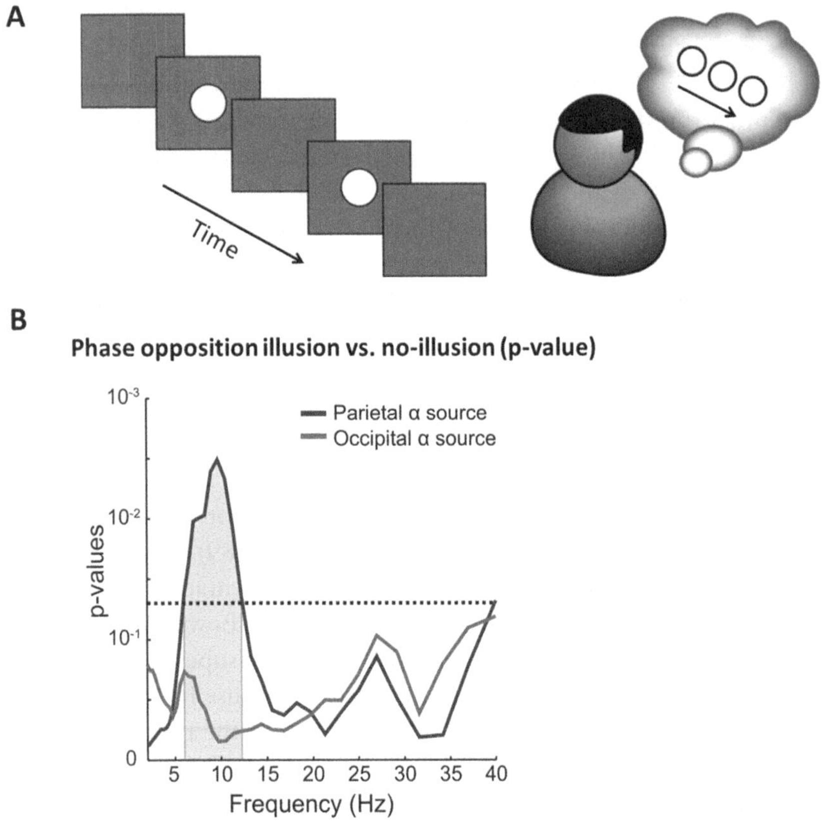

Fig. 17.3 The triple-flash illusion. (a) When two succeeding stimuli are shown with a specific delay in between (~100 ms), they can be perceived either as two or as three stimuli. (b) This effect is linked to oscillations in the impulse response function (IRF) of the brain to a visual stimulus. If the delay between two stimuli matches the period of the individual IRF of the observer, their superposition leads to the perception of an illusory third flash stimulus. However, three flashes are only perceived in 50% of the trials; this may be due to the pre-stimulus phase of ongoing parietal (dark gray) but not occipital (light gray) alpha sources. Gray-shaded area represents the frequencies that showed a significant pre-stimulus phase opposition effect between illusion and no-illusion trials (Figure panel (a) adapted from VanRullen, 2016; panel (b) adapted from Gulbinaite et al., 2017)

effect was the phase of pre-stimulus parietal (but not occipital) alpha oscillations (Gulbinaite et al., 2017). A given pre-stimulus phase would tend to favor the later perception of the third flash, whereas the opposite phase would prevent the illusory percept (Fig. 17.3b); as the stimulus onset was unpredictable and the pre-stimulus phases were uniformly distributed, the "optimal" phase for the triple-flash illusion could not happen more than 50% of the time on average. Thus, these findings are in line with the idea of rhythmic visual perception, where excitability fluctuates with the oscillatory phase of ongoing alpha oscillations, leading in the moments of high excitability to the perception of a third stimulus where only two were present.

While the two visual illusions that were presented above are evidence for *rhythmic* visual perception, the continuous wagon wheel illusion and the flash-lag effect both reflect *discrete* visual perception and will be presented in the following.

4 The Continuous Wagon Wheel Illusion

An illusion that has been intensively investigated by VanRullen and others is the continuous wagon wheel illusion (c-WWI). Most of us have already experienced this phenomenon while watching a movie, when the wheel of a forward moving car suddenly changes its direction while the car continues moving forward. This can be explained by the properties of the camera that filmed the object during the movement, and/or the properties of the screen or apparatus displaying the movie. When the sampling rate of the camera or the refresh rate of the screen is too slow to capture the temporal frequency of a moving object, the object will change its direction. Several studies showed that this illusion also exists in "the real world", under continuous light conditions (Purves, Paydarfar, & Andrews, 1996), or, for example, in a laboratory environment where observers are exposed to a rotating wheel stimulus on a fast-refresh computer screen (VanRullen, Reddy, & Koch, 2005). Just like in the movies, the perceived direction of the wheel spontaneously changes, while its physical movement continues in the same direction (Fig. 17.4); importantly, the illusion only emerges when participants are paying attention to the wheel

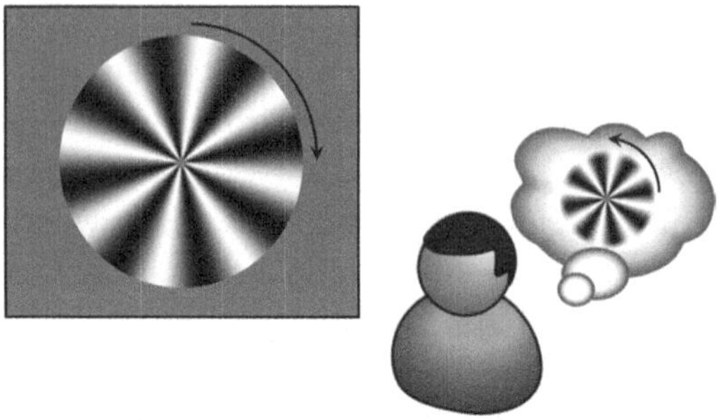

Fig. 17.4 The continuous wagon wheel illusion. The continuous wagon wheel illusion can be experienced on TV, when watching a turning wheel of a forward moving car which suddenly changes direction, although the car continues to move forward. This is due to temporal aliasing; either the sampling rate of the camera that filmed the object was too low as to capture the movement or it is due to a low refresh rate of the screen that shows the filmed object. This illusion can also be experienced under continuous light. In the brain, it has been suggested that the temporal aliasing is due to periodic sampling happening at ~13 Hz (Figure adapted from VanRullen, 2016)

stimulus (VanRullen et al., 2005). These findings could suggest that our visual perception of motion shares the discrete sampling principles of a camera, and thus provide more evidence for discrete visual perception. And indeed, in an EEG study VanRullen and colleagues found evidence for an oscillatory mechanism underlying the c-WWI (VanRullen et al., 2006): they showed that brain oscillations at ~13 Hz recorded over the right parietal regions were correlated with the perception of the illusion. The power in this frequency band decreased just before the illusion occurred and increased before the wheel was again rotating in the "right" direction (VanRullen et al., 2006). They hypothesized that the parietal 13 Hz effect could reflect a potential candidate for the neural mechanism underlying discrete perception, dividing incoming visual information into individual snapshots. Given the fact that the illusion only occurs when observers pay attention to the wheel stimulus (VanRullen et al., 2005), and given the topography of the 13 Hz effect being located over the right parietal areas, VanRullen further hypothesized that the right intraparietal sulcus (IPS), known to be implicated in attention processes (e.g. Corbetta & Shulman, 2002), could be the anatomical source of the illusion. They thus tested, in a second study, the causal relationship between the right IPS and the strength of the illusion (VanRullen, Pascual-Leone, & Battelli, 2008). They used repetitive transcranial magnetic stimulation (rTMS) to inhibit the right IPS before participants were exposed to an annulus version of the wheel stimulus. They showed that inhibiting this brain region with rTMS had a weakening effect on the illusion (VanRullen et al., 2008). These findings were concordant with those of a subsequent functional Magnetic Resonance Imaging (fMRI) study by Reddy and colleagues that confirmed the implication of the right IPS in the emergence of the c-WWI (Reddy, Rémy, Vayssière, & VanRullen, 2011). We think that the c-WWI is probably one of the most powerful illusions to reveal the discrete character of visual perception, whose periodicity potentially relies on the ongoing alpha oscillations.

The above presented studies all converge to the notion that the c-WWI represents evidence for discrete visual perception; however, alternative explanations that are not in line with this theory have also been suggested. For instance, Kline and colleagues as well as Holcombe and colleagues purported that the illusion is provoked by erroneously activated low-level motion detectors (Holcombe, Clifford, Eagleman, & Pakarian, 2005; Kline, Holcombe, & Eagleman, 2004), and/or by motion adaptation processes (Holcombe & Seizova-Cajic, 2008; Kline & Eagleman, 2008). However, we think that these explanations are incomplete, since they do not address the following experimental findings: (i) the illusion is maximal at a specific temporal frequency (~13 Hz) for both first- and second-order stimuli (VanRullen et al., 2005), while the temporal frequency tuning of motion detectors shows a high variability between first- and second-order motion (Hutchinson & Ledgeway, 2006); (ii) attentional manipulation has been shown to influence the magnitude of the illusion, its spatial extent and temporal frequency (VanRullen et al., 2006; VanRullen et al., 2005; VanRullen & MacDonald, 2012)—attentional manipulation could also have an effect on the amount of motion adaptation (Chaudhuri, 1990; Rezec, Krekelberg, & Dobkins, 2004), however probably not on the frequency tuning of low-level motion detectors; (iii) manipulating the stimulus' contrast or eccentricity can decrease the motion aftereffects while the magnitude of the illusion increases and vice versa (VanRullen, 2007), which effectively dissociates motion adaptation from the c-WWI; (iv) lastly, the brain regions that are involved in the c-WWI lie in the right parietal lobe, which would suggest a higher level mechanism to be implicated in the c-WWI, rather than the adaptation of low-level motion detectors.

5 The Flash-Lag Effect

The flash-lag effect (FLE) is another illusion that could be tied to discrete visual perception. In this illusion, a brief visual flash is misperceived to lag behind a moving object that is presented next to it (Eagleman & Sejnowski, 2000, 2007; Nijhawan, 1994, 2002) (Fig. 17.5). It has been suggested that the FLE reflects conscious updating, a process that happens late in the hierarchy of visual perception and tracks changes in the visual scene around us (Eagleman & Sejnowski, 2000; Moore & Enns, 2004; Watanabe & Yokoi, 2006). Conscious updating does seem to happen in a continuous stream, since we are usually not aware of discrete updating events. Since extensive work in our laboratory and by others had revealed evidence for a discrete character of visual perception, like the detection of low-threshold visual stimuli in the visual periphery (Busch et al., 2009; Busch & VanRullen, 2010; Dugué et al., 2011; Mathewson et al., 2009; Scheeringa et al., 2011), Chakravarthi and VanRullen decided to investigate the neural correlates of the flash-lag illusion in an EEG study (Chakravarthi & VanRullen, 2012). Here, observers were exposed to a computer screen that showed a clock stimulus with a moving hand. The flash-lag illusion was produced when the rim of the clock turned from gray to red for a very short time interval (100 ms). Participants then had to indicate at what "time" this flash occurred by moving the hand on a clock that was presented after each trial. The difference between the actual position of the hand and the estimated position was the measure of the flash-lag for each trial. Replicating previous findings (Linares, Holcombe, & White, 2009), Chakravarthi and VanRullen found that the variability of the flash-lag duration was very high (SD ~50 ms) and investigated whether this could reflect the fact that conscious updating is not a stationary, continuous process, but instead that it is more efficient at some moments in time than at others, which could point to a periodicity in this high-level process of visual perception. Indeed, they found, on the single-trial level, strong correlations between the duration of the flash-lag illusion and the pre-stimulus phase of occipital theta (5–10 Hz; −50 ms) oscillations as well as post-

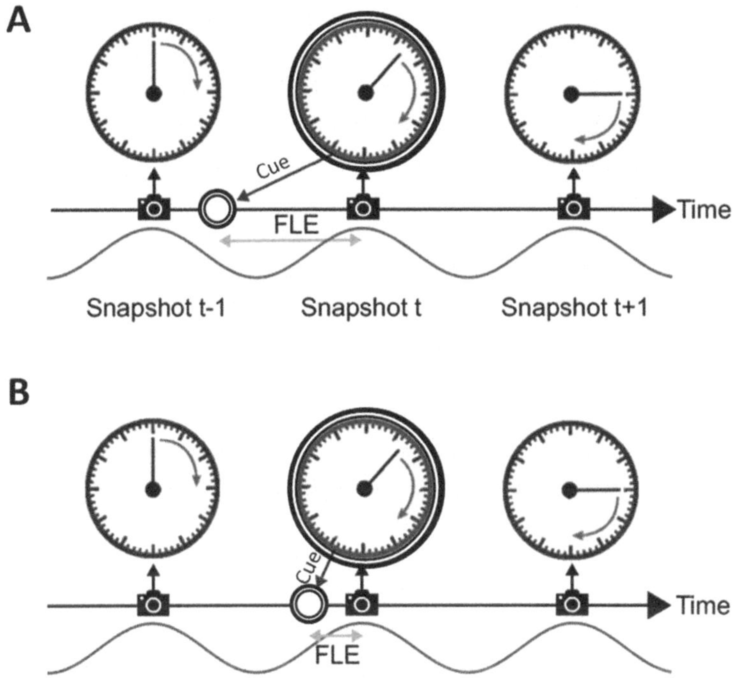

Fig. 17.5 The flash-lag effect. The flash lag can be experienced when a moving object (here the hand of the clock) and a brief flash stimulus (e.g. changing color of the rim of the clock, here indicated by the black circle surrounding the clock stimulus) are observed together. When observers are asked to report the location of the hand at the moment of the flash, the moving object tends to lag behind its actual location. (a) When the flash is presented at an early phase of the oscillation (here: black circle), the lag would be bigger than if the flash is presented at a later phase (b) (Theoretical explanation of the flash-lag effect by Schneider, 2018; figure adapted from Chota & VanRullen, 2019)

stimulus phase of fronto-central high alpha/low beta oscillations (12–20 Hz; 60 ms). In addition, they found that up to 10 ms of the ~50 ms trial-to-trial variability in the flash-lag illusion could be explained by the phase of these ongoing oscillations. These 20% of total variability might not seem a lot; however, one should keep in mind that this study was based on single-trial EEG data, which typically shows a very low signal-to-noise ratio. For every participant, there was therefore an optimal phase for the flash to arrive, where conscious updating would occur optimally and the duration of the lag would be minimal, whereas the opposite phase would result in sub-optimal conscious updating, and a maximum lag. These findings reveal that conscious updating is a periodic process, fluctuating with the phase of ongoing theta and high alpha/low beta oscillations; thus, the study further contributes to the growing body of evidence for a discrete nature of visual perception.

Recently, Schneider published a highly related theoretical explanation of the flash-lag, which purports that this illusion can be explained by discrete sampling in the visual system (Schneider, 2018). Schneider assumes that if our visual perception is divided into individual perceptual moments of a specific duration, the variability in the flash-lag duration can be explained by the specific phase of a perceptual moment where the flash is presented. According to Schneider, the position of a moving object is always recorded at the end of a perceptual moment. Thereby, a flash at the beginning of a perceptual moment would lead to a lag of duration (Fig. 17.5a), whereas a flash that occurs at the end of a perceptual moment would lead to a very short or no lag between the flashed and the moving object, since the time of the flash and the recording of the position of the moving object would be aligned (Fig. 17.5b). In a behavioral study, Chota and VanRullen further investigated the relationship between the phase of a perceptual moment (which they assume to be the phase of visual alpha oscillations) and the duration of the flash-lag (Chota & VanRullen, 2019). To control the phase at which a flash occurred, they visually entrained the brain with a flickering annulus at 10 Hz that was surrounding a clock stimulus with a handle (cf. Chakravarthi & VanRullen, 2012). The flash was then presented at different phases of the entrained oscillation, and the duration of the flash-lag was measured on every trial. Chota and VanRullen could show that the flash-lag duration was indeed modulated periodically by the entrained 10 Hz oscillation; furthermore, the entrained oscillations showed strong phase coherence between subjects. Therefore, they interpret their findings as causal evidence for the discrete perception theory (Chota & VanRullen, 2019).

While the above presented illusions could either be classified into evidence for *rhythmic* or *discrete* visual perception, it is less clear for the last two illusions that will be presented in this chapter (the fluttering heart illusion and the saltation effect); these can be interpreted either as the correlates of rhythmic or discrete visual perception.

6 The Fluttering Heart Illusion

Another visual illusion that shows a regular jitter or flicker, similar to the flickering wheel illusion, was discovered already 150 years ago by Helmholtz who named it "the fluttering heart illusion". He observed that when colored shapes (e.g. hearts) were superposed on top of a colored paper and both were moved together, the shapes (hearts) seemed to "dance about on the paper" (Helmholtz, 1867). Several studies have investigated this illusion, which first seemed to result from interactions of the rod and cone systems, which show differences in temporal and spatial processing (von Grünau, 1976). More recently, Arnold and Johnston gained new insight into this illusion by conducting a behavioral study (Arnold & Johnston, 2003). They could show that the illusion was strongest with a low luminance contrast between moving fore- and background, and that the illusory flicker recurred regularly. Given this invariable periodicity of the jitter, they suggested that it reflects a dynamic neural process, whereby the two misaligned motion streams are periodically re-aligned (Arnold & Johnston, 2003). In a later study, observers estimated the illusory jitter frequency at around ~10 Hz (Amano, Arnold, Takeda, & Johnston, 2008). The hypothesis that this 10 Hz jitter reflects a dynamic neural process has been revisited by Minami and Amano in 2017 (Minami & Amano, 2017), who recorded brain activity via Magnetoencephalography (MEG), while observers were experiencing the illusion in a red square stimulus with a superposed isoluminant green bar in its center. The illusion was perceived when the stimulus was moving

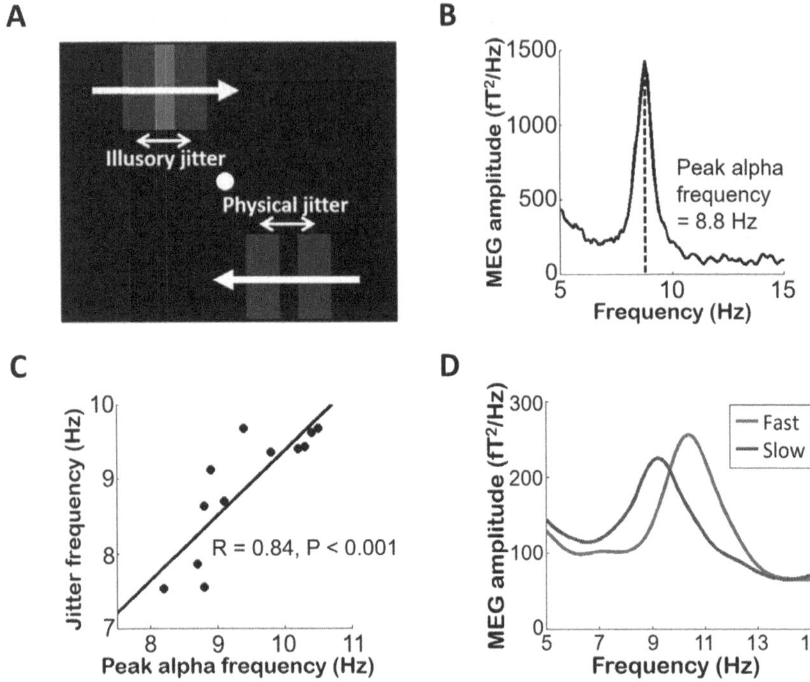

Fig. 17.6 The fluttering heart illusion. (a) A stimulus composed of a square (upper left stimulus, in darker gray, red in experiment) and an isoluminant bar centered on top of it (in lighter gray, green in experiment) is moving horizontally. Illusory jitter can be perceived in its center while the stimulus is moving. Participants reported whether the frequency of a physical jitter in the center of another horizontally moving square stimulus (lower right; red in experiment; jittering bar was of same luminance as the background) was faster than the perceived illusory jitter. (b) Power spectrum of MEG data while participants had their eyes closed. The frequency that showed the maximum amplitude within the alpha frequency band (8–13 Hz) was defined as peak alpha frequency (PAF). (c) Perceived jitter frequency correlates with the individual PAF at rest. (d) Sorting trials into "slow" and "fast" perceived illusory jitter revealed that slow jitter implied slow alpha and vice versa (Adapted from Minami & Amano, 2017)

horizontally (Fig. 17.6a). Since observers estimated the illusory jitter to lie in the alpha frequency band, Minami and Amano first identified the individual peak alpha frequency (PAF) from MEG data recorded during the resting state with eyes closed (cf. Fig. 17.6b) and correlated it with the illusory jitter frequency. They found that the PAF of each participant correlated with the perceived jitter frequency (Fig. 17.6c). Even at the individual subject level, trials in which the perceived illusory jitter frequency was lower (respectively, higher) than average displayed a lower-than-average (resp. higher-than-average) pre-stimulus PAF (Fig. 17.6d). They could further show that the perceived jitter frequency could be modulated by entraining brain oscillations via amplitude modulation transcranial current stimulation (AM-tACS) which differed slightly in frequency (i.e. slower entrained oscillations leading to slower perceived jitter frequency and vice versa) (Minami & Amano, 2017). Whether the hypothesis about a periodic correction of the mismatch between the two moving stimuli (Arnold & Johnston, 2003) is interpreted as evidence for discrete or rhythmic visual perception depends on whether motion is defined as temporal judgment or not. This ambiguity is also true for the saltation effect, which will be presented below.

7 The Saltation Effect

The last visual illusion that we will present here is the "saltation effect". The illusion comes about when a Gabor grating (Fig. 17.7a) is moving in one direction while its carrier grating is either stationary or moving in the opposite direction (Fig. 17.7b). Although the Gabor grating is moving continuously, the observer has the impression that it "jumps" from one location to the other in a rhythmic manner (Fig. 17.7c). Recently, Nakayama et al. published a behavioral study investigating this illusion (Nakayama et al., 2018). They found that the perceived saltation had a relatively constant frequency for every subject (range: 3–8 Hz; Fig. 17.7d), which was not modulated by stimulus characteristics, such as the speed of the moving grating or of the underlying window. The authors therefore claimed that this illusory saltation percept is potentially the result of an underlying periodic neuronal process acting in the theta frequency band (3–8 Hz). Although this study is showing purely behavioral results and its electrophysiological correlates still remain to be investigated, periodic processes in the theta EEG frequency band have

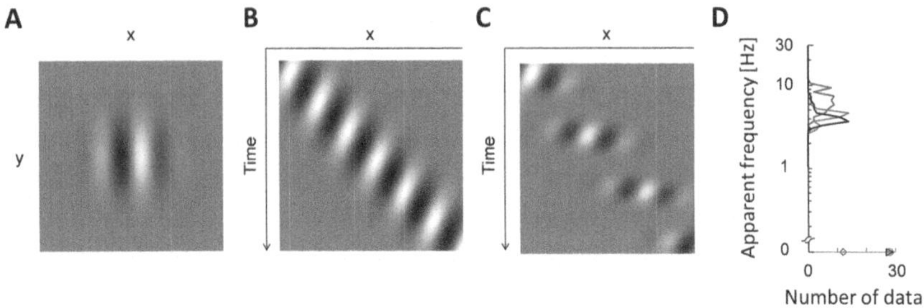

Fig. 17.7 The saltation effect. (a) Conventional Gabor grating. (b) In this space-time representation, the spatial window of the Gabor pattern is moving continuously in one direction while its carrier grating is moving in the opposite direction. (c) These opposing movements give rise to an illusory saltation of the Gabor grating at a regular frequency. (d) The apparent flicker frequency for all observers lies within the theta frequency band (3–8 Hz) (Adapted from Nakayama et al., 2018)

previously been revealed to exist in visual perception (Busch & VanRullen, 2010). Nakayama and colleagues conclude that whereas other studies of visual illusions revealed perceptual consequences of rhythmic or discrete perception in the alpha frequency band (e.g. Minami & Amano, 2017; Sokoliuk & VanRullen, 2013), this is the first illusion to show a perceptual correlate in the theta band. Periodicities in visual perception have been reported in both the theta and alpha frequency bands and have been interpreted, respectively, as attentional and sensory sampling rhythms (VanRullen, 2016). The saltation effect could thus be more related to attentional sampling, as discussed by the authors (Nakayama et al., 2018).

8 Conclusions

All the visual illusions presented here share a tight relationship with the ongoing brain oscillations in lower frequency bands (theta, alpha, beta). Furthermore, they exemplify periodicity in visual perception: the flickering wheel illusion and the triple-flash illusion both reflect evidence for *rhythmic* visual perception, where the illusory percept emerges from fluctuations in excitability. The continuous wagon wheel illusion and the flash-lag effect on the other hand are evidence for *discrete* visual perception, where the phase of ongoing brain oscillations determines the timing of visual perception. The fluttering heart illusion as well as the saltation effect are visual illusions that could be interpreted to be correlates of both, *rhythmic* or *discrete* visual perception. Both illusions likely arise because of a progressive mismatch between two motion signals. The flutter or saltation emerges when this mismatch gets corrected periodically. Whether these illusions reflect *discrete* or *rhythmic* visual perception depends on whether or not motion is considered as a temporal judgment.

References

Amano, K., Arnold, D. H., Takeda, T., & Johnston, A. (2008). Alpha band amplification during illusory jitter perception. *Journal of Vision, 8*(2008), 1–8.

Arnold, D. H., & Johnston, A. (2003). Motion-induced spatial conflict. *Nature, 425*(September), 181–184.

Baumgarten, T. J., Schnitzler, A., & Lange, J. (2015). Beta oscillations define discrete perceptual cycles in the somatosensory domain. *Proceedings of the National Academy of Sciences, 112*(39), 12187–12192.

Bowen, R. W. (1989). Two pulses seen as three flashes: A superposition analysis. *Vision Research, 29*(4), 409–417.

Busch, N. A., Dubois, J., & VanRullen, R. (2009). The phase of ongoing EEG oscillations predicts visual perception. *The Journal of Neuroscience, 29*(24), 7869–7876.

Busch, N. A., & VanRullen, R. (2010). Spontaneous EEG oscillations reveal periodic sampling of visual attention. *Proceedings of the National Academy of Sciences, 107*(37), 16048–16053.

Chakravarthi, R., & VanRullen, R. (2012). Conscious updating is a rhythmic process. *Proceedings of the National Academy of Sciences, 109*(26), 10599–10604.

Chaudhuri, A. (1990). Modulation of the motion aftereffect by selective attention. *Nature, 344*, 60–62.

Chota, S. & VanRullen, R. (2019. Visual Entrainment at 10 Hz causes periodic modulation of the Flash Lag Illusion. Frontiers in neuroscience, 13.

Corbetta, M., & Shulman, G. L. (2002). Control of goal-directed and stimulus-driven attention in the brain. *Nature Reviews Neuroscience, 3*(3), 201–215.

Dewan, E. M. (1967). Occipital alpha rhythm eye position and lens accommodation. *Nature, 214*(5092), 975.

Dubois, J., & VanRullen, R. (2011). Visual trails: Do the doors of perception open periodically? *PLoS Biology, 9*(5).

Dugué, L., Marque, P., & VanRullen, R. (2011). The phase of ongoing oscillations mediates the causal relation between brain excitation and visual perception. *Journal of Neuroscience, 31*(33), 11889–11893.

Eagleman, D. M., & Sejnowski, T. J. (2000). Flash lag effect: Differential latency, not postdiction: Response. *Science, 290*, 5494.

Eagleman, D. M., & Sejnowski, T. J. (2007). Motion signals bias localization judgments: A unified and Frohlich illusions. *Journal of Vision, 7*(2007), 1–12.

Exner, S. (1875). Experimentelle Untersuchung der einfachsten psychischen Processe. *Archiv Für Die Gesamte Physiologie Des Menschen Und Der Tiere, 11*(*1*), 403–432.

Fenwick, P. B., & Walker, S. (1968). The effects of eye position on the alpha rhythm. *Electroencephalography and clinical neurophysiology, 25*(5), 508.

Gulbinaite, R., İlhan, B., & VanRullen, R. (2017). The triple-flash illusion reveals a driving role of alpha-band reverberations in visual perception. *The Journal of Neuroscience, 37*(30), 7219–7230.

Helmholtz, H. (1867). *Handbuch der physiologischen Optik* (Vol. 9). Voss. *Handbuch Der Physiologischen Optik, 9.*

Holcombe, A. O., Clifford, C. W. G., Eagleman, D. M., & Pakarian, P. (2005). Illusory motion reversal in tune with motion detectors. *Trends in Cognitive Sciences, 9*(12), 559–560.

Holcombe, A. O., & Seizova-Cajic, T. (2008). Illusory motion reversals from unambiguous motion with visual, proprioceptive, and tactile stimuli. *Vision Research, 48*(17), 1743–1757.

Hutchinson, C. V., & Ledgeway, T. (2006). Sensitivity to spatial and temporal modulations of first-order and second-order motion. *Vision Research, 46*(3), 324–335.

James, W. (1890). *The Principles of psychology* (Vol. 1). Macmillan.

Jensen, O., & Mazaheri, A. (2010). Shaping functional architecture by oscillatory alpha activity: Gating by inhibition. *Frontiers in Human Neuroscience, 4.*

Kline, K., Holcombe, A. O., & Eagleman, D. M. (2004). Illusory motion reversal is caused by rivalry, not by perceptual snapshots of the visual field. *Vision Research, 44*(23), 2653–2658.
Kline, K. A., & Eagleman, D. M. (2008). Evidence against the temporal subsampling account of illusory motion reversal. *Journal of Vision, 8*(4), 13.
Kristofferson, A. B. (1967). Attention and psychophysical time. *Acta Psychologica, 27*, 93–100.
Linares, D., Holcombe, A. O., & White, A. L. (2009). Where is the moving object now? Judgments of instantaneous position show poor temporal precision (SD = 70 ms). *Journal of Vision, 9*(13), 9–14.
Mathewson, K. E., Gratton, G., Fabiani, M., Beck, D. M., & Ro, T. (2009). To see or not to see: Prestimulus phase predicts visual awareness. *Journal of Neuroscience, 29*(9), 2725–2732.
Mathewson, K. E., Lleras, A., Beck, D. M., Fabiani, M., Ro, T., & Gratton, G. (2011). Pulsed out of awareness: EEG alpha oscillations represent a pulsed-inhibition of ongoing cortical processing. *Frontiers in Psychology, 2*.
Minami, S., & Amano, K. (2017). Illusory jitter perceived at the frequency of alpha oscillations. *Current Biology, 27*(15), 2344–2351.e4.
Moore, C. M., & Enns, J. T. (2004). Object updating and the flash-lag effect. *Psychological Science, 15*(12), 866–871.
Mulholland, T., & Evans, C. R. (1965). An unexpected artefact in the human electroencephalogram concerning the alpha rhythm and the orientation of the eyes. *Nature, 207*(992), 36.
Nakayama, R., Motoyoshi, I., & Sato, T. (2018). Discretized theta-rhythm perception revealed by moving stimuli. *Scientific Reports, 8*(1), 2–10.
Nijhawan, R. (1994). Motion extrapolation in catching. *Nature, 370*(6487), 256–257.
Nijhawan, R. (2002). Neural delays, visual motion and the flash-lag effect. *Trends in Cognitive Sciences, 6*(9), 387–393.
Purves, D., Paydarfar, J. A., & Andrews, T. J. (1996). The wagon wheel illusion in movies and reality. *Proceedings of the National Academy of Sciences, 93*(April), 3693–3697.
Reddy, L., Rémy, F., Vayssière, N., & VanRullen, R. (2011). Neural correlates of the Continuous Wagon Wheel Illusion: A functional MRI study. *Human Brain Mapping, 32*(2), 163–170.
Rezec, A., Krekelberg, B., & Dobkins, K. R. (2004). Attention enhances adaptability: Evidence from motion adaptation experiments. *Vision Research, 44*(26), 3035–3044.

Scheeringa, R., Mazaheri, A., Bojak, I., Norris, D. G., & Kleinschmidt, A. (2011). Modulation of visually evoked cortical fMRI responses by phase of ongoing occipital alpha oscillations. *Journal of Neuroscience, 31*(10), 3813–3820.

Schneider, K. A. (2018). The flash-lag, Fröhlich and related motion illusions are natural consequences of discrete sampling in the visual system. *Frontiers in Psychology, 9*(July), 1–8.

Sokoliuk, R., & VanRullen, R. (2013). The flickering wheel illusion: When rhythms make a static wheel flicker. *Journal of Neuroscience, 33*(33), 13498–13504.

Stroud, J. M. (1956). The fine structure of psychological time. *Information Theory in Psychology: Problems and Methods, 1*, 174–207.

VanRullen, R., Reddy, L., & Koch, C. (2006). The continuous wagon wheel illusion is associated with changes in electroencephalogram power at ~ 13 Hz. *Journal of Neuroscience, 26*(2), 502–507.

VanRullen, R. (2007). The Continuous Wagon Wheel Illusion depends on, but is not identical to neuronal adaptation. *Vision Research, 47*(16), 2143–2149.

VanRullen, R. (2016). Perceptual cycles. *Trends in Cognitive Sciences, 20*(10), 723–735.

VanRullen, R., & MacDonald, J. S. P. (2012). Perceptual echoes at 10 Hz in the human brain. *Current Biology, 22*(11), 995–999.

VanRullen, R., Pascual-Leone, A., & Battelli, L. (2008). The continuous wagon wheel illusion and the "When" pathway of the right parietal lobe: A repetitive transcranial magnetic stimulation study. *PLoS ONE, 3*(8), 1–5.

VanRullen, R., Reddy, L., & Koch, C. (2005). Attention-driven discrete sampling of motion perception. *Proceedings of the National Academy of Sciences, 102*(14), 5291–5296.

Varela, F. J., Toro, A., John, E. R., & Schwartz, E. L. (1981). Perceptual framing and cortical alpha rhythm. *Neuropsychologia, 19*(5), 675–686.

von Grünau, M. W. (1976). The "Fluttering Heart" and spatio-temporal characteristics of color processing-III. *Vision, 16*, 397–401.

Watanabe, K., & Yokoi, K. (2006). Object-based anisotropies in the flash-lag effect. *Psychological Science, 17*(8), 728–735.

Zeki, S. (1991). Cerebral akinetopsia (visual motion blindness). *Brain, 114*, 811–824.

18

Time Markers and Temporal Illusions

Valtteri Arstila

1 Introduction

Three notions of time are commonly distinguished in the literature of the temporal illusions. The *event time* is the time when an external event occurs. The *brain time of experience* (brain time for short) is the time when the neural processes that realize the experience of the event occur (i.e., neural correlates of experiences). The *subjective time of experience* (subjective time for short) refers to the time in which an event is experienced to occur. Much of the discussion on the temporal illusions in recent years has been about the relationship between the last notion of time and the other two.

In this chapter, I approach this topic by focusing on temporal order illusions. At the heart of the discussion is the *thesis of temporal isomorphism*, according to which the distinction between subjective time and

V. Arstila (✉)
Department of Philosophy & Turku Institute for Advanced Studies,
University of Turku, Turku, Finland
e-mail: valtteri.arstila@utu.fi

V. Arstila et al. (eds.), *The Illusions of Time*,
https://doi.org/10.1007/978-3-030-22048-8_18

brain time is merely conceptual; in reality, subjective time and brain time are isomorphic. The aim is not to argue for the thesis of temporal isomorphism, but to explicate how it figures in the empirical accounts of temporal illusions.[1] Since similar positions in philosophy mostly concern theories about the temporal structure of consciousness, rather than about temporal illusions, the related philosophical literature is not so relevant to this volume.[2]

In what follows, I begin by summarizing the parts of Daniel Dennett and Marcel Kinsbourne's *Time and the Observer* (1992) that directly relate to the thesis. It is the most discussed exposition of the thesis of temporal isomorphism; the arguments in the paper are frequently cited as the reasons to reject the thesis. After describing this theoretical background, I turn to two debates in which the thesis figures prominently: the perception of asynchronous visual chances within one stimulus and temporal order perception between the occurrences of two stimuli. The explanatory frameworks in both debates contrast the *brain-time view*, which at minimum subscribes to the thesis, with the *event-time view*, according to which subjective time mirrors (or aims to mirror) the event time rather than the brain time. However, I argue that both views have been understood in different ways in these debates. For example, the two versions of the event-time view concur with two alternatives Dennett and Kinsbourne present as the thesis of temporal isomorphism. As a result, one version of the event-time view is compatible with the brain-time view's claim that subjective time and brain time are isomorphic.

---

[1] It is worth noting, however, that I argued for the thesis in Arstila (2015, 2016b). I also take it that the thesis is widely endorsed in science. For instance, Wolfgang Köhler (1947, p. 62) already argued that "experienced order in time is always structurally identical with a functional order in the sequence of correlated brain processes." More recently, a version of the thesis has figured in the explanations of perceptual simultaneity (Kopinska & Harris, 2004), duration estimation and reproduction (Reutimann, Yakovlev, Fusi, & Senn, 2004; Wittmann, van Wassenhove, Craig, & Paulus, 2010), spatiotemporal illusions (Arstila, 2015; Whitney & Murakami, 1998; Whitney, Murakami, & Cavanagh, 2000), and timing of visual feature changes (Arnold & Wilcock, 2007; Moutoussis & Zeki, 1997).

[2] For the discussion of the thesis as such in different contexts, see, for example, Grush (2007), Kelly (2005), Kiverstein & Arstila (2013), Mölder (2014), and Power (2010). For the discussion of the thesis complemented with additional qualifications, see, for example, Grush (2009), Lee (2014), and Phillips (2014).

2 The Thesis of Temporal Isomorphism and Alternative Views

In *Time and the Observer*, Dennett and Kinsbourne contrasted the Cartesian theater model of consciousness with the Multiple Drafts model of consciousness. Both models hold that feature detections and discriminations of the features of stimuli are done at different parts of the cortex. That is, the processing of the features is localized. According to the Cartesian theater model, we become conscious of the localized discriminations when they are represented at the hypothetical Cartesian theater. This is a place "where 'it all comes together,' and the discriminations in all modalities are somehow put into registration and 'presented' for subjective judgment." (Dennett & Kinsbourne, 1992, p. 183) The subjective time of the experiences is determined by the time the localized discriminations are (re)presented at the Cartesian theater. This means that the subjective time order of experienced events is the same as the order in which experiences of the events occur. In short, the Cartesian theater model entails the thesis of temporal isomorphism.

Dennett and Kinsbourne's objection to the Cartesian theater model makes use of three temporal anomalies—the backward metacontrast masking effect, the color phi phenomenon, and the cutaneous rabbit illusion—each of which brings about incorrect reports of the stimuli.[3] As an example, consider the ordinary phi phenomenon. Here, subjects are presented with spatially separated static stimuli (flashes) with a short temporal gap between them, and they report seeing that something moves between the locations of the presented stimuli. The temporally anomalous part of the illusion is that the apparent motion can be processed in the brain only after the second stimulus is processed—or so Dennett and Kinsbourne argue (for the opposing view, see Arstila, 2016a)—because, otherwise the motion would not have direction. Thus the order of neural processing (first stimulus, second stimulus, apparent motion) does not

[3] At the latter part of the paper, they also argue that if Benjamin Libet's results suggesting the delay of conscious intention and the backward referrals of time are sound, then the model of consciousness they advocate would explain the results better than the Cartesian theater model of consciousness.

appear to match with the reported order of experiences (first stimulus, apparent motion, second stimulus).

The described discrepancy raises the question of whether our report is correct or not. If the *Stalinesque revision* took place, then the report is correct. We see the motion before the second stimulus, because the order of localized discriminations is tampered with before they reach the Cartesian theater. In this case, the order in which localized discriminations are processed is different from the order in which they arrive at the theater. If the *Orwellian revision* takes place, then our experience of the second stimulus precedes our experience of motion, but we report them in the incorrect order because our memory of the order of events is tampered with. In the two possible scenarios, only the Stalinesque revision concurs with the thesis of temporal isomorphism.

According to Dennett and Kinsbourne, the Cartesian theater model necessitates that there is a way to distinguish which scenario is correct. Their main claim in the paper is that there is no such way. Consequently, they maintain that the Cartesian theater model is incorrect, and, with it, the thesis of temporal isomorphism is questionable as well. Dennett and Kinsbourne further argue that there is no reason that an experience of stimulus S_1 preceding stimulus S_2 needs to be accomplished by, first, a neural state realizing an experience of S_1 and then a neural state realizing an experience of S_2. The temporal order is not determined by the order in which the localized discriminations are represented at the Cartesian theater; it is determined by the temporal content of the individual representings of the events.

Dennett and Kinsbourne present two possibilities for what it means, in practice, to separate the time in the content of experience (experienced time) and the time of the vehicle of experience (the neural events realizing the experience). It is worth noting, however, that they do not explicate either possibility in detail. Instead, they provide analogies meant to illustrate the ideas behind them.

According to the first possibility, which I refer to as *the timestamp view*, the temporal properties of experience utilize timestamps, a kind of abstract temporal label, linked to the experienced contents. Timestamps have temporal information as their content. Dennett and Kinsbourne's analogy for timestamps is postmarks on letters: the postmarks represent the order in which letters are sent, not the order in which they are received. What matters for the subjective order of experiences are the postulated timestamps that accompany the experiences, not the order in which localized discriminations are represented ("received") at the hypothetical Cartesian theater.

Dennett and Kinsbourne call the second possibility *the content-sensitive settling*. Their analogy for this possibility comes from matching the auditory and visual tracks of a film. Such matching can be done by determining abrupt changes in both tracks and trying to find the best match between the changes, even when we do not know much about the film or even understand the spoken language.

Dennett and Kinsbourne (1992, p. 189) prefer the content-sensitive settling view over the timestamp view because they regard the first "a cheaper, less foolproof but biologically more plausible." Nonetheless, it is the latter position that is commonly attributed to them in the discussions of temporal illusions. This is possibly because part of their argument requires the timestamp view. For example, the content-sensitive settling cannot explain the experiences of features that are not in the stimuli, like an experience of motion in the phi phenomenon, because it only adjusts the relative order of experienced events. Moreover, the content-sensitive settling does not specify the stage of processing in which the settling occurs. For instance, it could take place before localized discriminations reach the Cartesian theater or even before localized discriminations. These possibilities mean that the content-sensitive settling is compatible with the Cartesian theater model and the thesis of temporal isomorphism.[4]

To summarize the relevant points briefly, the Cartesian theater model of consciousness maintains that we are conscious of localized discriminations of stimulus features only when they are represented in the Cartesian theater. Concurring with the thesis of temporal isomorphism, the subjective time order mirrors the brain time order, where brain time is the order of localized discriminations represented in the Cartesian theater. Since localized discriminations, formed in the various regions of the cortex,

[4]On this topic, it is interesting to note that even though many of the almost 30 commentators of Dennett and Kinsbourne's paper had reservations about the Multiple Drafts model, only one of them challenged the view that subjective time and brain time can come apart.

remain unconscious if they are not represented in the Cartesian theater, conscious and unconscious perceptions can have different temporal properties and structures. Finally, after rejecting the Cartesian theater and the thesis of temporal isomorphism, Dennett and Kinsbourne propose that the subjective time of experiences is determined by the temporal content of localized discriminations. They presented two ways in which this could happen: the timestamp view and the content-sensitive settling.

3 Perceptual Asynchrony in Visual Perception

One experimental research area in which the brain-time and the event-time views have figured in prominently is the alleged perceptual asynchrony of changes of visual features. The central finding in question is that when a moving stimulus changes its color and direction of movement simultaneously and repeatedly, observers report that the change in color occurs 60–100 milliseconds before the change in direction. This effect takes a little while to occur as it is not reported in the first changes of the features or if the features change only once (Nishida & Johnston, 2002).

The brain-time view, sometimes also called the brain-time account and the brain-time theory of perception (Arnold, 2005, 2010), explains the findings by maintaining that colors and changes in color are processed faster than motion and motion reversals (Moutoussis & Zeki, 1997, 2002). As Semir Zeki and Andreas Bartels (1999, p. 234) put it, their "result implies that the color processing system reaches its perceptual endpoint before the motion processing system." The difference in perceptual latency is correctly reflected in the judgments of the temporal features of the events. One implication of Zeki and Bartels' claim is that the difference between the cases when the changes are perceived as asynchronous and those in which they are not (e.g., the stimuli are not shown repeatedly) is due to some sort of adaptation to the asynchronous changes, rather than our reports changing from being incorrect to being correct. Since the subjective order of the changes is the same as the order in which the experiences of changes are neurally processed, the brain-time view subscribes to the thesis of temporal isomorphism.

The brain-time view is attributed to Zeki and his colleagues, who were first to report the described findings. Zeki took the findings, in concurrence with the known functional properties of the visual cortex, to support his theory of consciousness (Zeki, 2007; Zeki & Bartels, 1999). This is significant because the subsequent discussion of perceptual asynchrony in vision and the brain-time view has been understood in terms of Zeki's theory of consciousness, which differs from the Cartesian theater model of consciousness.

For the topic at hand, the most important difference between these models of consciousness is that, in Zeki's theory, the activity of the parts of the visual cortex that process a visual feature can be a neural correlate of the conscious experience. In Dennett and Kinsbourne's terms, Zeki argues that when localized discrimination is finished, we can become conscious of the detected visual feature without it being represented in the Cartesian theater. Thus, there is no separation between the time when the processing of localized discrimination is finished and the time when we experience the result.

Citing Dennett and Kinsbourne, Alan Johnston and Shin'ya Nishida (2001, R428) argue that the brain-time view faces "some thorny philosophical problems," at the root of which is the commitment to the thesis of temporal isomorphism. Moreover, they maintain that "a brain time mechanism is poorly designed in the sense that processing delay is added to event time estimation." (Nishida & Johnston, 2010, p. 286) And indeed, if the subjective time is inherently linked to the brain time, then the subjective timing of events is influenced by the neural latencies. This makes deducing the event time, the actual time of external events, practically impossible. After all, to do so, we would need to take into account the neural latencies, which differ both within one sense modality (e.g., Allik & Kreegipuu, 1998; Maunsell et al., 1999) and between sense modalities, and yet neurons are spatiotemporally limited in the sense that they do not have the information about the various latencies at their disposal (Scharnowski, Rees, & Walsh, 2013).[5]

Nishida and Johnston's (2002, 2010) method of avoiding this problem is based on two ideas. First, they distinguish between two types of temporal changes. Determining the first-order temporal changes, called *transitions*, requires that the stimulus is compared in two points in time. For instance, we can detect the change in color if we compare the color of a stimulus at times t_1 and t_2. Determining the second-order temporal changes, called *turning points*, requires that the stimulus is compared in three points in time. For instance, they argue, the detection of motion reversal demands that we compare the spatial position of the stimulus at times t_1, t_2, and t_3. Consequently, the implication is that even if a transition and a turning point were to co-occur at t_2, determining the time of turning points takes longer than determining the time of transitions. Nishida and Johnston's results as regards synchrony between different types of transitions and turning points provide support for this claim. Given that the changes in color and in the direction of motion are different types of temporal changes, simultaneous changes in them would not be judged to be simultaneous.

Nishida and Johnston's (2002, p. 365) idea suggests that the reports of perceptual asynchrony are "a result of an error of subjective temporal judgments" rather than asynchrony in the experiences of changes in color and motion. As such, it is compatible with the thesis of temporal isomorphism; the thesis concerns only the relationship between subjective time and brain time. It is also compatible with Dennett and Kinsbourne's view that subjective judgments occur when localized discriminations have been presented at the Cartesian theater. That is, their position too allows that our temporal judgments of experiences can be incorrect.

The second idea that Nishida and Johnston put forward is that temporal judgments are based on time markers—markers that are themselves based on early neural activity. The resulting view, which they call the *time-marker view*, distances them from the way in which the relationship

[5] Obviously, if one rejects, as many proponents of the brain-time view reject, the assumption that a temporal comparator tries to deduce the event time from the brain time, then this problem does not emerge.

between brain time and subjective time was envisaged in the Cartesian theater model. By time markers (also temporal markers), they mean "temporally localized representations of salient temporal features" (2002, p. 360), which are based on "the temporal pattern of the neural activity elicited by [external] events" (2002, p. 366). The claim that the relevant neural activity takes place very early on in the processing hierarchy has the consequence that the processing of time markers is dissociated from the finished localized discriminations and conscious experiences, and especially the timing thereof. Since temporal order judgments utilize time markers rather than the brain time of experience, their view resembles Dennett and Kinsbourne's timestamp view in this respect.

Nishida and Johnston's view as regards time markers also resembles a more concrete suggestion, namely Benjamin Libet's notion of a cortical time marker. Briefly, Libet argued that conscious experience of a tactile stimulus requires cortical processing for 500 milliseconds and that subjectively the stimulus is felt to occur already around the time when the processing is commenced. That is, the subjective time of experience and the brain time of experience come apart—the subjective time of skin sensations are "referred back" to the time when the processing commenced rather than when the neural processing of the stimulus becomes "adequate" to bring about the sensations. Libet hypothesized that such separation occurs because the first components of the evoked potential (caused by direct stimulation of the skin) are cortical time markers that determine the subjective time of experience.

The time-marker view allows Nishida and Johnston mostly to avoid the problem they thought the brain-time view suffered from. If subjective time is linked to the time markers grounded on the early neural activation caused by the external events, then subjective time reflects the time when the processing of a feature is commenced rather than finished. In this case, the differences in neural latencies and the processing times in the later processing stages do not influence the temporal judgments. This, in turn, means that subjective time and temporal judgments mirror event time rather than brain time. This also applies to the Libet's hypothesis: since the first evoked potential in the cortex occurs very fast after the presentation of the stimulus (in some of Libet's experiments, 15 milliseconds after the stimulation), the neural activity used for temporal coding mirrors the timing of external events certainly more closely than if all neural latencies would be taken into account. In short, all three argue for the event-time view.

4 The Views and Temporal Order Judgments

The second topic in which notions of the brain-time view and the event-time view figure in is explanations of the temporal order perception of separate stimuli. The relevant phenomena requiring explanation are when two simultaneous stimuli are perceived as non-simultaneous, and when two non-simultaneous stimuli are perceived as simultaneous. For example, a dim flash shown at the same time as a bright flash is perceived as occurring after the bright flash. Or, when a toe is touched tens of milliseconds before a nose, the two are perceived as occurring at the same time. Similar illusions also happen with multimodal stimuli; for example, a flash and a sound can appear simultaneous even though they are separated by a hundred milliseconds.

One straightforward explanation for these results appeals to the latency differences: the simultaneous stimuli are perceived as non-simultaneous because the neural processing of them took different amounts of time. In accordance with this explanation, the empirical evidence suggests that a bright stimulus is processed faster than a dimmer stimulus, and it takes longer time for a signal from a toe to reach the cortex than from a nose.[6] In this context, this explanation has been called the brain-time view.[7] Again, the view concurs with the thesis of temporal isomorphism.

The brain-time view has been taken to be challenged by the evidence that we adapt to non-simultaneous multimodal stimuli. One interpretation

[6] See Holcombe (2015) for a thorough discussion of the evidence supporting this explanation in various unimodal and cross-modal temporal order tasks.

[7] Since this position emphasizes the importance of simple latency differences, it has also been called as the simple brain-time view (Yarrow & Arnold, 2015). In the debates concerning spatiotemporal illusions, it is referred to as the latency difference hypothesis, the simple differential latency hypothesis, and the simple differential lag model (Arnold, Ong, & Roseboom, 2009; Chappell & Hine, 2004; Whitney & Murakami, 1998; Whitney et al., 2000).

of this evidence is that the adaptation is achieved by the brain physically aligning the timing of the signals it receives (e.g., Harrar & Harris, 2008; Harris, Harrar, Jaekl, & Kopinska, 2010; Navarra, García-Morera, & Spence, 2012).[8] In effect, the suggested adaptation is what Dennett and Kinsbourne's content-sensitive settling does—the non-simultaneity of the sudden occurrences of stimuli serves as the time-markers based on which the adaptation process takes place. If the brain recalibrates the sensory signals in this manner, then the explanation of temporal order perception is no longer a matter of simple latency differences. Instead, the possibly adaptive nature of temporal order perception increases the accuracy of the subjective time in relation to the actual order of events. Since, in this case, the subjective time reflects the event time—thus resembling the event-time view of the previous section—this position has also been called the event-time view.

Hence, in this context, the crucial difference between the brain-time view and the event-time view concerns the possible existence of an adaptive mechanism that temporally aligns the sensory signals. This difference means that, departing from how the event time is implemented in the time-marker view, the event-time view is in theory compatible with the thesis of temporal isomorphism. This is because the thesis is only concerned with the relationship between subjective time and brain time, and thus allows some adaptation to the asynchronous stimuli to occur before the perceptual end point is reached. The two are also likely to be compatible in practice; the adaptation has been postulated to take place in early sensory pathways, before experiences are formed (e.g., Heron, Roach, Hanson, McGraw, & Whitaker, 2012; Maravita, Bolognini, Bricolo, Marzi, & Savazzi, 2008). If correct, then the subjective time of experience and the neural time of experience could be isomorphic. An interesting consequence of this compatibility is that the event-time view as understood in typical temporal order judgments counts as a brain-time view in the perceptual asynchrony literature.

Alex Holcombe (2015) has argued that experience does not comprise a single ordered timeline because grouping and integrating stimuli is more important than ordering events in a timeline. He makes this

[8] The evidence for the adaptation to a non-simultaneous multimodal stimulus can also be explained by the shifts in decision criteria (Yarrow, Jahn, Durant, & Arnold, 2011). Since such shifts happen at the post-experiential decision level, this explanation concurs with the brain-time view.

comment in relation to the brain-time and the event-time views as they are understood in the context of the temporal order judgments of separate stimuli. Holcombe's claim suggests another difference between the two versions of the brain-time view (and possibly the event-time views too): While Zeki's brain-time view can be understood as comprising several different timelines, one for each localized discriminator, the brain-time view in the context of simultaneity and temporal order judgment tasks implies a single ordered timeline (similarly to the Cartesian theater model).

However, it is not obvious that Holcombe's claim is correct. On the positive side, he is correct to emphasize the importance of grouping and integrating stimuli, and the (relative) independence of these mechanisms from temporal order mechanisms. There is evidence, for example, that grouping and binding do not require subjective synchrony of the grouped and bound features (Fahle & Koch, 1995; Guttman, Gilroy, & Blake, 2007). In fact, studies on audiovisual perception even suggest that subjective simultaneity is not optimal for subjective integration of the stimuli. The integration of two stimuli is better when they deviate from the point of subjective simultaneity determined in the temporal order tasks (Freeman et al., 2013; Ipser, Karlinski, & Freeman, 2018). One interpretation for these results is that there is no ordered single timeline.

Another reason why one could think that there is no single ordered timeline is that we can utilize the temporal information that is implicitly present in our experiences to make temporal judgments as well. That is, many processes utilize and preserve temporal information by transforming it to the non-temporal content of experiences; such information, rather than explicit time markers, is also utilized when we make temporal judgments. Moreover, on some occasions, this information brings about the temporal resolution that is a magnitude smaller than the 30–40 milliseconds threshold, which is the typical temporal order threshold. For example, when the asynchrony between the stimuli is 2–5 milliseconds, visual stimuli can bring about the experiences of apparent motion and the direction of motion can be used to determine which stimulus was presented first (Sweet, 1953; Wehrhahn & Rapf, 1992; Westheimer & McKee, 1977). In the auditory modality, loudness, pitch, and other micropatterns can be used to determine the order of stimuli (Babkoff &

Sutton, 1963; Efron, 1973; Fostick & Babkoff, 2013). For instance, when two auditory stimuli are presented a few milliseconds apart, the first stimulus sounds louder than the second stimulus. Interestingly, at least in this case, subjects are better at determining the temporal order of two auditory stimuli when they are separated by less than four milliseconds than when they are separated by ten milliseconds (Babkoff, 1975).

Despite the evidence above, there are two reasons to resist Holcombe's claim. First, the brain-time view and the event-time view concern different sorts of timelines as the evidence. Whereas the ordered timeline that the both views bring about is the subjective timeline—how experiences are temporally ordered by the order in which they reach their perceptual endpoint—the examples above pertain to the temporal information used by pre-experiential processes. In Dennett and Kinsbourne's terms, the evidence concerns localized discriminators, not the temporal properties of the Cartesian theater.

Second, similarly to the evidence, the temporal order of events is only implicitly represented in the brain-time view and the event-time view. However, whereas the temporal information is preserved by non-temporal content in the examples above (e.g., in the direction of motion), it is represented by the temporal order of experiences in the views. The broader issue here is that such representation does not yet amount to the explicit representation of temporal order. Instead, the latter needs to be separately represented.[9] This claim concurs with Dennett and Kinsbourne's definition of the Cartesian theater, according to which it is the place where discriminations were "'presented' for subjective judgment," meaning that the order of experiences is not the same as the judged order of experiences. Moreover, neither of the views makes claims about the mechanisms for reporting temporal properties. For this reason, these mechanisms can utilize different information or have different decision criteria as suggested by simultaneity judgments, temporal order judgments, and reaction time studies (see, e.g., Jaśkowski, 2014; Steglich & Neumann, 2000; Sternberg & Knoll, 1973; Tappe, Niepel, & Neumann, 1994) without

[9] Paraphrasing William James' remark—agreed by most philosophers and motivating the lively debate concerning the temporal structure of consciousness—the succession of experience is not the experience of succession.

this pertaining to the nature of the ordered timeline as implied by the views. Overall, the plausibility of a single ordered timeline appears to boil down to the details of what we mean by such a timeline.

5 Conclusions

This chapter focused on the views about the relationships between subjective time, brain time, and event time. In particular, the focus was on the thesis of temporal isomorphism—an event is perceived to occur at the time neural signals the event evokes reach the processes responsible for consciousness—in the debates concerning temporal order illusions. It was argued that the view that is often associated with the thesis, namely the brain-time view, comes in two versions, and they differ in two respects.

First, the two versions differ regarding the processes responsible for consciousness. This disagreement is fundamentally over the neural correlates of consciousness that supposedly match the subjective time of experience. In the first version, the perceptual endpoint is reached when the process that determines the features of the perceived event finishes. The second version does not comment on the perceptual endpoint. Hence, unlike in Zeki's theory, the processes that determine the features of the perceived event can precede the processes responsible for consciousness. As a result, this version is open to the more commonly held view that conscious states are somehow unified. Second, the versions differ on their position toward the processes preceding consciousness. The first version is compatible with there being unconscious processes that aim to increase the match between subjective time and event time; the brain-time view, as understood in the temporal order judgment literature, rules out such unconscious processes.

The brain-time view is contrasted with the event-time view(s), whose crux is the idea that subjective time tracks or aims to track event time. It was argued that this idea has also been implemented in two very different ways. In Nishida and Johnston's time-marker view, which can be thought as inspired by Dennett and Zeki's timestamp view, subjective time utilizes time markers. The processing of time markers happens early in the processing hierarchy and is largely dissociated from the processing of the content of experiences (localized discriminations).

Hence, this implementation of the event-time view rejects the thesis of temporal isomorphism. In the context of typical temporal order judgment literature, in turn, the event-time view refers to a view in which subjective time aims to track event time by temporally aligning the propagating sensory signals. In practice, this is what Dennett and Zeki called content-sensitive settling. Because such adaptation is thought to happen before experiences are formed, this view is compatible with the thesis of temporal isomorphism.[10]

References

Allik, J., & Kreegipuu, K. (1998). Multiple visual latency. *Psychological Science, 9*(2), 135–138.

Arnold, D. H. (2005). Perceptual pairing of colour and motion. *Vision Research, 45*(24), 3015–3026.

Arnold, D. H. (2010). Relative timing and perceptual asynchrony. In R. Nijhawan & B. Khurana (Eds.), *Space and time in perception and action* (pp. 254–277). Cambridge, UK: Cambridge University Press.

Arnold, D. H., Ong, Y., & Roseboom, W. (2009). Simple differential latencies modulate, but do not cause the flash-lag effect. *Journal of Vision, 9*, 1–8.

Arnold, D. H., & Wilcock, P. (2007). Cortical processing and perceived timing. *Proceedings of the Royal Society B: Biological Sciences, 274*(1623), 2331–2336.

Arstila, V. (2015). Keeping postdiction simple. *Consciousness and Cognition, 38*, 205–216.

Arstila, V. (2016a). Theories of apparent motion. *Phenomenology and the Cognitive Sciences, 15*(3), 337–358.

Arstila, V. (2016b). The time of experience and the experience of time. In B. Mölder, V. Arstila, & P. Øhrstrøm (Eds.), *Philosophy and psychology of time* (pp. 163–186). Cham: Springer International Publishing.

Babkoff, H. (1975). Dichotic temporal interactions: Fusion and temporal order. *Perception & Psychophysics, 18*(4), 267–272.

[10] I am deeply grateful to Sean Power, whose help and incisive comments greatly improved the paper.

Babkoff, H., & Sutton, S. (1963). Perception of temporal order and loudness judgments for dichotic clicks. *The Journal of the Acoustical Society of America, 35*(4), 574–577.
Chappell, M., & Hine, T. (2004). Events before the flash do influence the flash-lag magnitude. *Vision Research, 44*(3), 235–239.
Dennett, D. C., & Kinsbourne, M. (1992). Time and the observer. *Behavioral and Brain Sciences, 15*(2), 183–247.
Efron, R. (1973). Conservation of temporal information by perceptual systems. *Perception & Psychophysics, 14*(3), 518–530.
Fahle, M., & Koch, C. (1995). Spatial displacement, but not temporal asynchrony, destroys figural binding. *Vision Research, 35*(4), 491–494.
Fostick, L., & Babkoff, H. (2013). Different response patterns between auditory spectral and spatial temporal order judgment (TOJ). *Experimental Psychology, 60*(6), 432–443.
Freeman, E. D., Ipser, A., Palmbaha, A., Paunoiu, D., Brown, P., Lambert, C., … Driver, J. (2013). Sight and sound out of synch: Fragmentation and renormalisation of audiovisual integration and subjective timing. *Cortex, 49*(10), 2875–2887.
Grush, R. (2007). Time and experience. In T. Müller (Ed.), *Philosophie der Zeit* (pp. 1–18). Frankfurt: Klosterman.
Grush, R. (2009). The temporal content of perceptual experience. In J. Symons & P. Calvo (Eds.), *The Routledge companion to philosophy of psychology* (pp. 592–606). Abingdon: Routledge.
Guttman, S. E., Gilroy, L. A., & Blake, R. (2007). Spatial grouping in human vision: Temporal structure trumps temporal synchrony. *Vision Research, 47*(2), 219–230.
Harrar, V., & Harris, L. R. (2008). The effect of exposure to asynchronous audio, visual, and tactile stimulus combinations on the perception of simultaneity. *Experimental Brain Research, 186*(4), 517–524.
Harris, L. R., Harrar, V., Jaekl, P., & Kopinska, A. (2010). Mechanisms of simultaneity constancy. In R. Nijhawan & B. Khurana (Eds.), *Space and time in perception and action* (pp. 232–253). Cambridge, UK: Cambridge University Press.
Heron, J., Roach, N. W., Hanson, J. V. M., McGraw, P. V., & Whitaker, D. (2012). Audiovisual time perception is spatially specific. *Experimental Brain Research, 218*(3), 477–485.
Holcombe, A. (2015). The temporal organization of perception. In J. Wagemans (Ed.), *The Oxford handbook of perceptual organization* (pp. 820–839). Oxford: Oxford University Press.

Ipser, A., Karlinski, M., & Freeman, E. D. (2018). Correlation of individual differences in audiovisual asynchrony across stimuli and tasks: New constraints on temporal renormalization theory. *Journal of Experimental Psychology: Human Perception and Performance, 44*(8), 1283–1293.

Jaśkowski, P. (2014). What determines simultaneity and order perception ? In V. Arstila & D. Lloyd (Eds.), *Subjective time: The philosophy, psychology, and neuroscience of temporality* (pp. 379–408). Cambridge, MA: MIT Press.

Johnston, A., & Nishida, S. (2001). Time perception: Brain time or event time? *Current Biology, 11*(11), R427–R430.

Kelly, S. D. (2005). The puzzle of temporal experience. In A. Brook & K. Akins (Eds.), *Cognition and the brain: The philosophy and neuroscience* (pp. 1–26). Cambridge University Press.

Kiverstein, J., & Arstila, V. (2013). Time in mind. In A. Bardon & H. Dyke (Eds.), *Blackwell companion to the philosophy of time* (pp. 444–469). Oxford: Wiley-Blackwell.

Köhler, W. (1947). *Gestalt psychology: An introduction to new concepts in modern psychology*. New York: Liveright.

Kopinska, A., & Harris, L. R. (2004). Simultaneity constancy. *Perception, 33*(9), 1049–1060.

Lee, G. (2014). Temporal experience and the temporal structure of experience. *Philosopher's Imprint, 14*(3), 1–21.

Maravita, A., Bolognini, N., Bricolo, E., Marzi, C. A., & Savazzi, S. (2008). Is audiovisual integration subserved by the superior colliculus in humans? *NeuroReport, 19*(3), 271–275.

Maunsell, J. H. R., Ghose, G. M., Assad, J. A., Mcadams, C. J., Boudreau, C. E., & Noerager, B. D. (1999). Visual response latencies of magnocellular and parvocellular LGN neurons in macaque monkeys. *Visual Neuroscience, 16*(1), 1–14.

Mölder, B. (2014). Constructing time: Dennett and Grush on temporal representation. In V. Arstila & D. Lloyd (Eds.), *Subjective time: The philosophy, psychology, and neuroscience of temporality* (pp. 217–238). Cambridge, MA: MIT Press.

Moutoussis, K., & Zeki, S. (1997). A direct demonstration of perceptual asynchrony in vision. *Proceedings of the Royal Society B: Biological Sciences, 264*(1380), 393–399.

Moutoussis, K., & Zeki, S. (2002). The relationship between cortical activation and perception investigated with invisible stimuli. *Proceedings of the National Academy of Sciences, 99*(14), 9527–9532.

Navarra, J., García-Morera, J., & Spence, C. (2012). Temporal adaptation to audiovisual asynchrony generalizes across different sound frequencies. *Frontiers in Psychology, 3*.

Nishida, S., & Johnston, A. (2002). Marker correspondence, not processing latency, determines temporal binding of visual attributes. *Current Biology, 12*, 359–368.

Nishida, S., & Johnston, A. (2010). The time marker account of cross-channel temporal judgments. In R. Nijhawan & B. Khurana (Eds.), *Space and time in perception and action* (pp. 278–300). Cambridge, UK: Cambridge University Press.

Phillips, I. (2014). The temporal structure of experience. In V. Arstila & D. Lloyd (Eds.), *Subjective time: The philosophy, psychology, and neuroscience of temporality* (pp. 139–158). Cambridge, MA: MIT press.

Power, S. (2010). Complex experience, relativity and abandoning simultaneity. *Journal of Consciousness Studies, 17*(3–4), 231–256.

Reutimann, J., Yakovlev, V., Fusi, S., & Senn, W. (2004). Climbing neuronal activity as an event-based cortical representation of time. *The Journal of Neuroscience, 24*, 3295–3303.

Scharnowski, F., Rees, G., & Walsh, V. (2013). Time and the brain: Neurorelativity. The chronoarchitecture of the brain from the neuronal rather than the observer's perspective. *Trends in Cognitive Sciences, 17*(2), 51–52.

Steglich, C., & Neumann, O. (2000). Temporal, but not spatial, context modulates a masked prime's effect on temporal order judgment, but not on response latency. *Psychological Research, 63*(1), 36–47.

Sternberg, S., & Knoll, R. L. (1973). The perception of temporal order: Fundamental issues and a general model. In S. Kornblum (Ed.), *Attention and performance IV* (pp. 629–685). New York: Academic Press.

Sweet, A. L. (1953). Temporal discrimination by the human eye. *The American Journal of Psychology, 66*(2), 185–198.

Tappe, T., Niepel, M., & Neumann, O. (1994). A dissociation between reaction time to sinusoidal gratings and temporal-order judgment. *Perception, 23*(3), 335–347.

Wehrhahn, C., & Rapf, D. (1992). ON- and OFF-pathways form separate neural substrates for motion perception: Psychophysical evidence. *The Journal of Neuroscience, 12*(6), 2247–2250.

Westheimer, G., & McKee, S. (1977). Perception of temporal order in adjacent visual stimuli. *Vision Research, 17*(8), 887–892.

Whitney, D., & Murakami, I. (1998). Latency difference, not spatial extrapolation. *Nature Neuroscience, 1*(2), 656–657.

Whitney, D., Murakami, I., & Cavanagh, P. (2000). Illusory spatial offset of a flash relative to a moving stimulus is caused by differential latencies for moving and flashed stimuli. *Vision Research, 40*(2), 137–149.

Wittmann, M., van Wassenhove, V., Craig, A. D., & Paulus, M. P. (2010). The neural substrates of subjective time dilation. *Frontiers in Human Neuroscience, 4*(February), 1–9.

Yarrow, K., & Arnold, D. H. (2015). The timing of experiences: How far can we get with simple brain time models? In B. Mölder, V. Arstila, & P. Øhrstrøm (Eds.), *Philosophy and psychology of time* (pp. 187–201). Dordrecht, The Netherlands: Springer.

Yarrow, K., Jahn, N., Durant, S., & Arnold, D. H. (2011). Shifts of criteria or neural timing? The assumptions underlying timing perception studies. *Consciousness and Cognition, 20*(4), 1518–1531.

Zeki, S. (2007). A theory of micro-consciousness. In M. Velmans & S. Schneider (Eds.), *The Blackwell companion to consciousness* (pp. 580–588). Malden, MA: Blackwell Publishing.

Zeki, S., & Bartels, A. (1999). Toward a theory of visual consciousness. *Consciousness and Cognition, 8*(2), 225–259.

Index[1]

[1] Note: Page numbers followed by 'n' refer to notes.

V. Arstila et al. (eds.), *The Illusions of Time*,
https://doi.org/10.1007/978-3-030-22048-8

The manufacturer's authorised representative in the EU is Springer Nature Customer Service Centre GmbH, Europaplatz 3, 69115 Heidelberg, Germany. If you have any concerns regarding our products, please contact ProductSafety@springernature.com

Printed and bound by CPI Group (UK) Ltd, Croydon, CR0 4YY
30/09/2026
02225439-0011